Chemical, Mineralogical and Isotopic Studies of Diagenesis of Carbonate and Clastic Sediments

Chemical, Mineralogical and Isotopic Studies of Diagenesis of Carbonate and Clastic Sediments

Editors

Ihsan Al-Aasm

Howri Mansurbeg

MDPI • Basel • Beijing • Wuhan • Barcelona • Belgrade • Manchester • Tokyo • Cluj • Tianjin

Editors
Ihsan Al-Aasm
University of Windsor
Canada

Howri Mansurbeg
University of Windsor
Canada

Editorial Office
MDPI
St. Alban-Anlage 66
4052 Basel, Switzerland

This is a reprint of articles from the Special Issue published online in the open access journal *Minerals* (ISSN 2075-163X) (available at: https://www.mdpi.com/journal/minerals/special_issues/diagenesis_carbonate).

For citation purposes, cite each article independently as indicated on the article page online and as indicated below:

LastName, A.A.; LastName, B.B.; LastName, C.C. Article Title. *Journal Name* **Year**, *Volume Number*, Page Range.

ISBN 978-3-0365-0078-2 (Hbk)
ISBN 978-3-0365-0079-9 (PDF)

Cover image courtesy of Ihsan S. Al-Aasm and Howri Mansurbeg.

Contents

About the Editors . . . . **vii**

Ihsan S. Al-Aasm and Howri Mansurbeg
Editorial for Special Issue "Chemical, Mineralogical and Isotopic Studies of Diagenesis of
Carbonate and Clastic Sediments"
Reprinted from: *Minerals* **2020**, *10*, 1035, doi:10.3390/min10111035 . . . . **1**

**Irene Cantarero, David Parcerisa, Maria Alexandra Plata, David Gómez-Gras,
Enrique Gomez-Rivas, Juan Diego Martín-Martín and Anna Travé**
Fracturing and Near-Surface Diagenesis of a Silicified Miocene Deltaic Sequence:
The Montjuïc Hill (Barcelona)
Reprinted from: *Minerals* **2020**, *10*, 135, doi:10.3390/min10020135 . . . . **5**

María J. Herrero, Rafaela Marfil, Jose I. Escavy, Ihsan Al-Aasm and Michael Scherer
Diagenetic Origin of Bipyramidal Quartz and Hydrothermal Aragonites within the Upper
Triassic Saline Succession of the Iberian Basin: Implications for Interpreting the Burial–Thermal
Evolution of the Basin
Reprinted from: *Minerals* **2020**, *10*, 177, doi:10.3390/min10020177 . . . . **27**

Guoqiang Sun, Yetong Wang, Jiajia Guo, Meng Wang, Yun Jiang and Shile Pan
Clay Minerals and Element Geochemistry of Clastic Reservoirs in the Xiaganchaigou Formation
of the Lenghuqi Area, Northern Qaidam Basin, China
Reprinted from: *Minerals* **2019**, *9*, 678, doi:10.3390/min9110678 . . . . **45**

Bo Chen, Feng Wang, Jian Shi, Fenjun Chen and Haixin Shi
Origin and Sources of Minerals and Their Impact on the Hydrocarbon Reservoir Quality of the
PaleogeneLulehe Formation in the Eboliang Area, Northern Qaidam Basin, China
Reprinted from: *Minerals* **2019**, *9*, 436, doi:10.3390/min9070436 . . . . **63**

Marco Tortola, Ihsan S. Al-Aasm and Richard Crowe
Diagenetic Pore Fluid Evolution and Dolomitization of the Silurian and Devonian
Carbonates, Huron Domain of Southwestern Ontario: Petrographic, Geochemical and Fluid
Inclusion Evidence
Reprinted from: *Minerals* **2020**, *10*, 140, doi:10.3390/min10020140 . . . . **89**

Namam Salih, Howri Mansurbeg and Alain Préat
Geochemical and Dynamic Model of Repeated Hydrothermal Injections in Two Mesozoic
Successions, Provençal Domain, Maritime Alps, SE-France
Reprinted from: *Minerals* **2020**, *10*, 775, doi:10.3390/min10090775 . . . . **141**

**MerveÖzyurt, M. Ziya Kırmacı, Ihsan Al-Aasm, Cathy Hollis, Kemal Taslı
and Raif Kandemir**
REE Characteristics of Lower Cretaceous Limestone Succession in Gümüşhane, NE Turkey:
Implications for Ocean Paleoredox Conditions and Diagenetic Alteration
Reprinted from: *Minerals* **2020**, *10*, 683, doi:10.3390/min10080683 . . . . **165**

Leilei Yang, Linjiao Yu, Donghua Chen, Keyu Liu, Peng Yang and Xinwei Li
Effects of Dolomitization on Porosity during Various Sedimentation-Diagenesis Processes in
Carbonate Reservoirs
Reprinted from: *Minerals* **2020**, *10*, 574, doi:10.3390/min10060574 . . . . **191**

About the Editors

Ihsan Al-Aasm earned his B.Sc. and M.Sc. degrees from the University of Baghdad and his Ph.D. from the University of Ottawa. He worked at the Geological Survey of Iraq from 1978 to 1980, as a postdoctoral fellow at the University of Ottawa from 1984 to 1986, and then as an assistant professor (1986–1988) at the University of Ottawa. He joined the University of Windsor in 1988, and he is now a full professor at the School of the Environment. His principal area of research is petrologic and chemical attributes of carbonate and siliciclastic diagenesis, dolomitization, reservoir characterization, and environmental geochemistry. He has acted as an associate editor for *Journal of Sedimentary Research, Marine and Petroleum Geology, Canadian Journal of Earth Science, Journal of Paleogeography, Arabian J. of Geosciences*, and *Minerals*, and he holds memberships in AAPG, GAC, SEPM, and IAS. He has published over 140 papers in peer-reviewed national and international journals.

Howri Mansurbeg is a sedimentary geologist with expertise in diagenesis of siliciclastic and carbonate reservoirs. He obtained his Ph.D. from Uppsala University, Sweden, in 2007. His current research deals with hydrocarbon exploration in carbonate reservoirs in the Middle East and siliciclastic deposits in the North Sea, utilizing an integrated multidisciplinary approach. Howri has worked for several international oil companies in Europe and the US and has produced many confidential technical reports and research articles in the past 15 years. He is currently a professor at the General Directorate of Scientific Research Center, Salahaddin University, Erbil, the Kurdistan Region of Iraq. He is also an adjunct professor at the School of the Environment, University of Windsor, Windsor, Canada.

Editorial

Editorial for Special Issue "Chemical, Mineralogical and Isotopic Studies of Diagenesis of Carbonate and Clastic Sediments"

Ihsan S. Al-Aasm [1,*] and Howri Mansurbeg [1,2]

[1] School of the Environment, University of Windsor, Windsor, ON N9B 3P4, Canada;
   Howri.Mansurbeg@uwindsor.ca
[2] General Directorate of Scientific Research Center, Salahaddin University-Erbil, Erbil 44002,
   The Kurdistan Region of Iraq, Iraq
* Correspondence: alaasm@uwindsor.ca

Received: 10 November 2020; Accepted: 18 November 2020; Published: 20 November 2020

Diagenesis of carbonates and clastic sediments encompasses the biochemical, mechanical and chemical changes that occur in sediments after deposition and prior to low-grade metamorphism. Parameters which, to a large extent, control diagenesis in carbonate and clastic sediments include primary composition of the sediments, depositional facies, pore water chemistry, burial-thermal and tectonic evolution of the basin, and paleo-climatic conditions [1–6].

Diagenetic processes involve a widespread chemical, mineralogical and isotopic modifications affected by original mineralogy of carbonate and clastic sediments. These diagenetic alterations will impose a major control on porosity and permeability and hence on hydrocarbon reservoirs, water aquifers as well as the presence of other important economic minerals [7–10].

This Special Issue "Chemical, Mineralogical and Isotopic Studies of Diagenesis of Carbonate and Clastic Sediments", is a collection of eight selected papers that show up to date detailed geochemical, geological and sedimentological data on the diagenesis.

The paper by Cantarero et al. [11], entitled "Fracturing and near-surface diagenesis of a silicified Miocene deltaic sequence: the Montjuic Hill (Barcelona)" provided petrographic and geochemical evidence for the diagenetic overprints within the deltaic sequence investigated. These diagenetic modifications were affected by fracturing and cementation of a variety of minerals, such as barite and silicates. The authors also discussed the sources and nature of the diagenetic fluids that affected these rocks.

Studies by Yang et al. [12], entitled "Effect of dolomitization on porosity during various sedimentation-diagenesis processes in carbonate reservoirs", present how complex diagenetic processes, such as dolomitization, can affect porosity development in carbonate reservoirs. Multiphase fluid flow and solute transport simulation was employed to investigate dolomitization and its effect on porosity evolution from deep carbonates in Tarim Basin, northwest China. The numerical modeling applied in this study quantified dolomitization and other diagenetic processes and their relationships to fluid composition and hydrodynamic characteristics in the basin.

Tortola et al. [13], in their paper entitled "Diagenetic pore fluid evolution and dolomitization of the Silurian and Devonian carbonates, Huron Domain of southwestern Ontario: petrographic, geochemical and fluid inclusion evidence", investigated the nature of diagenetic fluids in part of Michigan Basin using sedimentologic and geochemical tracers. In both age groups, the authors identified three types of dolomite replacement matrix: RD1 (precipitated at shallow burial conditions), RD2 and RD3 (formed at intermediate burial conditions). In addition, a coarse crystalline ferron saddle dolomite cement (formed at intermediate burial conditions) filling fractures and vugs has been documented in the Silurian successions. Early- and late-stage calcite cement have been distinguished in both groups of formations including isopachous, syntaxial overgrowth, dogtooth, drusy and blocky calcite. The authors linked the different types of cements to the fluid-rock interactions, which are associated to the tectonic evolution of the basin.

A unique presence of diagenetic bipyramidal quartz and aragonite was discussed thoroughly by Herrero et al. [14], in their paper entitled "Diagenetic origin of bipyramidal quartz and hydrothermal aragonites within the Upper Triassic Saline succession of the Iberian Basin: implications for interpreting the burial-thermal evolution of the basin". The authors applied several analytical techniques to investigate the nature of these minerals, such as petrography, SEM, Raman and fluid inclusions. They were able to relate the formation of these minerals to tectonic and thermal evolution of the Iberian Basin. Fluid composition and timing of migration were suggested based on mineralogical and geochemical evidence.

Sun et al. [15], in their paper entitled "Clay minerals and element geochemistry of clastic reservoirs in the Xiaganchaigou Formation of the Lenghuqi area, northern Qaidam Basin, China", discussed the formation of clay minerals and their diagenetic modification in Oligocene sandstones deposited in the Qaidam Basin in China. They used mineralogical and geochemical techniques to characterize the clay minerals and assigned depositional, paleoclimatic and diagenetic environments to their formation.

In a paper by Ozyurat et al. [16], entitled "REE characteristics of Lower Cretaceous limestone succession in Gumushane, NE Turkey: implications for ocean paleoredox conditions and diagenetic alteration", the effect of paleoredox conditions on diagenetic alteration of carbonates was investigated using REE and other geochemical proxies. Water–rock interactions between seawater and the surrounding basaltic rocks have an important role on the distribution and behavior of REE in the studied limestones. This paper provided an excellent example on the validity of using REE and other geochemical tracers to investigate paleoocean redox conditions.

Diagenetic alteration of the Paleogene sandstone of the Lulehe Formation in northern Qaidam Basin in China was the subject of detailed mineralogical and geochemical investigation by Chen et al. [17], in their paper entitled "Origin and sources of minerals and their impact on the hydrocarbon reservoir quality of the Paleogene Lulehe Formation in the Eboliang area, northern Qaidam Basin, China". The effects of diagenetic processes, such as cementation, dissolution and compaction on reservoir quality were discussed in this paper. These processes resulted in heterogeneity of the reservoir.

Hydrothermal fluid flow and resultant diagenetic alteration of Mesozoic successions from Maritime Alps, SE France was the focus of a paper by Salih et al. [18], entitled "Geochemical and dynamic model of repeated hydrothermal injections in two Mesozoic successions, Provencal Domain, Maritime Alps, SE-France". The authors used field, petrographic and geochemical evidence to evaluate dolomitization in the studied stratigraphic sections. They attributed the formation of these dolomites to episodic fracturing related to the flow of multiple fluxes of hydrothermal fluids.

The papers in this Special Issue demonstrate the interplay between mineralogical and chemical changes in carbonates and clastic sediments and diagenetic processes, fluid flow, tectonics, mineral reactions at variable scales and environments from a variety of sedimentary basins. Quantitative analyses of diagenetic reactions in these sediments, using a multitude of techniques, are essential for understanding pathways of in different diagenetic environments. These papers offer exciting new analytical and modeling techniques for understanding diagenesis of sedimentary rocks in a variety of sedimentary basins under different tectonic and hydrodynamic regimes.

Author Contributions: I.S.A.-A. wrote this editorial; proof-read by H.M. All authors have read and agreed to the published version of the manuscript.

Acknowledgments: We thank the authors of the articles included in this Special Issue and the organizations that have financially supported the research in the areas related to this topic. We thank the Editorial Board for their suggestions which improved the quality of this editorial.

Conflicts of Interest: The author declares no conflict of interest.

References

1. Moore, C.H. *Carbonate Reservoirs*; Elsevier: Amsterdam, The Netherlands, 2001; 444p.
2. Tucker, M.E.; Wright, P.V. *Carbonate Sedimentology*; Blackwell Sci. Publ.: Hoboken, NJ, USA, 1990; 482p.
3. Morad, S.; Al Suwaidi, M.; Mansurbeg, H.; Morad, D.; Cerini, A.; Paganoni, M.; Al-Aasm, I. Diagenesis of a limestone reservoit (Lower Cretaceous), Abu Dhabi, UAE: Comparison between the anticline crest and flanks. *Sedimentol. Geol.* **2019**, *380*, 127–142. [CrossRef]
4. Seibel, M.J.; James, N.P. Diagenesis of Miocene, incised valley-filling limestones; Provence, Southern France. *Sedimentol. Geol.* **2017**, *347*, 21–35. [CrossRef]
5. Goldstein, R.H.; Franseen, E.K. Meteoric calcite cementation: Diagenetic response to relative fall in sea-level and effect on porosity and permeability, Las Negras area, southeastern Spain. *Sedimentol. Geol.* **2017**, *348*, 1–18.
6. Burley, S.D.; Wordon, R. *Sandstone Diagenesis: Recent and Ancient*; Blackwell: Hoboken, NJ, USA, 2009.
7. James, N.P.; Jones, B. *Origin of Carbonate Rocks*; John Wiley & Sons: Hoboken, NJ, USA, 2015.
8. Swart, P.K. The geochemistry of carbonate diagenesis: The past, present and future. *Sedimentology* **2015**, *62*, 1233–1304. [CrossRef]
9. Al-Aasm, I. Origin and characterization of hydrothermal dolomite in the Western Canada Sedimentary Basin. *J. Geochem. Exp.* **2003**, *78–79*, 9–15. [CrossRef]
10. Morad, D.; Nader, F.H.; Gasparrini, M.; Morad, S.; Rossi, C.; Marchionda, E.; Al Darmaki, F.; Martines, M.; Hellevang, H. Comparison of the diagenetic and reservoir quality evolution between the anticline crest and flank of an Upper Jurassic carbonate gas reservoir, Abu Dhabi, UAE. *Sediment. Geol.* **2018**, *367*, 96–113. [CrossRef]
11. Cantarero, I.; Parcerisa, D.; Plata, M.A.; Gomez-Gras, D.; Gomez-Rivas, E.; Martin-Martin, J.D.; Trave, A. Fracturing and near-surface diagenesis of a silicified Miocene deltaic sequence: The Montjuic Hill (Barcelona). *Minerals* **2020**, *10*, 135. [CrossRef]
12. Yang, L.; Yu, L.; Chen, D.; Liu, K.; Yang, P.; Li, X. Effect of dolomitization on porosity during Various sedimentation-diagenesis processes in carbonate reservoirs. *Minerals* **2020**, *10*, 574. [CrossRef]
13. Tortola, M.; Al-Aasm, I.S.; Crowe, R. Diagenetic and pore fluid evolution and dolomitization of the Silurian and Devonian carbonates, Huron Domain of southwestern Ontario: Petrographic, geochemical and fluid inclusion evidence. *Minerals* **2020**, *10*, 140. [CrossRef]
14. Herrero, M.; Marfil, R.; Escavy, J.I.; Al-Aasm, I.; Scherer, M. Diagenetic origin of bipyramidal quartz and hydrothermal aragonites within the Upper Triassic Saline succession of the Iberian Basin: Implications for interpreting the burial-thermal evolution of the basin. *Minerals* **2020**, *10*, 177. [CrossRef]
15. Sun, G.; Wang, Y.; Guo, J.; Wang, M.; Jiang, Y.; Pam, S. Clay minerals and element geochemistry of clastic reservoirs in the Xiaganchaigou Formation of the Lenghuqi area, northern Qaidam Basin, China. *Minerals* **2020**, *9*, 678. [CrossRef]
16. Ozyurt, M.; Kirmaci, M.Z.; Al-Aasm, I.; Hollis, C.; Tasli, K.; Kandemir, R. REE characteristics of Lower Cretaceous limestone succession in the Gumushane, NE Turkey: Implications of ocean paleoredox conditions and diagenetic alteration. *Minerals* **2020**, *10*, 683. [CrossRef]
17. Chen, B.; Wang, F.; Shi, J.; Chen, F.; Shi, H. Origin and sources of minerals and their impact on the hydrocarbon reservoir quality of the Paleogenelulehe Formation in the Eboliang area, northern Qaidam Basin, China. *Minerals* **2019**, *9*, 436. [CrossRef]
18. Salih, N.; Mansurbeg, H.; Preat, A. Geochemical and dynamic model of repeated hydrothermal injections in two Mesozoic successions, Provencal Domain, Maritime Alps, SE-France. *Minerals* **2020**, *10*, 775. [CrossRef]

Publisher's Note: MDPI stays neutral with regard to jurisdictional claims in published maps and institutional affiliations.

Article

Fracturing and Near-Surface Diagenesis of a Silicified Miocene Deltaic Sequence: The Montjuïc Hill (Barcelona)

Irene Cantarero [1,*], David Parcerisa [2], Maria Alexandra Plata [1], David Gómez-Gras [3], Enrique Gomez-Rivas [1], Juan Diego Martín-Martín [1] and Anna Travé [1]

[1] Departament de Mineralogia, Petrologia i Geologia Aplicada, Facultat de Ciències de la Terra, Universitat de Barcelona (UB), C/Martí i Franqués s/n, 08028 Barcelona, Spain; alexa854@gmail.com (M.A.P.); e.gomez-rivas@ub.edu (E.G.-R.); juandiegomartin@ub.edu (J.D.M.-M.); atrave@ub.edu (A.T.)

[2] Departament d'Enginyeria Minera, Industrial i TIC, Escola Politècnica Superior d'Enginyeria de Manresa, Universitat Politècnica de Catalunya, Av. Bases de Manresa 63-71, 08242 Manresa, Spain; dparcerisa@epsem.upc.edu

[3] Departament de Geologia, Universitat Autònoma de Barcelona, 08193 Bellaterra (Cerdanyola del Vallès), Spain; David.Gomez@uab.cat

* Correspondence: i_cantarero@ub.edu; Tel.: +34-934031165

Received: 29 November 2019; Accepted: 31 January 2020; Published: 4 February 2020

Abstract: Near-surface diagenesis has been studied in the Langhian siliciclastic rocks of the Montjuïc Hill (Barcelona Plain) by means of petrographical (optical and cathodoluminescence) and geochemical (electron microprobe, $\delta^{18}O$, $\delta^{13}C$, $\delta^{34}S$ and $^{87}Sr/^{86}Sr$) analyses. In the hill, these rocks are affected by strong silicification, but the same unit remains non-silicified at depth. The results reveal that fracturing took place after lithification and during uplift. Fracture cementation is clearly controlled by the previous diagenesis of the host rock. In non-silicified areas, cementation is dominated by calcite, which precipitated from meteoric waters. In silicified areas, fractures show multiepisodic cementation produced firstly by barite and secondly by silica, following the sequence opal, lussatite, chalcedony, and quartz. Barite precipitated only in fractures from the mixing of upflowing seawater and percolating meteoric fluids. The presence of silica stalactites, illuviation, and geopetal structures, and $\delta^{18}O$ values indicate that silica precipitation occurred in the vadose regime from low-temperature percolating meteoric fluids, probably during a glacial period. Moreover, the presence of alunite suggests that silica cement formed under acidic conditions. Karst features (vugs and caverns), formed by arenisation, reveal that silica was derived from the dissolution of surrounding silicified host rocks.

Keywords: silicification; meteoric diagenesis; fractures; deltaic sequence; karst; glacial period

---

1. Introduction

Depending on the depositional environment, interstitial waters in deltaic depositional systems can be of meteoric or marine origin, or a mixture of both [1] but can be modified due to relative sea-level fluctuations [1–5]. However, other external factors not related to the depositional system, such as fracturing, can also control the type and distribution of fluids. Fractures can act as conduits for fluids, which can be external fluids with different chemical and/or thermal properties or as barriers, causing compartmentalization of the system. Thus, the quality and heterogeneity of deltaic reservoirs depend on the depositional facies, diagenesis, and the presence of fractures [6–11]. Accordingly, the systematic study of cements, in both host rocks and fractures, can provide valuable information about the origin of fluids, the degree of fluid-rock interaction, the paleofluid flow paths, and the role of fractures for fluid transport [12–18].

Common early diagenetic processes in sandstones are burrowing and boring, cementation, often as concretions (typically of carbonate minerals, iron oxides, and phosphates), and soil formation [5,19]. Quartz and silica minerals are some of the most abundant cements in sandstones. Despite that, silicification processes are usually related to burial diagenesis and hydrothermalism [18,20–22], silicification can also develop in the form of chert or silcrete in near-surface conditions at relatively low temperatures [23–26]. Recent studies have demonstrated that different silicification events, starting with low temperature and followed by high-temperature silicification, had strongly modified the petrophysical properties of pre-salt carbonate reservoirs of the Campos and Santos Basin (offshore Brazil) and in the Kwanza Basin (offshore Angola) [27,28].

In the Catalan Coastal Ranges (NE Spain), previous studies show that the main diagenetic processes linked to Cenozoic and Mesozoic deformation events are calcite cementation and dolomitization ([15,29–37]). Only two locations have been found to show extensive silicification, both related to the Neogene extension: the Camp dels Ninots (Caldes de Malavella) and the Montjuïc Hill (in the city of Barcelona). In the Camp dels Ninots, silicification is represented by opaline chert nodules developed in maar-diatreme lake sediments, that is, in Pliocene sediments deposited in a lake that formed in a volcanic crater caused by a phreatomagmatic eruption. This silicification was produced by hydrothermal fluids associated with Neogene NW–SE normal faults and Pliocene volcanism [38]. On the contrary, the Montjuïc hill is a horst delimited by Neogene NE–SW normal faults and the silicification of its deltaic sequence has been attributed to low-temperature meteoric fluids [39–42].

In this contribution, we focus on the complex fracture diagenesis developed in previously silicified sandstones of a Miocene deltaic sequence in the Montjuïc Hill. Specifically, the aims of this study are fivefold: (i) to petrologically and geochemically characterize the successive generations of fracture-filling cements; (ii) to determine the composition and origin of fluids that circulated along the faults; (iii) to unravel the paleofluid pathways; (iv) to infer possible paleoclimatic implications from silica cements in fractures; and, (v) to determine the source of silica in fractures.

2. Geological Setting

The Catalan Coastal Ranges (CCR) in the NE of Spain, constitutes the northwestern edge of the Valencia Trough, which opened during the Neogene extensional event (Oligocene–Middle Miocene, [43]) (Figure 1A). During this extension, the CCR acquired a well-developed horst and graben structure limited by listric faults striking NE–SW and NNE–SSW with a detachment level at 12–16 km [44,45] (Figure 1B). This structure was also segmented by later faults trending NW–SE to NNW–SSE. The Neogene extensional event has been divided into a syn-rift stage (Aquitanian–Late Burdigalian), an early post-rift stage (Langhian–Serravallian), and a late post-rift stage (late Serravallian–Pliocene) [46]. Two small compressional episodes are recognized during the Neogene extension: one between the late Langhian–Serravalian (early post-rift) and the other during the Messinian (late post-rift) [45].

One of the grabens that formed at this time is the Barcelona Plain, a ~40 km long and ~2–10 km wide graben situated in the central sector of the CCR, and mainly filled with Miocene continental-transitional siliciclastic deposits and Quaternary fluvio-deltaic deposits. Its northern boundary is limited by the Collserola-Montnegre Horst, which is mainly formed by Paleozoic rocks consisting of Cambro-Ordovician shales and phyllites, Silurian black shales and phyllites, Devonian carbonates, Carboniferous Culm facies, and late Hercynian leucogranites, tonalites, and granodiorites [47]. The Montjuïc Hill is a tilted horst formed within the Miocene sediments filling the Barcelona Plain. It is bounded to the SE by the Morrot fault, a NE–SW normal fault that dips to the SE with a minimum vertical throw of 215 m [41] (Figure 2). It has been postulated that this block is also bounded to the NW by another NE–SW normal fault that dips to the NW [48,49].

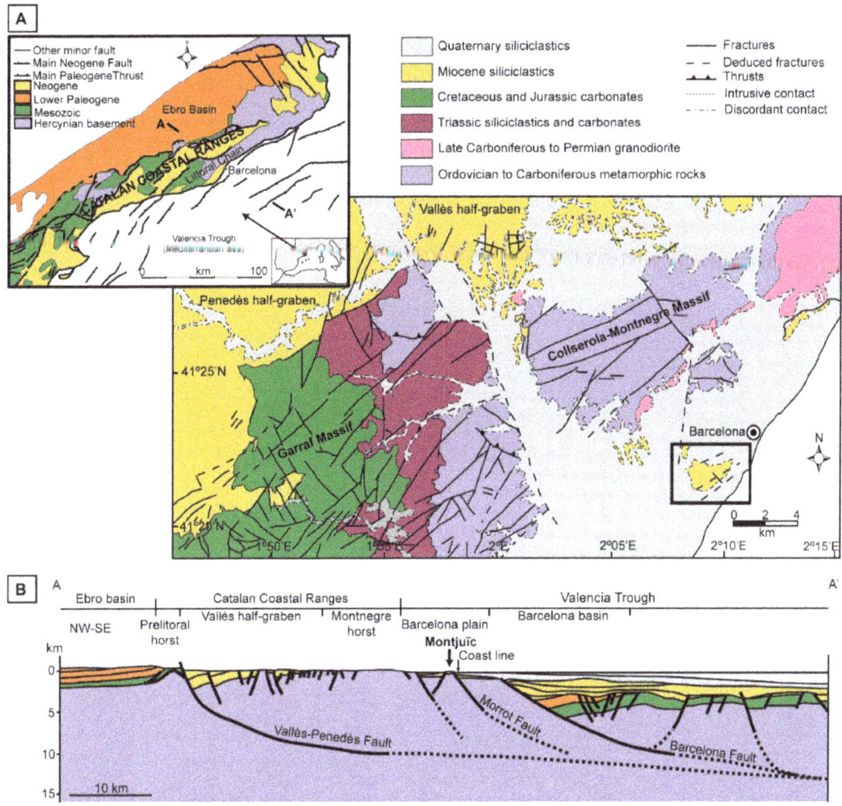

Figure 1. Geological setting. (**A**) Location and schematic map of the Catalan Coastal Ranges and magnification of the Littoral Chain around the study area. The Montjuïc Hill is marked with a black square. (**B**) Cross-section of the CCR indicated in (**A**) [45].

The Montjuïc block is formed by alternating sandstones and conglomerates units and minor lutitic beds interpreted as a prograding deltaic sequence of Serravallian age [40,41]. However, recent biostratigraphy studies place the planktonic zonations N9 and N10 of [50] in the Langhian [51]. Four lithostratigraphic units from base to top have been defined [40] (Figure 2): (1) The basal Morrot unit, formed by 80 m of conglomerates and sandstones, interpreted as delta plain deposits; (2) The Castell unit, represented by 100 m of siltstones and mudstones and well-cemented conglomerates and sandstones, interpreted as delta-front deposits; (3) The Miramar unit, defined by 15 m of marls, interpreted as prodelta deposits, and finally, (4) The Mirador unit, formed by 20 m of massive conglomerates and sandstones, interpreted as proximal delta-front deposits. The uplift of the hill took place between the Pliocene and the Quaternary, according to the presence of faulted Pliocene rocks [52].

Host Rock Petrology and Diagenesis

The petrology and diagenesis of sandstones and conglomerates of Montjuïc have been studied in detail before [39–42]. A brief summary is given in this subsection.

Siliciclastic rocks range from lithic to arkosic arenites/rudites to greywackes. Detrital grains are well-rounded and are composed of quartz (35%), potassium feldspar (9%), rock fragments (20%, mainly quartzites and schists) and minor muscovite, zircon, tourmaline, and biotite. The shale matrix, composed of illite-mica, is frequently transformed into opal and microquartz with variable amounts of remaining clay minerals. Porosity is low (1–3%), primary as intergranular or secondary due to the dissolution of altered feldspar and shale matrix. Detrital fragments indicate that the source area most probably was the Collserola Massif (Figure 1), where Paleozoic rocks crop out.

Lithification of the Montjuïc sandstones was mostly due to the authigenesis of silica minerals. Silicification of these sandstones only occurred on high ground, as demonstrated by numerous boreholes around the mountain where these units appear non-silicified. On high ground, specifically in the Morrot area, the presence of some lenticular remnants of non-silicified rocks allows the observation that silicification fronts preferentially used coarse-grained sediments and fractures. Consequently, two main diagenetic facies with characteristic associations of authigenic minerals are identified: (i) the non-silicified and (ii) silicified facies. Non-silicified facies are constituted by ochre-colored fine-grained sandstones. In these facies, cementation is scarce and generally forms minor feldspar overgrowths around detrital k-feldspar as well as layers or nodules of calcite spar cement mainly filling interparticle porosity. Additionally, they present moldic porosity filled with calcite cement. Silicified facies are characterized by red to purple-colored coarse-grained sandstones containing opal, microquartz and quartz overgrowths as well as other minor authigenic minerals such as Ti and Fe oxides and alunite. Sparitic calcite cement filling residual porosity of silicified sandstones has also been described. Alunite and opal often appear at the boundary between silicified and non-silicified facies.

According to the aforementioned authors [39–42], the geological setting and the lack of compaction suggest that cementation occurred at shallow conditions by oxidizing groundwaters.

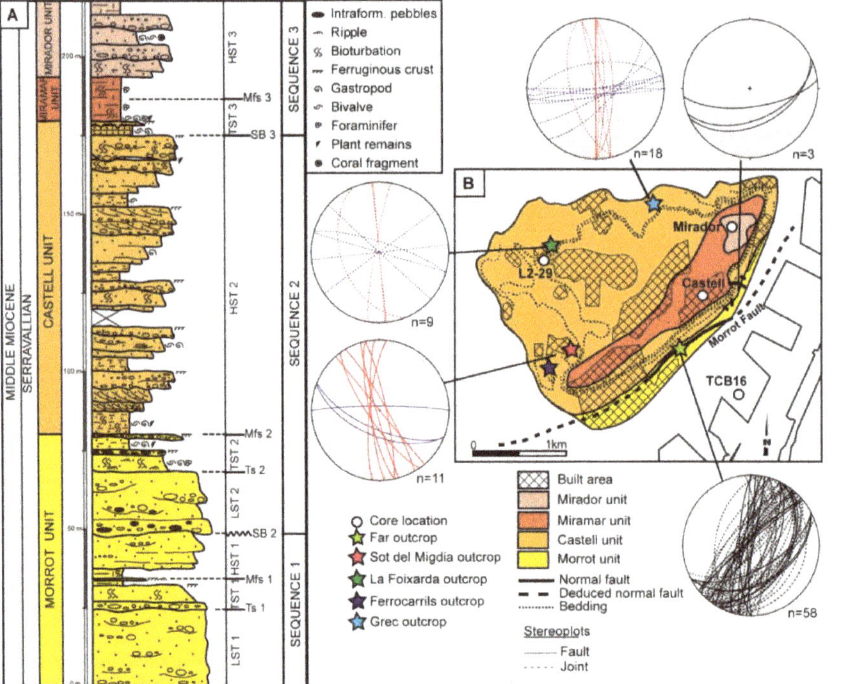

Figure 2. (**A**) General stratigraphic section of Montjuïc Hill [40]. (**B**) Arrangement of the lithostratigraphic units in the Montjuïc Hill [41] and Schmidt lower-hemisphere stereoplot projections of the different generations of fractures found in the study area. The three generations of fractures are indicated with different colors: set 1 (black), set 2 (blue), and set 3 (red).

3. Methodology

Five outcrops (Far, Sot del Migdia, La Foixarda, Ferrocarrils, and Grec) were chosen because of the good exposure of Miocene rocks affected by fracturing to study the fluid flow events related to fracturing occurring in previously silicified rocks of the Montjuïc Hill. Moreover, access to four boreholes (L2-29, Castell, Mirador, and TCB16) allowed us to have a more representative lateral and vertical distribution of fracture-filling cements. Fractures were measured in the field and crosscutting relationships were established. Structural data were plotted and analyzed in lower hemisphere Schmidt stereoplots (Figure 2D). The different fracture-filling minerals developed in the different fracture systems were sampled in outcrops and boreholes for petrographic observations and geochemical analyses.

Thirty samples were collected from the outcrops and forty-eight samples were obtained from the four different boreholes. Thin sections were studied using optical and cathodoluminescence microscopes. A Technosyn Cold Cathodoluminescence Model 8200 MkII (Technosyn Limited, Cambridge, UK) operating at 16–19 kV and 350 µA gun current were used to establish the different calcite cement generations. Some thin sections were also examined under ESEM Quanta 200 FEI, XTE 325/D8395 scanning electronic microscope (FEI Europe B.V., Eindhoven, The Netherlands).

X-ray diffraction of bulk rock and oriented aggregates have been performed with a Bragg-Brentano PAnalytical X'Pert PRO MPD alpha 1(PANalytical B.V., Almelo, The Netherlands) operating at 1.5406 Å, 45 kV, and 40 mA in order to establish the mineralogy of fine-grained fracture fillings.

After the petrographic study, geochemical analysis of the different fracture-filling cements and host rock intergranular and moldic cements (elemental geochemistry, stable isotopes, and $^{87}Sr/^{86}Sr$ ratios) were carried out to characterize the different fluids (i.e., origin, pathways, fluid-rock interaction, temperature).

Carbon-coated thin sections were used for elemental analyses of carbonate and barite cements with a CAMECA model SX-50 microprobe (CAMECA SAS, Gennevilliers, France). It was operated using 15 nA of current intensity, 20 kV of acceleration voltage, and a beam diameter of 10 µm. The detection limits for carbonates are 99 ppm for Na, 554 ppm for Ca, 491 ppm for Mg, 223 ppm for Fe, 158 ppm for Mn, and 141 ppm for Sr. Detection limits (d.l.) for barites are 105 pp for Ca, 298 ppm for Co, 793 ppm for Sr, 1516 ppm for Ba, and 805 ppm for Pb. The precision of major elements is about 0.64% (at 2σ level).

Microsamples for oxygen, carbon, sulfur, and strontium isotope analysis were powdered with a 4 mm-diameter micro drill.

Collected samples for carbon and oxygen stable isotopes in carbonates were reacted with 100% phosphoric acid at 70 °C for two minutes in an automated Kiel Carbonate Device attached to a Thermal Ionization Mass Spectrometer Thermo Electron MAT-252 (Thermo Fisher Scientific, Bremen, Germany). The International Standard NBS-18 and the internal standard RC-1, traceable to the International Standard NBS-19, were used for calibration [53,54]. The results are expressed in δ‰ VPDB standard (Vienna Pee Dee Belemnite). Standard deviation is ±0.02‰ for $\delta^{13}C$ and ±0.05‰ for $\delta^{18}O$.

Oxygen isotopes were analyzed in ten samples of chalcedony and quartz cement and speleothems. The CO_2 to be analyzed was obtained by laser fluorination, using a CO_2 laser of 25 W and ClF_3 as a reactant, following the method of [55]. The obtained CO_2 was analyzed in a Dual Inlet SIRA-II mass spectrometer (VG-Isotech, Cheshire, UK). The results are expressed in δ‰ VSMOW standard (Vienna Standard Mean Ocean Water). These analyses were performed at the stable isotopic service of the Universidad de Salamanca.

Sulphur and oxygen isotopes were analyzed in ten barite samples to establish the origin of the precipitating fluid. The CO and SO_2 gases produced from the barites were analyzed on a continuous-flow elemental analyzer Thermo Delta Plus XP mass spectrometer (Thermo, Bremen, Germany), with a TC/EA pyrolyser for oxygen and a Finnigan MAT CHNS 1108 analyzer (Finnigan, Bremen, Germany) for sulfur. Results were calibrated with NBS-127, SO-5, and SO-6 international standards [56] and the internal standard YCEM (+12.78‰ CDT). The analytical error is ±0.4‰ CDT (Canyon Diablo Troilite) for $\delta^{34}S$ and ±0.5‰ VSMOW for $\delta^{18}O$.

The $^{87}Sr/^{86}Sr$ ratios of three barite samples were obtained. The Sr contained in the powder was obtained by means of chromatography using a SrResinTM (Triskem International). After evaporation, samples were loaded onto a Re filament with 1 µL of 1M phosphoric acid and 2 µL of Ta_2O_5. Isotopic ratio measurements were carried out in a TIMS-Phoenix mass spectrometer (Isotopx, Cheshire, UK) with a dynamic multicollector during 10 blocks of 16 sweeps each, with an ^{88}Sr beam intensity of 3V. Isotopic ratios were corrected from ^{87}Rb interferences and normalized using a value of $^{88}Sr/^{86}Sr$ = 0.1194 to correct the possible mass fractionation during loading and analysis of the sample. During sample analysis, the isotopic standard NBS-987 was measured seven times obtaining a mean of 0.710247 and a standard deviation of 0.000009. These values have been used for the correction of the analyzed values in the samples. Analytical error in the $^{87}Sr/^{86}Sr$ ratio is 0.01%. The standard deviation is 0.000003. These analyses were performed at the CAI (Centro de Apoyo a la Investigación) of Geochronology and Isotopic Geochemistry from the Universidad Complutense de Madrid.

4. Results

4.1. Fracturing

The Montjuïc Hill is affected by intense fracturing produced during the Neogene post-rift period (Langhian to present). Fractures are well-developed within the silicified areas whereas they disappear into the non-silicified areas. Three fracture sets have been distinguished according to their orientations, kinematics, and crosscutting relationships (Figures 2B and 3).

Figure 3. Outcrop photographs of the different fracture sets. (**A**) Interpreted outcrop view of the subsidiary fault zone of the Morrot fault in the far outcrop. Red areas highlight main fault planes and black arrows indicate the orientation and displacement direction of slickenlines. Rock hammer for scale. (**B**) Outcrop view perpendicular to the image shown in A. Sigmoidal shapes and alteration halos produced by iron oxides are observed. (**C**) Outcrop view of the second fracture set characterized by straight and discrete planes coated by orange and purple iron oxides. Ferrocarrils outcrop. (**D**) Horsetail ending of a fault from the third fracture set partially filled with silica. Pen for scale. Ferrocarrils outcrop.

The first set of fractures is represented by NE–SW to NNE–SSW trending joints and normal and strike-slip faults with highly variable dips (Figure 2B). This array has been mainly observed in the southeastern part of the hill. In the far outcrop, a 20 m thick fault zone parallel and subsidiary to the Morrot fault is observed (Figure 3A,B). Individual planes are discrete and undulating, forming an arrangement of sigmoidal shapes that indicate dip-slip movement of the fault zone (Figure 3A,B). Fault planes show both dip-slip and strike-slip slickenlines, the latter being probably the result of the re-activation of the previous normal faults during the Messinian. Fault planes are cemented by silica and are surrounded by red to purple alteration halos towards the host rocks caused by the precipitation of iron oxides (Figure 3B). The second set of fractures is defined by E–W to WNW–ESE trending normal faults and joints dipping between 60 and 85° to both N and S (Figure 2B). They are characterized by straight to slightly undulating discrete planes (Figure 3C). Finally, the third set is constituted by N–S to NNW–SSE trending normal faults and joints with high dips (80–90°) to the E and W (Figure 2B). Commonly, these faults show horsetail endings (Figure 3D). Some fault planes of the second and third sets appear enlarged by dissolution and are cemented by silica minerals or filled with breccias cemented by silica, whereas other planes are coated by orange to purple iron oxides (Figure 3C,D).

4.2. Fracture Fillings

The fractures of the Montjuïc Hill contain different fillings depending on whether they are located in the non-silicified area, silicification front or in the silicified area.

Fractures in the non-silicified area are cemented by euhedral equant calcite crystals, ranging from 90 μm to 1 mm in size, that display a drusy texture (Figure 4A,B). Two calcite cement generations have been recognized. The first one, Cc1, shows a zoned bright orange to brown cathodoluminescence, similar to host rock intergranular cement, whereas the second generation, Cc2, has a uniform bright orange luminescence similar to that of cement filling molds (Figure 4B,C). The change between the two calcite cement generations in fractures is gradual.

Fractures within the silicification front are filled with clays with isolated grains of quartz and feldspar from the host rock (Figure 4D). This clayey filling has white or yellow to brownish color and it consists of sericite, kaolinite, goethite, and minor smectite and alunite.

In the silicified area, fractures display multi-episodic cementation produced by: (1) barite and (2) several silica varieties.

Barite cementation has been only identified in core L2-29 (Figure 2B) and in the Sot del Migdia outcrop (Figure 2B). Specifically, it is found along the uppermost 40 m below the present-day surface. It usually forms crusts up to 1 cm thick but in the L2-29 core, it completely fills a 30 cm-wide fracture. Barite cementation is characterized by the first stage of up to 300 μm-long prismatic crystals that grow decussated from the fracture wall and the second stage of up to 1 mm-long prismatic crystals that grow perpendicular to the fracture wall (Figure 5A). In some samples, in addition to barite, millimeter-sized cubic pyrite crystals and 20 to 50 μm long prismatic jarosite crystals are present. Pyrite crystallized either before the first barite generation or coevally during the last growth stage of this barite. Barite crystals of the second generation commonly have their margins replaced by chalcedony (Figure 5B). The presence of prismatic crystals with low relief next to quartz grains implies the complete replacement of barite by quartz.

With regard to silica cementation, the sequence opal → lussatite → chalcedony →quartz is recognized (Figure 5C), representing an increase in crystallinity and purity towards the inner part of fractures. Purity is defined by the presence of cations other than silica (mostly in low crystallinity silica varieties) but also by the presence of clay-sized particles of other minerals coming from the sandstone matrix. Three successive phases of opal precipitation have been identified according to the amount of lithic fragments, the content in titanium oxides, and the different degree of recrystallization to microquartz (Figure 5C,D). In some of the thickest veins, opal shows millimeter-scale cracks and the silicic clasts develop both quartz rims and corrosion pits (Figure 5D,E). Lussatite has a fibrous texture, arranged in pseudospherulites, and usually forms single 20 to 35 μm-thick bands (Figure 5C,F).

Chalcedony shows multiple precipitation episodes defined by 30 to 600 µm-thick bands constituted by fibers with radial disposition, spherulites or semispherulites (Figure 5F). It presents two varieties, chessboard chalcedony and chalcedonite. Finally, quartz precipitates in the remaining porosity as euhedral crystals (Figure 5F). Quartz crystals normally range between 20–70 µm in size but in some moldic porosity of the Castell cores (Figure 2B), they can reach 2 mm. This silica precipitation sequence locally alternates with precipitation of goethite (Figure 5G,H).

Breccias and internal sediment, cemented by the mineralogical sequence described above, are also present filling the fractures in the silicified area. Breccias are made of angular to subrounded clasts up to 3 cm in size. Sediments show a brownish color and were deposited as fine laminae at the base of fractures and on top of breccia clasts (Figure 6A,B). Geopetal structures are usually present (Figure 6C). These sediments comprise opal, smectite, iron oxides, barite, alunite and K-feldspar, and variable contents of host rock grains (0–60%).

In La Foixarda, Grec and Sot del Migdia outcrops, silica crusts, and stalactites are observed coating the fault planes (Figures 2B and 6D,G). Moreover, in the Grec outcrop flowstone speleothems are also present (Figure 6D). At least three generations of crusts are present and consist of 1 mm thick wavy layers of opal and chalcedony with different amounts of iron oxides inclusions (Figure 6E). Most stalactites are small (up to 7 mm long) and present several morphologies. Some of them are cylindrical, up to 2 mm in diameter, and formed by several concentric layers (Figure 6F). They do not show a drip-water channel through their length. Other stalactites display discontinuous and thin wavy "curtains" (Figure 6G), whilst some fracture planes are coated with mammillary microstalactites. The color of stalactites varies between grey, yellow and orange, probably due to the influence of small amounts of iron compounds.

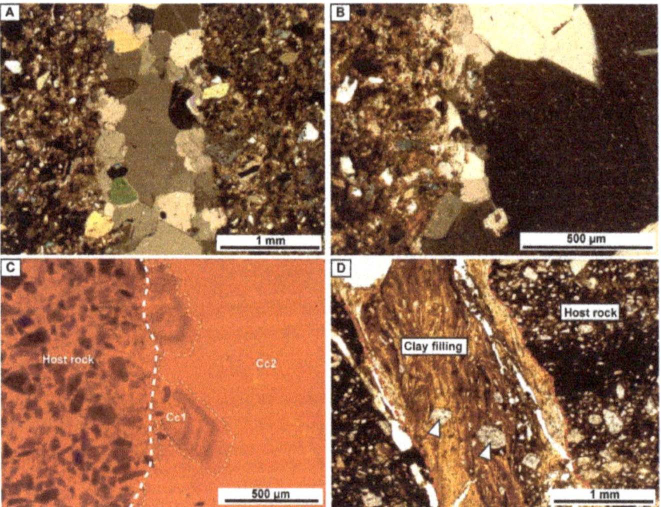

Figure 4. Fracture fillings from the non-silicified area and silicification front. (**A**) Fracture from the non-silicified area cemented by a drusy mosaic of equant calcite crystals (XPL). (**B**,**C**) Cross-polarized and cathodoluminescence images of a fracture from the non-silicified area filled with two generations of calcite cement. The fracture wall is marked with a white thick dashed-line. Note the gradual transition from the bright orange-brown zoned luminescence (Cc1) to the uniform bright orange luminescence (Cc2). The white thin dashed-line shows the limit between the two calcite generations. (**D**) Fracture from the silicification front, affecting reddish fine-grained sandstones due to the presence of iron oxide cement, filled with brownish clays and grains from the host rock (white arrows) (PPL). The limit between the host rock and the clayey filling has been marked with a red dashed-line.

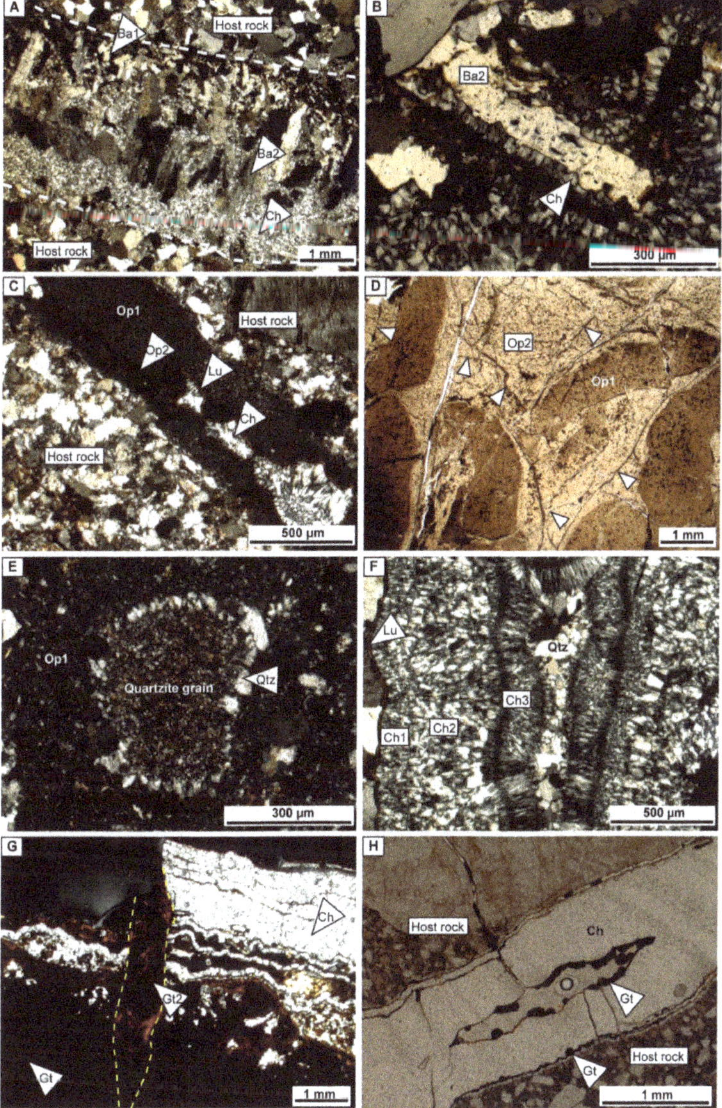

Figure 5. Cements in the fractures in the silicified area of the Castell Unit. (**A**) Fracture cemented by two generations of barite (Ba1 and Ba2) and a later generation of chalcedony (Ch) (XPL). Fracture walls are marked with a white dashed-line. (**B**) Detail of the replacement of barite (Ba2) by chalcedony (Ch) (XPL). (**C**) Silica varieties cementing a fracture according to their purity and crystallinity: two opal generations (Op1 and Op2), lussatite (Lu) and chalcedony (Ch) (XPL). (**D**) Fracture filled by two generations of opal affected by millimeter-scale cracks (white arrows). (**E**) Quartzite clast contained in a fracture cemented by opal that has developed a quartz rim (Qtz). (**F**) Fracture cemented by the silica sequence, lussatite, three generations of chalcedony and quartz (XPL). (**G**) Fracture filled by goethite (Gt) (and minor barite), alternation of goethite and chalcedony and finally pure chalcedony. The sequence is later cut by a fracture (yellow dashed- line) cemented by iron oxides (Gt2) (XPL). (**H**) Fracture cemented by several episodes of chalcedony and later goethite botryoids (PPL).

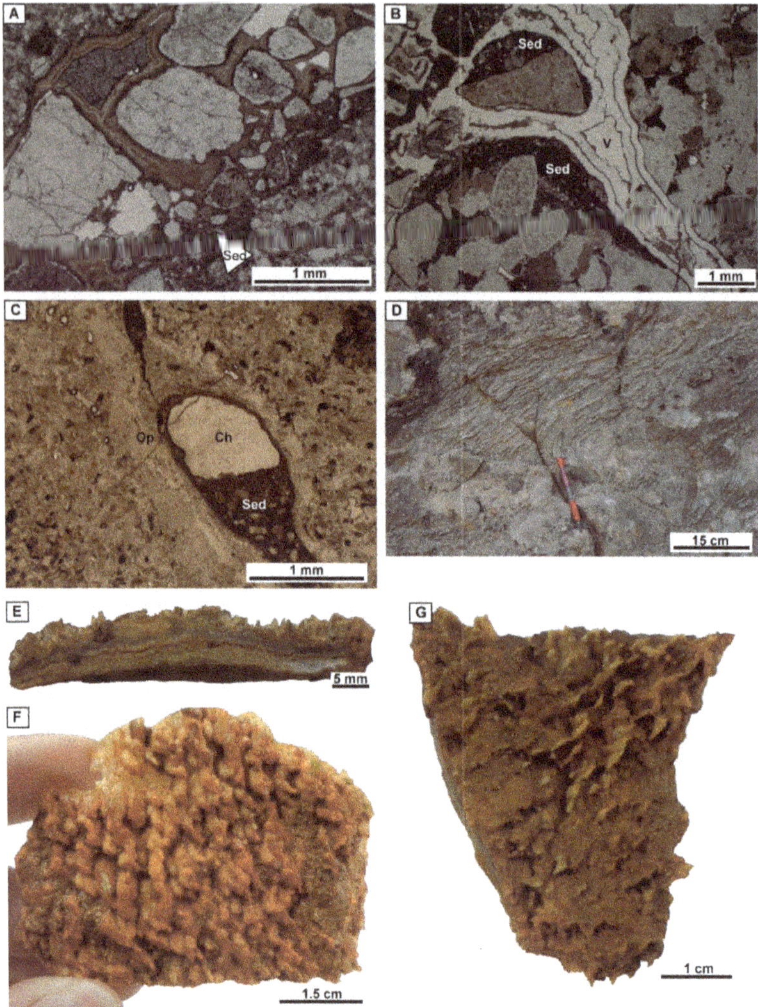

Figure 6. (**A**) Fracture filled with sediment (Sed) (deposited at the fracture bottom) and a microbreccia cemented by opal and lussatite (PPL). (**B**) Fracture filled with sediment and a microbreccia cemented by chalcedony. Some of the clasts contain earlier sediment fragments. V: void (PPL). (**C**) Geopetal structure filled with sediment at the base and chalcedony cement (Ch) at the top developed after opal cementation (Op) (PPL). (**D**) Cascade speleothems on a fault plane of the Grec outcrop. (**E**) Hand sample of several crust generations that pre-date stalactite formation. (**F**) Hand sample of cylindrical microstalactites. (**G**) Hand sample of small silica curtains.

4.3. Geochemistry

4.3.1. Elemental and Isotopic Composition of Barite

EPMA analyses revealed a large variation in barite composition, with Ba contents varying between 58.2% and 60.24%, Sr contents between 800 ppm and 1.36%, Ca contents between 100 and 700 ppm and Pb contents from below the detection limit to 700 ppm. Strontium and barium show a good negative correlation ($R^2 = 0.832$, $n = 16$), which reflects the isomorphous substitution of these elements in the barite-celestite solid solution (Figure 7A). Barites from the second generation that were replaced by silica show lower Sr contents than those that were not replaced.

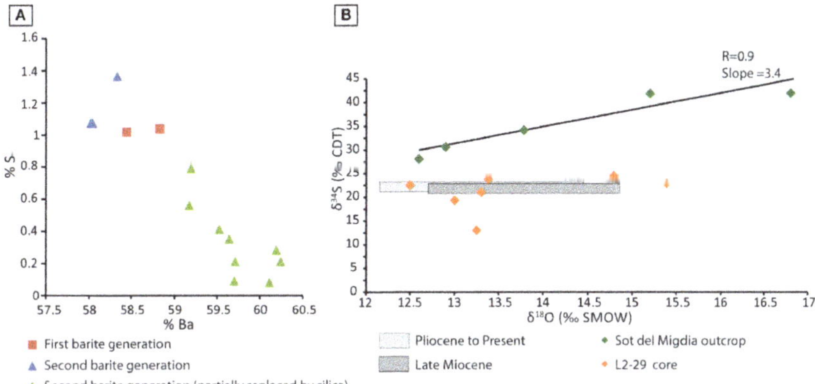

Figure 7. (**A**) Barite compositions plotted in a %Sr vs %Ba binary diagram. (**B**) δ^{34}S vs δ^{18}O binary diagram. Grey boxes represent the isotopic composition of seawater for the Late Miocene and from Pliocene to present according to [57].

Barite cement has δ^{34}S values ranging between +19.4‰ to +24.7‰ CDT in the core L2-29, whereas more δ^{34}S-enriched values, from +28.2‰ to +42‰ CDT, have been measured in the Sot del Migdia outcrop (Table 1). These values do not show covariation with δ^{18}O data, which ranges in both areas from +12.5‰ to +16.8‰ SMOW (Figure 7B). Barite has ^{87}Sr/^{86}Sr ratios varying between 0.712446 and 0.715218 (Table 1).

Table 1. Sulphur, oxygen, and strontium isotopes of barite cement.

Sample	δ^{34}O (‰CDT)	δ^{18}O (‰SMOW)	^{87}Sr/^{86}Sr	Location and Description
S29-12 A	23.8	13.4		L2-29 core at 35.20 m (±jarosite)
S29-12 B	22.6	12.5		L2-29 core at 35.20 m (+iron oxides)
S29-1	19.4	13	0.712446	L2-29 core at 37.85 m (+chalcedony)
S29-2	21.1	13.3		L2-29 core at 37.85 m (+chalcedony)
S29-14 B	22.7	15.4		L2-29 core at 39 m (+chalcedony)
S29-20 A	24.7	14.8		L2-29 core at 47 m. Next to fracture wall (+ch)
S29-20 B	13.1	13.2		L2-29 core at 47 m. Far from fracture wall (+ch)
MTJ-32 A	42	16.8	0.712939	Sot del Migdia. Next to fracture wall
MTJ-32 B	28.2	12.6		Sot del Migdia. Central part of the fracture
MTJ-32 C	34.2	13.8	0.715218	Sot del Migdia. Far from fracture wall
FMP-21a	30.7	12.9		Sot del Migdia
FMP-21b	41.9	15.2		Sot del Migdia

4.3.2. Elemental and Isotopic Composition of Calcite Cements

Calcite cement Cc1, characterized by zoned cathodoluminescence, has Mg contents from below the detection limit to 2584 ppm, Mn and Fe from below the detection limit up to 7079 and 2028 ppm, respectively, and low Sr content (from below the detection limit to 665 ppm) (Table 2). The elemental composition of this calcite cement in fractures and intergranular porosity prior to silicification is very similar, although the latter reaches higher Fe contents (up to 24,189 ppm). Calcite cement Cc1 is characterized by δ^{18}O values between −10.5‰ and −6.7‰VPDB and δ^{13}C values between −5.8‰ and −1.7‰VPDB (Figure 8). Calcite Cc2, which cements fractures and molds, has higher Mn contents, from 1300 to 14,272 ppm, than cement Cc1. It has low Mg contents (average below the detection limit), Fe from below the detection limit to 3100 ppm, and very low Sr contents. Na is below the detection limit in both cement generations. Cement Cc2 has δ^{18}O values between −18.2‰ and −12.1‰VPDB and δ^{13}C values between −5.5‰ and −3.5‰VPDB (Figure 8).

Table 2. Elemental composition of calcite cements (<d.l.: below detection limit).

Cement Generation	Statistics	ppm Mg	ppm Ca	ppm Mn	ppm Fe	ppm Sr	ppm Na
Cc1 Fractures *n* = 23	min	<d.l.	382,700	<d.l.	<d.l.	<d.l.	<d.l.
	max	2584	403,564	7079	2028	665	<d.l.
	average	712	392,395	3452	645	285	<d.l.
Cc2 Fractures *n* = 12	min	<d.l.	383,978	1300	<d.l.	<d.l.	<d.l.
	max	2500	392,893	14,272	3100	400	<d.l.
	average	<d.l.	388,025	8288	669	228	<d.l.
Cc1 Intergranular cement *n* = 17	min	<d.l.	377,13[illegible]	1[illegible]	<d.l.	<d.l.	<d.l.
	max	1269	398,220	5885	24,189	1060	<d.l.
	average	611	390,352	3124	4263	351	<d.l.

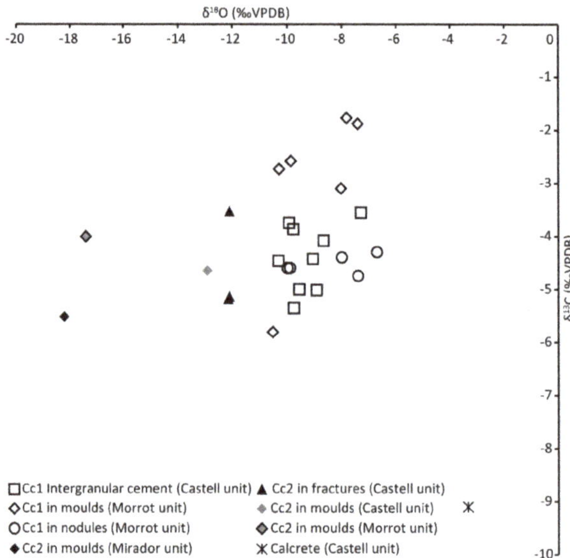

Figure 8. $\delta^{18}O$ vs $\delta^{13}C$ plot of calcite cements in fractures, molds and intergranular positions in the different stratigraphic units of the Montjuïc Hill.

4.3.3. Isotopic Composition of Silica Cements

Oxygen isotope measurements for chalcedony and quartz cements and speleothems yield similar values regardless of mineralogy. $\delta^{18}O$ values range from +9.2‰ to +25.2‰ SMOW (Table 3). The heaviest values are from speleothems of La Foixarda and El Far outcrops, which show a progressive enrichment in $\delta^{18}O$ (up to 11 units) towards the outer layers.

Table 3. Oxygen isotope values of silica cements and speleothems.

Sample	$\delta^{18}O$ (‰SMOW)	Location and Description
MTJ-31-1	16.4	Ferrocarrils. Chalcedony botryoids in fracture
MTJ-37-1	12.4	La Foixarda. Chalcedony stalactite nucleus
MTJ-37-2	23.4	La Foixarda. Chalcedony stalactite cover
MTJ-38-1	25.2	La Foixarda. Outer stalactitic layer
MTJ-38-2	22.3	La Foixarda. Inner stalactitic layer
MTJ-3A	18.8	El Far. Chessboard and fibrous chalcedony crosscut by iron oxides
S29-14A	9.2	L2-29 core. Chalcedony associated with barite in fracture
S29-17A	10.3	L2-29 core. Radial fibrous chalcedony
SR9B-03A	10	Castell core. Quartz in a vug

5. Discussion

The petrological study presented here has allowed us to establish the cementation sequence within the fractures of the Montjuïc Hill. In the silicified areas, the assemblage is barite followed by silica ± iron oxides, whereas in the non-silicified areas, cementation consists solely of calcite precipitation.

5.1. Origin of Calcite Cements

Calcite cements within the host rock and fractures record the diagenetic history previous to silicification.

Calcite cement Cc1 fills intergranular porosity, usually as nodules, moldic porosity, and fractures. The exclusive presence of this cement in the Morrot and Castell units probably indicates that its precipitation was prior to the sedimentation of the upper units and thus, formed relatively early. This is consistent with the formation of the calcite nodules, a feature most frequently associated with early diagenesis [2,58]. The oxygen isotope signature of this calcite cement, which ranges between −10.5‰ and −6.7‰ VPDB, is consistent with precipitation from meteoric water [59,60]. The zoned CL pattern may reflect variations in the oxidation state produced by oscillations of the water table level [61,62]. Several authors have proposed that carbonate shells are the main source of calcite cement in the concretions [63,64]. In this study, this observation is supported by the presence of shell molds and $\delta^{13}C$ values, which are consistent with a mixture of light carbon from soil-derived CO_2 [65] and heavier carbon derived from Miocene marine shells [66].

Calcite cement Cc2 has been identified in the Morrot, Castell, and Mirador units and yields more $\delta^{18}O$-depleted values, thus reflecting a variation in precipitation temperature or in water composition with respect to Cc1. If we consider that this oxygen depletion is due to burial, assuming a Miocene meteoric water composition of −7‰ SMOW [31] and applying the equation of [67], the precipitation temperature of this cement would be between 40 and 78 °C. With a regional geothermal gradient of 30 °C/km [68], these temperatures would imply a burial depth between 800 and 2100 m, which does not match the regional geology because late Miocene deposits do not exceed 500 m in thickness [43,69] and Plio-Pleistocene sediments were never deposited on top of the Montjuïc Hill [52]. The mechanical compaction of only the ductile clasts and the lack of chemical compaction [39–42] also support the hypothesis of the shallow burial of these rocks. Precipitation of calcite Cc2 from brine or from mixed meteoric and marine waters, both characterized by more $\delta^{18}O$-enriched values than meteoric water, would imply higher precipitation temperatures [67]. This leaves two options to explain the low $\delta^{18}O$ values at such shallow burial depths. Firstly, cement Cc2 precipitated from hydrothermal fluids, which have already been reported in several locations along the Catalan Coastal Ranges related to the Neogene extension [37,70,71], or secondly, the geothermal gradient may have been locally higher due to an increase of the lithospheric thinning towards the offshore, a fact that has been also suggested before [43,72,73].

5.2. Origin of Barite Formation in Fractures

Barite is only found in the fractures of the uppermost 40 m below the present-day surface, indicating their precipitation from at least one superficial fluid. Barites from the L2-29 core have $\delta^{18}O$ and $\delta^{34}S$ values consistent with the isotopic composition of seawater from the Late Miocene to present (from +12.2‰ to +14.9‰ SMOW and from +20.8‰ to +23.1‰ VCDT, according to [57]). However, barites from the Sot del Migdia outcrop show heavier $\delta^{34}S$ values than Late Miocene to present seawater. $\delta^{34}S$-enriched values indicate that the sulfate anion resulted from the microbial sulfate-reduction of seawater [74]. Such processes induced by microbial activity normally take place near the sediment-water interface but bioturbation, or fractures in our case, generate pathways for downward sulfate penetration [75–77]. In fact, the heaviest values have been obtained from outcrop rather than borehole samples (Table 1). Sulfate-reduction modifies both the sulfur and the oxygen signature, which could account for the oxygen values lying outside the range for seawater [78,79].

According to [79], the $\delta^{34}S/\delta^{18}O$ ratio can vary at different sites between 3.5 and 1.4 due to kinetic isotope effects during sulfate reduction. In Figure 7B, barite samples from Sot del Migdia outcrop show a strong covariant trend ($R = 0.9$) with a slope of 3.4. The presence of jarosite in close association with barite and pyrite may indicate local H_2S reoxidation, which would lower the pH in order to precipitate this mineral. However, seawater could not be the superficial fluid controlling barite precipitation close to the surface. After the marine-transitional deposition of the Montjuïc sediments during the Langhian, host rock diagenesis occurred under meteoric conditions [42]. The ensuing marine transgression, recorded in the marine sediments within the Barcelona Plain, occurred during the Pliocene and did not affect these rocks as the uplift of the mountain commenced at this time [52]. Thus, as barite precipitation is limited to the fractures, seawater seemingly had to flow up through the faults. As observed by [80], seawater is not enough to precipitate barite and an external Ba-rich, sulfate-depleted fluid is also required. This fluid had to control the precipitation of barite close to the surface. Taking into account the uplift of the block and the proposed silicification model of the host rocks by means of meteoric groundwaters moving from the horst to the basin [39–42], such meteoric water is the most probable external fluid. The high $^{87}Sr/^{86}Sr$ ratios of barite (0.712446–0.715218), which are more radiogenic than those obtained from bivalves of the Morrot unit (0.708965) and seawater (0.709000, according to [66]), supports the meteoric character of these fluids [81]. K-feldspars and K-micas are the most common source of barium in sedimentary basins [80] and also of radiogenic strontium [82], and are important components within the basement granodiorites and phyllites and the Miocene siliciclastics filling the basin.

5.3. Origin and Age of Silicification in Fractures

The change from sulfate to silica precipitation indicates a change in water chemistry as the concentration of silica is low in sulfate-supersaturated waters [83,84]. Thus, considering the model of host rock silicification and barite precipitation, either the entrance of silica-rich meteoric groundwaters had to increase or the upflow through fractures of sulfate-rich waters had to stop in order to decrease silica solubility and start the process of fracture silicification [85].

The presence of stalactites demonstrates that these silica-rich fluids percolated downwards through the vadose regime. In addition, illuviation structures, as capping structures on top of breccia clasts formed by laminated and sorted sediment, and geopetal structures are indicative of water percolation near the surface. The presence of several generations of stalactites, illuviation structures together with the silica sequence, from opal to quartz, and corrosion gulfs indicates successive episodes of precipitation, dissolution, and redeposition of silica [86–88]. The cracks in opal cements can be interpreted as shrinkage cracks formed during such episodes. Both shrinkage cracks and quartz rims between host rock grains and the opaline matrix reveal initial precipitation of opal as a hydrated silica gel [85,87,89]. All these features and processes are very similar to those described in pedogenic silcretes [85,89–92], thus suggesting a shallow and low-temperature precipitation environment. The oxygen isotopic values of the outer layers of stalactites, which range between +22.3‰ and +25.2‰ VSMOW, are in agreement with precipitation from low-temperature meteoric waters. Nevertheless, more depleted values as those registered in the nucleus, the first layers of stalactites and minor fractures in boreholes, which all vary between +9.2‰ and +16.4‰ VSMOW, are not consistent with such an origin. One possible explanation is that contamination by other mineral phases could have occurred in samples collected close to the host rock or in very thin fractures, resulting in a different final isotopic signature. These can be silicate grains from the host rock, internal sediment deposited at the base of fractures or barite crystals, which grow close to the fracture wall and are disseminated between quartz and chalcedony crystals. Another option is that prior to meteoric percolation, an episode of hydrothermal fluids ascending through faults took place. However, this origin is difficult to reconcile with geotropic indications of a vadose environment. A third option is that the more depleted $\delta^{18}O$ values were linked to the depletion of meteoric waters during glacial periods [93] with the concomitant increase of the relative elevation and distance of the Montjuïc Hill to the coast.

The alternation between silica and iron oxides (Figure 5G) may reflect small fluctuations from reducing to oxidizing conditions, which favored the precipitation of iron derived from the weathering of iron-rich minerals [94]. Such physico-chemical fluctuations were probably related to oscillations of the water table level produced by episodic rainfall/evaporation or seasonality (wet and dry periods) [87,95]. These climatic variations have in turn been proposed to explain the dissolution-precipitation cycles [85,87]. Thus, the model of fracture silicification would be the same as that proposed for silcretes by [87]: during dry periods, silica dissolution occurs due to alkaline conditions, whereas from the onset of the wet season, the entry of the acid rainwater lows the pH producing silica precipitation. Moreover, the precipitation of alunite, in the fractures of the silicification front and in capping sediments of the silicified area, required acidic conditions. The coexistence of opal, alunite, and kaolinite in the sediments and in the silicification front points to a pH between 3.5 and 5.5 [96]. Such a low pH would have been caused by the oxidation of iron sulfides during the onset of the wet season, resulting in the formation of H_2SO_4 that reacted with kaolinite, smectite, and illite to form alunite [87,97]. Iron sulfide oxidation is inferred from the presence of pyrite pseudomorphs transformed to iron oxides within the host sandstones.

Silica speleothems in the Montjuïc area formed in the meteoric vadose environment from the Pliocene, when uplift of the hill commenced, to recent (Quaternary) times, whereas present-day speleothems in Montjuïc are composed of calcite. It was suggested by [25,26] that, in some cases, low-temperature silica cements formed during cold periods due to the decrease of silica solubility at low temperatures. Particularly, [98] report the case of Fontainebleau sandstone diagenesis where silica cements precipitated during glacial periods. Alternate precipitation of quartz and low-crystallinity silica varieties depending on the temperature gradients of meteoric waters was found by [95]. Although silicification driven by glacial climates has been primarily linked to groundwater silicifications [26,98], a decrease of silica solubility in vadose meteoric waters during cold periods may have contributed to silica precipitation in fractures within silicified sandstones in the Montjuïc Hill. This linkage with glacial periods would support the third option proposed for the interpretation of $\delta^{18}O$ signatures and can explain why silica speleothems do not form nowadays. In fact, colder periods have been identified in post-Tortonian calcite speleothems in the Barcelona Plain [99] and in Pleistocene calcite speleothems in Mallorca [100] and southern Spain [101].

5.4. Karst Evidence: Silica Dissolution

Silica dissolution occurs at the micro-scale, as vug porosity (Figure 9A), and at the macroscale, as irregular caverns, both within the silicified host rock. The caverns, up to 60 cm in size, are formed at the intersection between discontinuity surfaces such as bedding planes and fractures (Figure 9B), confirming that such discontinuities act as important pathways for fluids causing preferential dissolution [102,103]. Thus, most silica precipitating within fractures is provided by the surrounding sandstones and conglomerates. At the micro-scale, dissolution preferentially affects the sandstone matrix constituted by opal and clays, feldspar and mica grains, and syntaxial feldspar overgrowths (Figure 9A,C). At fracture walls, dissolution normally starts along crystal borders so that sand grains are finally removed and included in the fracture by flowing water (Figure 9C,D). This process of dissolution, erosion, and winnowing of loose material by flowing waters has been described as "arenisation" [102,104–108]. In the Montjuïc Hill, arenisation is clearly triggered by dissolution, which is the dominant process, and therefore vugs and caverns are real karst forms and not pseudokarst features, which would imply a minor role of dissolution in their formation [102,109].

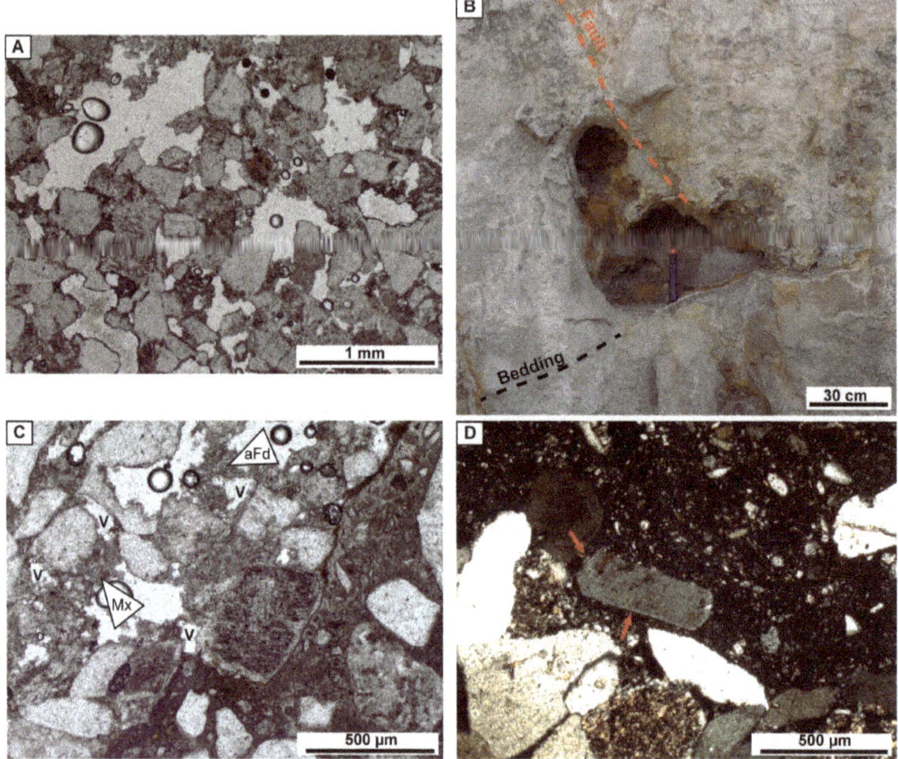

Figure 9. Evidence of in situ dissolution and arenisation. (**A**) Vuggy porosity. (**B**) Cavern porosity developed at the intersection between faulting and bedding. (**C**) Vuggy porosity (V) produced by the matrix (Mx) and altered feldspar (aFd) dissolutions in a sandstone sample. Note how the feldspar grain in the center of the image is about to be included in the fracture by dissolution around its borders. (**D**) Feldspar grain removed from the host rock and included in the fracture infilling by a process of dissolution around the grain boundaries (red arrows) and erosion (arenisation).

6. Conclusions

The fractures affecting the deltaic sequence of the Montjuïc Hill, formed after lithification and during the Montjuïc block uplift, show a complex diagenetic evolution controlled by the previous diagenesis of their host rocks.

Non-silicified areas are characterized by calcite cementation. Calcite cement Cc1 formed very early, prior to the sedimentation of the Miramar and Mirador units, and precipitated in molds, intergranular porosity (calcite nodules), and fractures from meteoric waters. Calcite cement Cc2, found in fractures and some molds, shows more δ^{18}O-depleted values, which are interpreted as a result of a temperature increase linked to a higher geothermal gradient during the Neogene extension, rather than to overburden.

Fractures within the silicification front are filled by clays (sericite, kaolinite, goethite and minor smectite, and alunite) with isolated grains of quartz and feldspar from the host rock.

In silicified areas, fractures show multiepisodic cementation produced by barite and silica. Barite is only found in fractures along the uppermost 40 m below the present-day surface and precipitated from the mixing of upflowing marine fluids, as suggested by δ^{18}O and δ^{34}S values, and percolating meteoric waters, which supplied barium and more radiogenic ^{87}Sr/^{86}Sr ratios. δ^{34}S values heavier than those of Late Miocene seawater in a covariant trend with $\delta^{18}O$ indicate that microbial sulfate reduction occurred in outcrop samples. Silica cements follow the sequence opal, lussatite, chalcedony, and quartz, all precipitated in the vadose zone from low-temperature percolating meteoric fluids as indicated by the presence of silica stalactites, illuviation, and geopetal structures and $\delta^{18}O$ values between +22.3‰ and +25.2‰VSMOW. Silica precipitation occurred in acidic conditions and under fluctuating redox conditions. A model of fracture silicification, similar to that of silcretes, and based on the alternation of dry and wet periods is proposed here. However, a decrease of silica solubility due to cold periods may also have contributed to silica precipitation, as indicated by $\delta^{18}O$ values between +9.2‰ and +16.4 ‰VSMOW and the present-day development of calcite speleothems and calcretes.

Dissolution occurs as vug porosity and caverns, which are generated by an arenisation process. Thus, silica precipitating within fractures was from the surrounding sandstones and conglomerates.

Author Contributions: Conceptualization, D.P. and D.G.-G; methodology, all authors; formal analysis, all authors; investigation, all authors; data curation, I.C., D.P., M.A.P.; writing—original draft preparation, I.C.; writing—review and editing, D.P., A.T., D.G.-G., E.G.-R., J.D.M.-M.; visualization, I.C.; supervision, D.P. and A.T.; funding acquisition, A.T. All authors have read and agreed to the published version of the manuscript.

Funding: This research was performed within the framework of DGICYT Spanish Projects CGL2015-66335-C2-1-R, PGC-2018-093903-B-C22, and Grup Consolidat de Recerca "Geologia Sedimentària" (2017-SGR-824).

Acknowledgments: We thank three anonymous referees for their constructive reviews, which helped to improve the quality of the manuscript. Electron microprobe, $\delta^{18}O$, $\delta^{13}C$, $\delta^{34}S$ analyses were performed at "Centres Científics i Tecnològics" of the Universitat de Barcelona. The oxygen isotopic analyses in silica were carried out at the Universidad de Salamanca and the strontium analyses at the Geochronology and Isotopic Geochemistry service of the Universidad Complutense de Madrid.

Conflicts of Interest: The authors declare no conflict of interest.

References

1. Karim, A.; Pe-Piper, G.; Piper, D.J.W. Controls on diagenesis of Lower Cretaceous reservoir sandstones in the western Sable Subbasin, offshore Nova Scotia. *Sediment. Geol.* **2010**, *224*, 65–83. [CrossRef]
2. Morad, S.; Ketzer, J.M.; DeRos, F. Spatial and temporal distribution of diagenetic alterations in siliciclastic rocks: implication for mass transfer in sedimentary basins. *Sedimentology* **2000**, *47*, 95–120. [CrossRef]
3. Ketzer, J.M.; Holz, M.; Morad, S.; Al-Aasm, I.S. Sequence stratigraphic distribution of diagenetic alterations in coal-bearing, paralic sandstones: Evidence from the Rio Bonito Formation (early Permian), southern Brazil. *Sedimentology* **2003**, *50*, 855–877. [CrossRef]
4. El-ghali, M.A.K.; Mansurbeg, H.; Morad, S.; Al-Aasm, I.; Ajdanlisky, G. Distribution of diagenetic alterations in fluvial and paralic deposits within sequence stratigraphic framework: Evidence from the Petrohan Terrigenous Group and the Svidol Formation, Lower Triassic, NW Bulgaria. *Sediment. Geol.* **2006**, *190*, 299–321. [CrossRef]
5. Worden, R.H.; Burley, S.D. *Sandstone Diagenesis: The Evolution of Sand to Stone*; Blackwell Publishing Ltd.: Hoboken, NJ, USA, 2009; ISBN 9781444304459.
6. Salem, A.M.; Ketzer, J.M.; Morad, S.; Rizk, R.R.; Al-Aasm, I.S. Diagenesis and Reservoir-Quality Evolution of Incised-Valley Sandstones: Evidence from the Abu Madi Gas Reservoirs (Upper Miocene), the Nile Delta Basin, Egypt. *J. Sediment. Res.* **2005**, *75*, 572–584. [CrossRef]
7. Mansurbeg, H.; Morad, S.; Salem, A.; Marfil, R.; El-ghali, M.A.K.; Nystuen, J.P.; Caja, M.A.; Amorosi, A.; Garcia, D.; La Iglesia, A. Diagenesis and reservoir quality evolution of palaeocene deep-water, marine sandstones, the Shetland-Faroes Basin, British continental shelf. *Mar. Pet. Geol.* **2008**, *25*, 514–543. [CrossRef]
8. Luo, J.L.; Morad, S.; Salem, A.; Ketzer, J.M.; Lei, X.L.; Guo, D.Y.; Hlal, O. Impact of diagenesis on reservoir-quality evolution in fluvial and lacustrine-deltaic sandstones: Evidence from jurassic and triassic sandstones from the Ordos Basin, China. *J. Pet. Geol.* **2009**, *32*, 79–102. [CrossRef]
9. Morad, S.; Al-Ramadan, K.; Ketzer, J.M.; De Ros, L.F. The impact of diagenesis on the heterogeneity of sandstone reservoirs: A review of the role of depositional fades and sequence stratigraphy. *Am. Assoc. Pet. Geol. Bull.* **2010**, *94*, 1267–1309. [CrossRef]

10. Khalifa, M.; Morad, S. Impact of structural setting on diagenesis of fluvial and tidal sandstones: The Bahi Formation, Upper Cretaceous, NW Sirt Basin, North Central Libya. *Mar. Pet. Geol.* **2012**, *38*, 211–231. [CrossRef]
11. Sun, N.; Zhong, J.; Hao, B.; Ge, Y.; Swennen, R. Sedimentological and diagenetic control on the reservoir quality of deep-lacustrine sedimentary gravity flow sand reservoirs of the Upper Triassic Yanchang Formation in Southern Ordos Basin, China. *Mar. Pet. Geol.* **2020**, *112*, 104050. [CrossRef]
12. Chan, M.A.; Parry, W.T.; Bowman, J.R. Diagenetic Hematite and Manganese Oxides and Fault-Related Fluid Flow in Jurass ic Sandstones, Southeastern Utah. *AAPG Bull.* **2000**, *84*, 1281–1310.
13. Ceriani, A.; Di Giulio, A.; Goldstein, R.H.; Rossi, C. Diagenesis associated with cooling during burial: An example from Lower Cretaceous reservoir sandstones (Sirt basin, Libya). *Am. Assoc. Pet. Geol. Bull.* **2002**, *86*, 1573–1591.
14. Andre, G.; Hibsch, C.; Fourcade, S.; Cathelineau, M.; Buschaert, S. Chronology of fracture sealing under a meteoric fluid environment: Microtectonic and isotopic evidence of major Cainozoic events in the eastern Paris Basin (France). *Tectonophysics* **2010**, *490*, 214–228. [CrossRef]
15. Cantarero, I.; Zafra, C.J.; Travé, A.; Martín-martín, J.D.; Baqués, V.; Playà, E. Fracturing and cementation of shallow buried Miocene proximal alluvial fan deposits. *Mar. Pet. Geol.* **2014**, *55*, 87–99. [CrossRef]
16. Cruset, D.; Cantarero, I.; Travé, A.; Vergés, J.; John, C.M. Crestal graben fluid evolution during growth of the Puig-reig anticline (South Pyrenean fold and thrust belt). *J. Geodyn.* **2016**, *101*, 30–50. [CrossRef]
17. Cruset, D.; Cantarero, I.; Vergés, J.; John, C.M.; Muñoz-López, D.; Travé, A. Changes in fluid regime in syn-orogenic sediments during the growth of the south Pyrenean fold and thrust belt. *Glob. Planet. Chang.* **2018**, *171*, 207–224. [CrossRef]
18. Yuan, G.; Cao, Y.; Gluyas, J.; Cao, X.; Zhang, W. Petrography, fluid-inclusion, isotope, and trace-element constraints on the origin of quartz cementation and feldspar dissolution and the associated fluid evolution in arkosic sandstones. *Am. Assoc. Pet. Geol. Bull.* **2018**, *102*, 761–792. [CrossRef]
19. Ulmer-Scholle, D.S.; Scholle, P.A.; Schieber, J.; Raine, R.J. *Memoir 109: A Color Guide to the Petrography of Sandstones, Siltstones, Shales and Associated Rocks*; Sweet, M.L., Ed.; The American Association of Petroleum Geologists: Tulsa, OK, USA, 2014; ISBN 978-0-89181-389-7.
20. Daza Brunet, R.; Bustillo Revuelta, M.Á. Exceptional silica speleothems in a volcanic cave: A unique example of silicification and sub-aquatic opaline stromatolite formation (Terceira, Azores). *Sedimentology* **2014**, *61*, 2113–2135. [CrossRef]
21. Khalifa, M.A.; Mansurbeg, H.; Morad, D.; Morad, S.; Al-Aasm, I.S.; Spirov, P.; Ceriani, A.; De Ros, L.F. Quartz and Fe-dolomite cements record shifts in formation-water chemistry and hydrocarbon migration in Devonian shoreface sandstones, Ghadamis Basin, Libya. *J. Sediment. Res.* **2018**, *88*, 38–57. [CrossRef]
22. Menezes, C.P.; Bezerra, F.H.R.; Balsamo, F.; Mozafari, M.; Vieira, M.M.; Srivastava, N.K.; de Castro, D.L. Hydrothermal silicification along faults affecting carbonate-sandstone units and its impact on reservoir quality, Potiguar Basin, Brazil. *Mar. Pet. Geol.* **2019**, *110*, 198–217. [CrossRef]
23. Hervig, R.L.; Williams, L.B.; Kirkland, I.K.; Longstaffe, F.J. Oxygen isotope microanalyses of diagenetic quartz: possible low temperature occlusion of pores. *Geochim. Cosmochim. Acta* **1995**, *59*, 2537–2543. [CrossRef]
24. Ullyott, J.S.; Nash, D.J. Micromorphology and geochemistry of groundwater silcretes in the eastern South Downs, UK. *Sedimentology* **2006**, *53*, 387–412. [CrossRef]
25. Thiry, M.; Milnes, A.; Brahim, M. Ben Pleistocene cold climate groundwater silicification, Jbel Ghassoul region, Missour Basin, Morocco. *J. Geol. Soc. London.* **2015**, *172*, 125–137. [CrossRef]
26. Thiry, M.; Milnes, A. Silcretes: Insights into the occurrences and formation of materials sourced for stone tool making. *J. Archaeol. Sci. Reports* **2017**, *15*, 500–513. [CrossRef]
27. Tritlla, J.; Esteban, M.; Loma, R.; Mattos, A.; Sanders, C.; Vieira De Luca, P.H.; Carrasco, A.; Gerona, M.; Herra, A.; Carballo, J. Where have most of the carbonates gone? Silicified Aptian pre-salt microbial (?) carbonates in South Atlantic basins (Brazil and Angola). In Proceedings of the Bathurst Meeting Mallorca 2019 16th International Meeting of Carbonate Sedimentologists, Palma de Mallorca, Spain, 9–11 July 2019.
28. Saller, A. Presalt lake evolution in the south Atlantic: the significance of microbial chert. In Proceedings of the Bathurst Meeting Mallorca 2019 16th International Meeting of Carbonate Sedimentologists Presalt, Palma de Mallorca, Spain, 9–11 July 2019.
29. Travé, A.; Calvet, F. Syn-rift geofluids in fractures related to the early-middle Miocene evolution of the Vallès-Penedès half-graben (NE Spain). *Tectonophysics* **2001**, *336*, 101–120. [CrossRef]

30. Travé, A.; Calvet, F.; Soler, A.; Labaume, P. Fracturing and fluid migration during Palaeogene compression and Neogene extension in the Catalan Coastal Ranges, Spain. *Sedimentology* **1998**, *45*, 1063–1082. [CrossRef]

31. Travé, A.; Roca, E.; Playà, E.; Parcerisa, D.; Gómez-Gras, D.; Martín-Martín, J.D. Migration of Mn-rich fluids through normal faults and fine-grained terrigenous sediments during early development of the Neogene Vallès-Penedès half-graben (NE Spain). *Geofluids* **2009**, *9*, 303–320. [CrossRef]

32. Parcerisa, D.; Gómez-Gras, D.; Travé, A. A model of early calcite cementation in alluvial fans: Evidence from the Burdigalian sandstones and limestones of the Vallès-Penedès half-graben (NE Spain). *Sediment. Geol.* **2005**, *178*, 197–217. [CrossRef]

33. Parcerisa, D.; Gómez-Gras, D.; Martín-Martín, J.D. Calcretes, oncolites, and lacustrine limestones in Upper Oligocene alluvial fans of the Montgat area (Catalan Coastal Ranges, Spain). *Geol. Soc. Am. Spec. Pap.* **2006**, *416*, 105–117.

34. Baqués, V.; Travé, A.; Benedicto, A.; Labaume, P.; Cantarero, I. Relationships between carbonate fault rocks and fluid flow regime during propagation of the Neogene extensional faults of the Penedès basin (Catalan Coastal Ranges, NE Spain). *J. Geochem. Explor.* **2010**, *106*, 24–33. [CrossRef]

35. Baqués, V.; Travé, A.; Roca, E.; Marín, M.; Cantarero, I. Geofluid behaviour in successive extensional and compressional events: a case study from the southwestern end of the Vallès-Penedès Fault (Catalan Coastal Ranges, NE Spain). *Pet. Geosci.* **2012**, *18*, 17–31. [CrossRef]

36. Baqués, V.; Travé, A.; Cantarero, I. Development of successive karstic systems within the Baix Penedès Fault zone (onshore of the Valencia Trough, NW Mediterranean). *Geofluids* **2014**, *14*, 75–94. [CrossRef]

37. Cantarero, I.; Travé, A.; Alías, G.; Baqués, V. Polyphasic hydrothermal and meteoric fluid regimes during the growth of a segmented fault involving crystalline and carbonate rocks (Barcelona Plain, NE Spain). *Geofluids* **2014**, *14*, 20–44. [CrossRef]

38. Miró, J.; Martín-Martín, J.D.; Ibáñez, J.; Anadón, P.; Oms, O.; Tritlla, J.; Caja, M.A. Opaline chert nodules in maar lake sediments from Camp dels Ninots (La Selva Basin, NE Spain). In Proceedings of the Geo-Temas, Huelva, Spain, 12–14 September 2016; Volume 16, pp. 387–390.

39. Gomez-Gras, D.; Parcerisa, D.; Bitzer, K.; Calvet, F.; Roca, E.; Thiry, M. Hydrogeochemistry and diagenesis of Miocene sandstones at Montjuic, Barcelona (Spain). *J. Geochem. Explor.* **2000**, *69*, 177–182. [CrossRef]

40. Gomez-Gras, D.; Parcerisa, D.; Calvet, F.; Porta, J.; Solé de Porta, N.; Civís, J. Stratigraphy and petrology of the Miocene Montjuïc delta: (Barcelona, Spain). *Acta Geològica Hispànica* **2001**, *36*, 115–136.

41. Parcerisa, D. *Petrologia i diagènesi en sediments de l'Oligocè superior i del Miocè inferior i mitjà de la depressió del Vallès i del Pla de Barcelona*; Universitat Autònoma de Barcelona: Barcelona, Spain, 2002.

42. Parcerisa, D.; Thiry, M.; Gomez-Gras, D.; Calvet, F. Tentative model for the silicification in Neogene Montjuic sandstones, Barcelona (Spain): authigenic minerals, geochemical environment and fluid flow. *Bull. La Soc. Geol. Fr.* **2001**, *172*, 751–764. [CrossRef]

43. Roca, E.; Sans, M.; Cabrera, L.; Marzo, M. Oligocene to Middle Miocene evolution of the central Catalan margin (northwestern Mediterranean). *Tectonophysics* **1999**, *315*, 209–233. [CrossRef]

44. Roca, E.; Guimerà, J. The Neogene structure of the eastern Iberian margin—Structural constraints on the crustal evolution of the Valencia trough (Western Mediterranean). *Tectonophysics* **1992**, *203*, 203–218. [CrossRef]

45. Gaspar-Escribano, J.M.; Garcia-Castellanos, D.; Roca, E.; Cloetingh, S. Cenozoic vertical motions of the Catalan Coastal Ranges (NE Spain): The role of tectonics, isostasy, and surface transport. *Tectonics* **2004**, *23*, TC1004. [CrossRef]

46. Calvet, F.; Travé, A.; Roca, E.; Soler, A.; Labaume, P. Fracturación y migración de fluidos durante la evolución tectónica neógena en el Sector Central de las Cadenas Costero Catalanas. *Geogaceta* **1996**, *20*, 1715–1718.

47. Julivert, M.; Duran, H. Paleozoic stratigraphy of the Central and Northern part of the Catalonian Coastal Ranges (NE Spain). *Acta Geol. Hisp.* **1990**, *25*, 3–12.

48. Llopis Lladó, N. Tectomorfología del macizo del Tibidabo y valle inferior del Llobregat. *Estud. Geográficos* **1942**, *3*, 321–383.

49. Roca, J.L.; Casas, A. *Gravimetria en zona urbana. Mapa Gravimétrico de la Ciudad de Barcelona; IV Asamblea Nacional de Geodesia y Geofísica. Presidencia de Gobierno. Comisión Nacional de Geodesia y Geofísica*; Asamblea Nacional de Geodesia y Geofisica: Barcelona, Spain, 1981.

50. Blow, W.H. Late Middle Eocene to Recent planktonic foraminiferal biostratigraphy. In *Proceedings of the First International Conference on Planktonic Microfossils, Geneva, Switzerland*; Bronnimann, P., Renz, H.H., Eds.; Brill Archive: Leiden, The Netherlands, 1969; pp. 199–421.

51. BouDagher-Fadel, M.K. *Biostratigraphic and Geological Significance of Planktonic Foraminifera*; UCLPress: London, UK, 2018; ISBN 978-1-910634-24-0.

52. Riba, O.; Colombo, F. *Barcelona: la Ciutat Vella i el Poblenou. Assaig de geologia urbana*; Institut d'Estudis Catalans i Reial Acadèmia de Ciències i Arts de Barcelona: Barcelona, Spain, 2009.

53. Friedman, I.; O'Neil, J.; Cebula, G. Two New Carbonate Stable-Isotope Standards. *Geostand. Newsl.* **1982**, *6*, 11–12. [CrossRef]

54. Hut, G. *Consultants Group Meeting on Stable Isotope Reference Samples for Geochemical and Hydrological Investigations*; International Atomic Energy Agency (IAEA): Vienna, Austria, 1987; p. 42.

55. Sharp, Z.D. A laser-based microanalytical method for the in situ determination of oxygen isotope ratios of silicates and oxides. *Geochim. Cosmochim. Acta* **1990**, *54*, 1353–1357. [CrossRef]

56. Halas, S.; Szaran, J. Improved thermal decomposition of sulfates to SO_2 and mass spectrometric determination of $\delta^{34}S$ of IAEA SO-5, IAEA SO-6 and NBS-127 sulfate standards. *Rapid Commun. Mass Spectrom.* **2001**, *15*, 1618–1620. [CrossRef]

57. Claypool, G.E.; Holser, W.T.; Kaplan, I.R.; Sakai, H.; Zak, I. The age curves of sulfur and oxygen isotopes in marine sulfate and their mutual interpretation. *Chem. Geol.* **1980**, *28*, 199–260. [CrossRef]

58. McBride, E.F.; Milliken, K.L.; Cavazza, W.; Cibin, U.; Fontana, D.; Picard, M.D.; Zuffa, G.G. Heterogeneous distribution of calcite cement at the outcrop scale in Tertiary sandstones, northern Apennines, Italy. *Am. Assoc. Pet. Geol. Bull.* **1995**, *79*, 1044–1063.

59. Moore, C.H. *Carbonate Diagenesis and Porosity. Developments in Sedimentology 46*; Elsevier Science Bv.: Amsterdam, The Netherlands, 1989; ISBN 0444874151.

60. Tucker, M.E.; Wright, P.V. *Carbonate Sedimentology*; Blackwell: Oxford, UK, 1990.

61. Rossi, C.; Canaveras, J.C. Pseudospherulitic fibrous calcite in paleo-groundwater, unconformity-related diagenetic carbonates (Paleocene of the Ager Basin and Miocene of the Madrid Basin, Spain). *J. Sediment. Res.* **1999**, *69*, 224–238. [CrossRef]

62. Parcerisa, D.; Gómez-Gras, D.; Travé, A.; Martín-Martín, J.D. Fe and Mn in calcites cementing red beds: A record of oxidation–reduction conditionsExamples from the Catalan Coastal Ranges (NE Spain). *J. Geochem. Explor.* **2006**, *89*, 318–321. [CrossRef]

63. Abdel-Wahab, A.; McBride, E.F. Anonymous Origin of giant calcite-cemented concretions, Temple Member, Qasr El Sagha Formation (Eocene), Faiyum Depression, Egypt. *J. Sediment. Res.* **2001**, *71*, 70–81. [CrossRef]

64. McBride, E.F.; Parea, G.C. Origin of highly elongate, calcite-cemented concretions in some Italian coastal beach and dune sands. *J. Sediment. Res.* **2001**, *71*, 82–87. [CrossRef]

65. Cerling, T.E. The stable isotopic composition of modern soil carbonate and its relationship to climate. *Earth Planet. Sci. Lett.* **1984**, *71*, 229–240. [CrossRef]

66. Veizer, J.; Ala, D.; Azmy, K.; Bruckschen, P.; Buhl, D.; Bruhn, F.; Carden, G.A.F.; Diener, A.; Ebneth, S.; Godderis, Y.; et al. 87Sr/Sr86, d13 C and d18 O evolution of Phanerozoic seawater. *Chem. Geol.* **1999**, *161*, 59–88. [CrossRef]

67. Craig, H. The measurements of oxygen isotope paleotemperatures. In *Stable Isotopes in Oceanographic Studies and Paleotemperatures*; Consiglio Nazionale delle, Richerche; Laboratorio di Geologia Nucleare; Tongiorgi, E., Ed.; Pisa: Rome, Italy, 1965; pp. 161–182.

68. Juez-Larré, J. Post late Paleozoic Tectonothermal Evolution of the Northeastern Margin of Iberia, Assessed by Fission-Track and (U-Th)/He Analyses. A Case History from the Catalan Coastal Ranges. Ph.D. Thesis, Vrije Universiteit, Amsterdam, The Netherlands, 2003.

69. Alonso, F.; Peón, A.; Villanueva, O.; Rosell, J.; Trilla, J.; Obrador, A. *Mapa y Memoria Explicativa de la Hoja nº 421 (Barcelona) del Mapa Geológico Nacional a Escala 1:50,000*; IGME: Madrid, Spain, 1977.

70. Carmona, J.M.; Bitzer, K.; López, E.; Bouazza, M. Isotopic composition and origin of geothermal waters at Caldetes (Maresme-Barcelona). *J. Geochem. Explor.* **2000**, *69–70*, 441–447. [CrossRef]

71. Cardellach, E.; Canals, À.; Grandia, F. Recurrent hydrothermal activity induced by successive extensional episodes: The case of the Berta F-(Pb-Zn) vein system (NE Spain). *Ore Geol. Rev.* **2002**, *22*, 133–141. [CrossRef]

72. Gallart, J.; Rojas, H.; Diaz, J.; Dañobeitia, J.J. Features of deep crustal structure and the onshore offshore transition at the Iberian flank of the Valencia Trough (Western Mediterranean). *J. Geodyn.* **1990**, *12*, 233–252. [CrossRef]

73. Vidal, N.; Gallart, J.; Dañobeitia, J.J. Contribution of the ESCI-Valéncia Trough Wide-Angle Data to a Crustal Transect in the NE Iberian Margin. *Rev. la Soc. Geol. Espana* **1997**, *8*, 417–429.

74. Seal, R.R. Sulfur Isotope Geochemistry of Sulfide Minerals. *Rev. Mineral. Geochem.* **2006**, *61*, 633–677. [CrossRef]

75. Fisher, I.S. Pyrite formation in bioturbated clays from the Jurassic of Britain. *Geochim. Cosmochim. Acta* **1986**, *50*, 517–523. [CrossRef]

76. Chowdhury, A.H.; Noble, J.P.A. Organic carbon and pyrite sulphur relationships as evidences of bottom water conditions of sedimentation, Albert Formation fine-grained lacustrine sediments, New Brunswick, Canada. *Mar. Pet. Geol.* **1996**, *13*, 79–90. [CrossRef]

77. Machel, H.G. Bacterial and thermochemical sulfate reduction in diagenetic settings—Old and new insights. *Sediment. Geol.* **2001**, *140*, 143–175. [CrossRef]

78. Habicht, K.S.; Canfield, D.E. Sulphur isotope fractionation in modern microbial mats and the evolution of the sulphur cycle. *Nature* **1996**, *382*, 342–343. [CrossRef]

79. Aharon, P.; Fu, B. Microbial sulfate reduction rates and sulfur and oxygen isotope fractionations at oil and gas seeps in deepwater Gulf of Mexico. *Geochim. Cosmochim. Acta* **2000**, *64*, 233–246. [CrossRef]

80. Hanor, J.S. Barite-Celestine Geochemistry and Environments of Formation. *Rev. Mineral. Geochem.* **2000**, *40*, 193–275. [CrossRef]

81. Palmer, M.R.; Edmond, J.M. Controls over the strontium isotope composition of river water. *Geochim. Cosmochim. Acta* **1992**, *56*, 2099–2111. [CrossRef]

82. Hanor, J.S. A model for the origin of large carbonate- and evaporite-hosted celestine (SrSO4) deposits. *J. Sediment. Res.* **2004**, *74*, 168–175. [CrossRef]

83. Marshall, W.L.; Warakomski, J.M. Amorphous silica solubilities-II. Effect of aqueous salt solutions at 25°C. *Geochim. Cosmochim. Acta* **1980**, *44*, 915–924. [CrossRef]

84. Marshall, W.L.; Chen, C.-T.A. Amorphous silica solubilities V. Predictions of solubility behavior in aqueous mixed electrolyte solutions to 300 °C. *Geochim. Cosmochim. Acta* **1982**, *46*, 289–291. [CrossRef]

85. Thiry, M.; Milnes, A.R.; Rayot, V.; Simon-Coinçon, R. Interpretation of palaeoweathering features and successive silicifications in the Tertiary regolith of inland Australia. *J. Geol. Soc. London.* **2006**, *163*, 723–736. [CrossRef]

86. Thiry, M.; Milnes, A.R. Pedogenic and groundwater silcretes at Stuart Creek opal field, South Australia. *J. Sediment. Petrol.* **1991**, *61*, 111–127.

87. Blanco, J.A.; Armenteros, I.; Huerta, P. Silcrete and alunite genesis in alluvial palaeosols (late Cretaceous to early Palaeocene, Duero basin, Spain). *Sediment. Geol.* **2008**, *211*, 1–11. [CrossRef]

88. Aubrecht, R.; Lánczos, T.; Gregor, M.; Schlögl, J.; Šmída, B.; Liščák, P.; Brewer-Carías, C.; Vlček, L. Sandstone caves on Venezuelan tepuis: Return to pseudokarst? *Geomorphology* **2011**, *132*, 351–365. [CrossRef]

89. Thiry, M.; Millot, G. Mineralogical Forms of Silica and their Sequence of Formation in Silcretes. *SEPM J. Sediment. Res.* **1987**, *57*, 343–352.

90. Milnes, A.R.; Thiry, M. Silcretes. In *Weathering, Soils and Palaeosols. Developments in Earth Surface Processes 2*; Martini, I.P., Chesworth, W., Eds.; Elsevier: Amsterdam, The Netherlands, 1992; pp. 349–377.

91. Thiry, M.; Simon-Coinçon, R. Tertiary paleoweatherings and silcretes in the southern Paris Basin. *Catena* **1996**, *26*, 1–26. [CrossRef]

92. Ullyott, J.S.; Nash, D.J.; Huggett, J.M. Cap structures as diagnostic indicators of silcrete origin. *Sediment. Geol.* **2015**, *325*, 119–131. [CrossRef]

93. Bantcev, D.V.; Ganyushkin, D.A.; Chistyakov, K.V.; Volkov, I.V.; Ekaykin, A.A.; Veres, A.N.; Tokarev, I.V.; Shtykova, N.B.; Andreeva, T.A. The Components of the Glacial Runoff of the Tsambagarav Massif from Stable Water Isotope Data. *Geosciences* **2019**, *9*, 297. [CrossRef]

94. Krumbein, W.C.; Garrels, R.M. Origin and Classification of Chemical Sediments in Terms of pH and Oxidation-Reduction Potentials. *J. Geol.* **1952**, *60*, 1–33. [CrossRef]

95. Meyer, R.; Pena dos Reis, R.B. Paleosols and alunite in continental Cenozoic of western Portugal. *J. Sediment. Res.* **1985**, *55*, 76–85.

96. Raymahashay, B.C. A geochemical study of rock alteration by hot springs in the Paint Pot Hill area, Yellowstone Park. *Geochim. Cosmochim. Acta* **1968**, *32*, 499–522. [CrossRef]
97. Wray, R.A.L. Alunite formation within silica stalactites from the Sydney Region, South-Eastern Australia. *Int. J. Speleol.* **2011**, *40*, 109–116. [CrossRef]
98. Thiry, M.; Schmitt, J.; Innocent, C.; Cojan, I. Sables et Grès de Fontainebleau: Que reste-t-il des faciès sédimentaires initiaux? In Proceedings of the 14ème Congrès Français de Sédimentologie, Paris, France, 4–8 November 2013; Association des Sédimentologistes Français: Paris, France, 2013; Volume 74, pp. 37–90.
99. Cantarero, I.; Lanari, P.; Vidal, O.; Alías, G.; Travé, A.; Baqués, V. Long-term fluid circulation in extensional faults in the central Catalan Coastal Ranges: P–T constraints from neoformed chlorite and K-white mica. *Int. J. Earth Sci.* **2014**, *103*, 165–188. [CrossRef]
100. Vesica, P.L.; Tuccimei, P.; Turi, B.; Fornós, J.J.; Ginés, A.; Ginés, J. Late Pleistocene Paleoclimates and sea-level change in the Mediterranean as inferred from stable isotope and U-series studies of overgrowths on speleothems, Mallorca, Spain. *Quat. Sci. Rev.* **2000**, *19*, 865–879. [CrossRef]
101. Jiménez de Cisneros, C.; Caballero, E.; Vera, J.A.; Durán, J.J.; Juliá, R. A record of pleistocene climate from a stalactite, Nerja Cave, southern Spain. *Palaeogeogr. Palaeoclimatol. Palaeoecol.* **2003**, *189*, 1–10. [CrossRef]
102. Wray, R.A.L. Quartzite dissolution: Karst or pseudokarst? *Cave Karst Sci.* **1997**, *24*, 81–86.
103. Sauro, F.; Lundberg, J.; De Waele, J.; Tisato, N.; Galli, E. Speleogenesis and speleothems of the Guacamaya cave, Auyan tepui, Venezuela. In Proceedings of the Karst and Caves in Other Rocks, Pseudokarst, 16th International Congress of Speleology, Brno, Czech Republic, 21–28 July 2013; pp. 298–304.
104. Martini, J.E.J. Karst in Black Reef Quartzite near Kaapsehoop, Eastern Transvaal. *Ann. S. Afr. Geol. Surv.* **1979**, *13*, 115–128.
105. Martini, J.E.J. Dissolution of quartz and silicate minerals. In *Speleogenesis: Evolution of Karst Aquifers*; Klimchouk, A.B., Ford, D.C., Palmer, A.N., Dreybrodt, W., Eds.; National Speleological Society: Huntsville, AL, USA, 2000; pp. 452–457.
106. Wray, R. A global review of solutional weathering forms on quartz sandstones. *Earth-Science Rev.* **1997**, *42*, 137–160. [CrossRef]
107. Piccini, L.; Mecchia, M. Solution weathering rate and origin of karst landforms and caves in the quartzite of Auyan-tepui (Gran Sabana, Venezuela). *Geomorphology* **2009**, *106*, 15–25. [CrossRef]
108. Sauro, F. Structural and lithological guidance on speleogenesis in quartz–sandstone: Evidence of the arenisation process. *Geomorphology* **2014**, *226*, 106–123. [CrossRef]
109. Wray, R.A.L. Phreatic drainage conduits within quartz sandstone: Evidence from the Jurassic Precipice Sandstone, Carnarvon Range, Queensland, Australia. *Geomorphology* **2009**, *110*, 203–211. [CrossRef]

Article

Diagenetic Origin of Bipyramidal Quartz and Hydrothermal Aragonites within the Upper Triassic Saline Succession of the Iberian Basin: Implications for Interpreting the Burial–Thermal Evolution of the Basin

María J. Herrero [1,*], Rafaela Marfil [1], Jose I. Escavy [2], Ihsan Al-Aasm [3] and Michael Scherer [4]

[1] Departamento de Mineralogía y Petrología, Facultad CC. Geológicas, Universidad Complutense de Madrid, 28040 Madrid, Spain; marfil@ucm.es

[2] Departamento de Ingeniería y Morfología del Terreno Escuela Técnica Superior de Ingenieros de Caminos, Canales y Puertos, Universidad Politécnica de Madrid, 28040 Madrid, Spain; ji.escavy@upm.es

[3] School of the Environment, University of Windsor Ontario, Windsor, ON N9B 3P4, Canada; alaasm@uwindsor.ca

[4] GG Consulting, 46282 Dorsten, Germany; mscherer1025@yahoo.com

* Correspondence: mjherrer@ucm.es

Received: 30 December 2019; Accepted: 12 February 2020; Published: 15 February 2020

Abstract: Within the Upper Triassic successions in the Iberian Basin (Spain), the occurrence of both idiomorphic bipyramidal quartz crystals as well as pseudohexagonal aragonite crystals are related to mudstone and evaporite bearing sequences. Bipyramidal-euhedral quartz crystals occur commonly at widespread locations and similar idiomorphic crystals have been described in other formations and ages from Europe, America, Pakistan, and Africa. Similarly, pseudohexagonal aragonite crystals are located at three main sites in the Iberian Range and are common constituents of deposits of this age in France, Italy, and Morocco. This study presents a detailed description of the geochemical and mineralogical characteristics of the bipyramidal quartz crystals to decipher their time of formation in relation to the diagenetic evolution of the sedimentary succession in which they formed. Petrographic and scanning electron microscopy (SEM) analyses permit the separation of an inner part of quartz crystals with abundant anhydrite and organic-rich inclusions. This inner part resulted from near-surface recrystallization (silicification) of an anhydrite nodule, at temperatures that were <40 °C. Raman spectra reveal the existence of moganite and polyhalite, which reinforces the evaporitic character of the original depositional environment. The external zone of the quartz contains no anhydrite or organic inclusions and no signs of evaporites in the Raman spectra, being interpreted as quartz overgrowths formed during burial, at temperatures between 80 to 90 °C. Meanwhile, the aragonite that appears in the same Keuper deposits was precipitated during the Callovian, resulting from the mixing of hydrothermal fluids with infiltrated waters of marine origin, at temperatures ranging between 160 and 260 °C based on fluids inclusion analyses. Although both pseudohexagonal aragonite crystals and bipyramidal quartz appear within the same succession, they formed at different phases of the diagenetic and tectonic evolution of the basin: bipyramidal quartz crystals formed in eo-to mesodiagenetic environments during a rifting period at Upper Triassic times, while aragonite formed 40 Ma later as a result of hydrothermal fluids circulating through normal faults.

Keywords: bipyramidal quartz; pseudohexagonal aragonite; Iberian Range; Upper Triassic; diagenesis; hydrothermal circulation

---

1. Introduction

Within the Upper Triassic sequences in the Iberian Basin, Spain (Figure 1a), and in other European countries, the occurrence of both bipyramidal quartz crystals (Figure 1b) and pseudohexagonal aragonite crystals (Figure 1c) is common within mudstone, evaporites, and marls that represent saline mudflats and salt pond facies. The aragonite crystals occur associated with outcrops of pre-Hettangian alkaline magmatism and related to normal faults dated as Middle-to-Upper Jurassic. Their formation resulted from the mixing of infiltrated marine waters with hydrothermal fluids that circulated synrift faults [1]. The quartz crystals, on the contrary, appear in a wider range of locations (Figure 1a) with no close relation to major faults or volcanic deposits. Instead, they appear to be related to facies rich in organic matter and have numerous saline mineral inclusions, indicating that the original sediments formed in environments under strong evaporitic conditions.

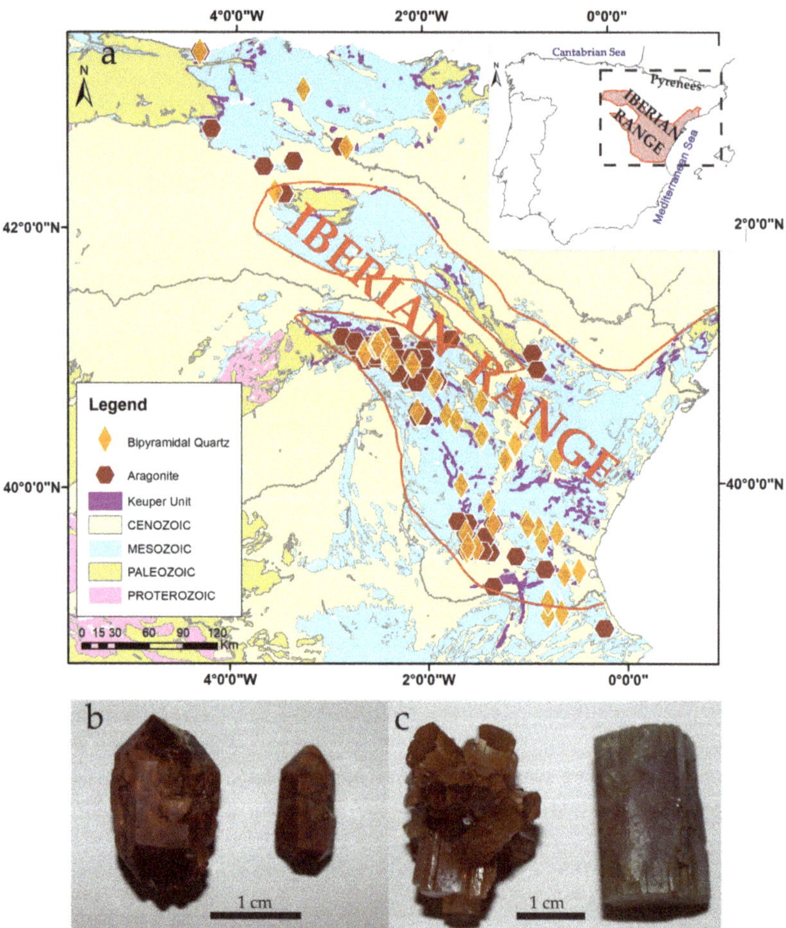

Figure 1. (a) Location of the bipyramidal quartz and aragonite crystals in the Iberian Range (Spain); (b) Bipyramidal quartz crystals that commonly are red-colored; (c) Pseudohexagonal aragonite prisms and rosettes, which show a wider color variation.

Euhedral quartz crystals, also named megaquartz crystals [2], have been described in the geological record within carbonate strata [3], normally as minor constituents [2,4–8]. They commonly appear related to sequences formed under evaporitic conditions [5,9–11] in dolomitic and salty sediments. It is, therefore, necessary to include the associated lithologies in the study. In the Iberian Peninsula, the occurrences of bipyramidal quartz are taken as indicators of the evaporitic Keuper (Upper Triassic) facies [12]. These are commonly known in Spain as "Jacintos de Compostela" [13].

Megaquartz crystals appear in Pleistocene and ancient sabkha dolomite accumulations from the Arabian Gulf, and their origin has been determined as resulting from dolomite and evaporite precursors replaced by silica in near-surface environments [5] during early diagenesis [2,14]. Such types of megaquartz crystals are found as well in Ukraine, Norway, Pakistan, and China [15]. The Herkimer diamonds [16], formed in a shallow marine Cambro–Ordovician succession, were interpreted as cements within cavities formed by dissolution produced from the flow of acidic water. Hexagonal crystals grew in the cavities very slowly, resulting in doubly terminated quartz crystals of exceptional clarity. The main factor controlling the carbonates silicification has been interpreted as fluctuations in pH around a value of 9, due to the inverse solubility relationship of silica and calcite [17,18]. Other factors are salinity changes in the pore fluids as well as the amount of host rock porosity and groundwater circulation [17].

Therefore, considering that authigenic quartz crystals and pseudohexagonal aragonites both appear within the same Keuper facies, the comparison of time, genesis, and factors controlling their formation should indicate variations in the tectonic evolution, i.e., rise and subsidence, and hence, in the thermal history of the basin.

2. Geological Setting

The Permo–Triassic Iberian Basin is a NW–SE trending rift transect that has been generally considered to be related to two main rifting cycles followed by post-rift periods of thermal subsidence [19–22]. The first rifting stage (Figure 2) occurred during the break-up of Pangea [23] between the early Permian and the Middle Triassic [24]. The second main cycle (Figure 2) spanned the late Jurassic to early Cretaceous and has been linked to the separation of Africa from Europe, and the simultaneous anticlockwise rotation of the Iberian Plate [25,26].

The current Iberian Range (Figure 1a) resulted from the tectonic inversion of the Iberian Basin and the up thrusting of basement rocks during the Pyrenean orogeny [27]. Basement uplifts, 5 to 30 km wide, are bounded by major NW–SE striking steeply dipping reverse faults, which were reactivated Variscan normal faults [28]. The Keuper succession (late Ladinian–early Carnian to Norian–Rhaetian) was formed during a post-rifting subsidence stage with minor tectonic activity [29]. The sedimentary sequence resulted from marine incursions during a transgressive period [30], first into narrow corridors and later surrounding the highest areas of the flanks [27]. The Keuper succession crops out with variable thicknesses and shows strong deformation due to its ductile character [31].

Five sequences (K1 to K5) compose the Spanish Keuper succession. K1, known as the "Lower Evaporitic Series", is a 40 to 380 m thick alternation of laminated gypsum and mudstone beds, with layers of sandstone, marl, and carbonates that represent lagoons and shallow coastal salinas [33,34]. K2 consists of *ca* 15 m of red and yellow mudstones with interbedded dolomite beds formed within fluvial and lacustrine environments. The third sequence, K3, consists of a *ca* 45 m succession of red mudstone that accumulated in a muddy intertidal setting [35]. Finally, the K4 and K5 units compose the "Upper Evaporitic Series", being K4 a 60 to 150 m-thick red mudstone sequence with abundant nodular gypsum deposited in coastal sabkhas, and K5 (up to 50 m thick) is mostly white-bedded gypsum, with interbedded mudstones deposited in evaporitic lagoons [36]. In the Iberian Peninsula, these saline deposits are the most abundant, covering 4394 km^2 [37,38]. The studied bipyramidal quartz and aragonite crystals are found in the topmost part of the evaporite-mudstone deposits of the K4 unit.

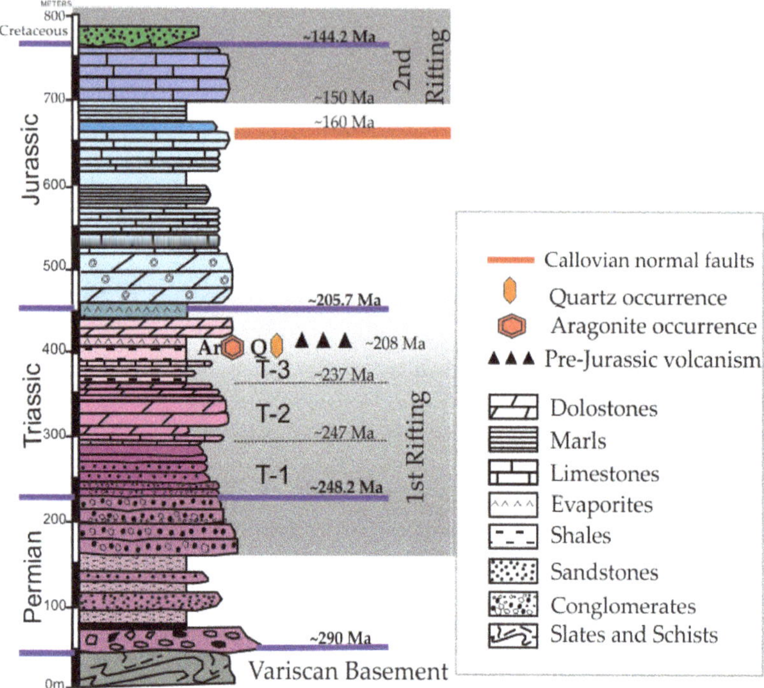

Figure 2. Upper Permian to Lower Cretaceous stratigraphic section in the Iberian Basin (lithology, thickness, and ages) and position of the occurrences of aragonite (Ar) and authigenic quartz (Q). Dates of the pre-Hettangian volcanism and the 1st and 2nd Rifting periods are indicated [32] (T-1: Buntsandstein; T-2: Muschelkalk; T-3: Keuper; Modified from [1].

The overlying Imón Formation represents an epicontinental carbonate platform [39] of pre-Hettangian age. Different magmatic episodes occurred from the Middle and Upper Triassic to Jurassic times [23]. The magmas were of tholeiitic and alkaline affinities, being pre-Hettangian volcanism commonly associated with the Imón Formation (Figure 2), characterized by sub-volcanic tabular bodies (sills) that intruded into the K4 and K5 units of the Keuper [40].

3. Materials and Methods

Analyses of 30 representative bipyramidal quartz crystals from selected outcrops located in the Iberian Range in the K4 unit of the Keuper Facies (Figure 2) were performed. Thin sections were prepared to distinguish between quartz and inclusions of different minerals. Sequential relationships were established between different phases. For petrographic studies, samples were cut with an oil-cooled diamond disc saw (Struers Discoplan-TS; Struers ApS, Ballerup, Denmark).

Bulk mineralogy of the powdered quartz crystals was determined by powder X-ray diffraction (XRD), employing a Philips PW 1720 diffractometer with Cu (Kα) radiation and a Bruker D8 Advance diffractometer equipped with a Sol-X detector and Cu (Kα) radiation (Bruker, Billerica, MA, USA). The mineralogical composition of crystalline phases was estimated following the Chuns [41] method and using Bruker software (EVA) at CAI de Técnicas Geológicas, Universidad Complutense de Madrid (UCM).

The occurrence and the textural relationship between quartz crystals and their inclusions were studied using a scanning electron microscope (SEM) coupled with energy-dispersive spectrometry

(EDS). For the SEM analyses, it has been used the model JEOL JSM 6400 equipped with an Oxford energy–dispersive X-ray micro analyzer (JEOL Limited, Tokyo, Japan) both in secondary electron (SE) and backscattered electron modes (BSE) was used. SEM-cathodoluminescence microscopy has been used to distinguish between (1) quartz (as several phases of quartz overgrowths), and (2) inclusions within quartz crystals [42]. SEM-cathodoluminescence (SEM-CL) colors in quartz have been related to the presence of activator trace elements, lattice order, and crystallization temperature [42,43]. SEM-CL observations were done on carbon-coated polished thin sections using a JEOL JSM-840 electron microscope equipped with a Gatan ChromaCL2 housed at the CAI de Técnicas Geológicas, UCM.

Microprobe mineral analyses on polished quartz crystal surfaces were performed to obtain additional data to determine the geochemical composition and decipher diagenetic processes. The chemical composition of inclusions in quartz crystals was analyzed by ion microprobe analyses (EPMA) using a JEOL JXA–8900 model. Operating conditions were characterized by voltage 15 kV, current intensity 20 nA, and electron beam diameter 5 μm. Detection limits were *ca* 150 ppm for Ca, 100 ppm for Mg, 250 ppm for Mn, 300 ppm for Fe, and 250 for Sr and Ba. Analytical totals were normalized to 100% moles of $CaCO_3$, SO_3, $MgCO_3$, $FeCO_3$, TiO_2, BaO, Al_2O_3, and MnO for comparison. These analyses were performed at the Centro Nacional de Microscopía electrónica (CNME).

Raman spectral analysis of samples was performed on an XRD confocal Raman Thermo Fischer Microscope at the Spectroscopy CAI, UCM. Excitation was provided by the 532 nm line of a diode laser.

4. Results

4.1. Petrographic Analyses

The quartz crystals appear as individual euhedral crystals or as clusters of crystals. Detailed petrographic analyses were used to characterize their textures.

Single crystals were either double terminated or had a well-developed single termination and well-defined crystal faces. Single crystals ranged from 0.0625 mm to 2.6 cm lengths (Figure 1b), while cluster appeared as groupings of quartz crystals that converged in a nucleus (Figure 1b). Petrographic analyses revealed that the center of the crystal had numerous evaporate mineral inclusions (Figure 3a) resembling anhydrite laths (Figure 3b,c) and radial-pattern inclusion filled by organic matter (Figure 3d,e). These inclusions appeared in a circular distribution (Figure 3c). Towards the edges, the crystal was zoned and free of inclusions (Figure 3a,f–h). Generally, all the inclusions with prismatic or needle shapes appeared to be filled by anhydrite, and all of them had the same extinction which appeared to be originated by the replacement of one single anhydrite crystal. Variations in the number of inclusions permitted the determination of a discontinuous zoning that formed bands (Figure 3a,d). There were at least 5 bands of zonation (Figure 3a), considered as potential crystal overgrowths.

Optical images (Figure 3c,d) showed subangular to round quartz grains (red line) in the center of the quartz crystals. Clusters of crystals embedded in a carbonate (calcite) poikilotopic cement occurred as well. There were also dispersed crystals of pyrite and chalcopyrite.

4.2. Geochemical Analyses

Microprobe analyses (Table 1) showed that the composition of quartz crystals was almost totally SiO_2, with a mean value of 98.14% ($n = 53$). The second highest component was Al_2O_3 with average percentages of 0.11%. Minor contents of Fe_2O_3, MnO, CaO, and MgO were present. Microprobe measurements of Al_2O_3 along a transect from the center to the edges of the quart generally varied from lower contents in the inner part of the crystal (Figure 4a–d) and a progressive rising towards the outer part (Figure 4b). Geochemical analysis performed within the inclusions ($n = 15$) indicated that they were dominantly composed of *ca* 52.81% of CaO and *ca* 0.8% of SO_3, with minor amounts of FeO (0.39%), MgO (0.2%), and SiO_2 (0.36%). Other inclusions near the detection limit appeared filled by anhydrite and organic matter.

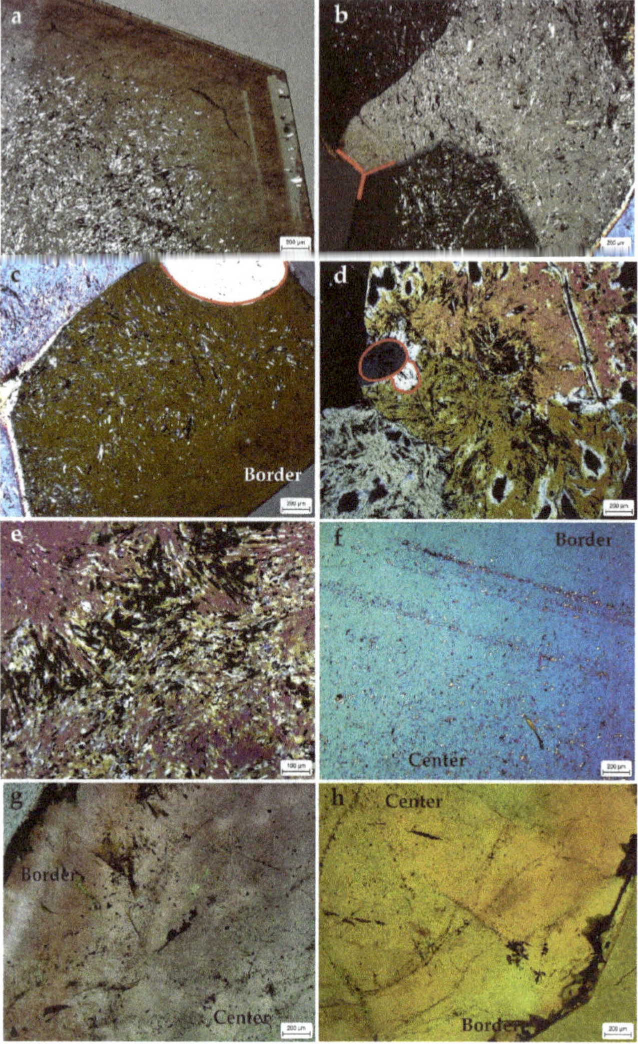

Figure 3. Euhedral quartz with hexagonal crystal shapes: (**a**) Cross-polarized light of a quartz crystal with abundant inclusions in the center and idiomorphic overgrowth towards the edges of the crystal; (**b**) Cross-polarized image of the quartz crystals, with crystal triple junctions (in red) that are clear and free of inclusions; (**c**) Cross-polarized image of another crystal where the pattern of the inclusion can be observed; (**d**) Cross-polarized image of the center of the crystal with abundant inclusions that change to bands free of inclusions towards the outer part; (**e**) Detail of the center of the crystal and the texture of the organic-rich inclusion; (**f**) Patches of elongate inclusions filled with organic matter and anhydrite laths in the central part of the crystal; (**g**) Plane-polarized light where the concretional morphology of the interior of the quartz marked by lineation of the organic matter inclusions. The external part of the crystal is free of inclusion; (**h**) Quartz appeared as idiomorphic crystals, although the inclusions-rich inner part revealed a dissolution surface in the interior of the crystal. To avoid the destruction of inclusions, thin sections were kept thicker and consequently changed the quartz interference colors in polarized light.

Table 1. Microprobe analyses of quartz crystals (the values are in wt%). Numbers of measurement points are valid for Figure 4.

	Point	SiO_2	Al_2O_3	FeO	MnO	MgO	CaO	Na_2O	K_2O	TiO_2	SO_3	BaO
BO17	1	99.52	0.29	0.12	0	0.01	0	0	0.02	0.01	0.03	0
BO17	2	99.67	0.24	0.04	0	0	0.01	0	0.01	0	0.01	0.02
BO17	3	99.72	0.27	0.01	0	0	0	0	0	0	0	0
BO17	4	99.78	0.05	0	0.06	0.02	0	0	0.02	0	0	0.07
BO17	5	99.92	0	0	0	0	0	0	0.02	0.01	0.04	0.01
BO17	6	99.83	0.02	0	0.02	0	0.03	0	0.01	0	0.01	0.08
BO17	7	99.89	0.02	0	0.01	0.01	0.03	0	0.01	0.02	0.01	0
BO17	8	99.68	0.10	0.07	0	0	0.03	0	0	0.03	0	0.09
BO17	9	99.85	0.04	0.03	0.03	0	0	0.02	0.01	0	0.02	0
BO17	10	99.84	0.04	0.02	0	0.01	0.01	0.01	0.01	0	0.03	0.03
BO17	11	99.88	0.04	0	0	0	0.03	0.01	0.02	0.01	0.01	0
BO17	12	99.78	0.04	0.05	0	0	0	0.01	0.01	0.01	0	0.10
BO17	13	99.89	0.04	0.03	0	0	0	0.01	0	0.02	0.01	0
BO17	14	99.8	0.10	0.02	0	0.01	0.01	0	0.02	0.02	0.02	0
BO17	15	99.8	0.11	0.01	0.02	0	0	0.01	0	0	0	0.05
BO17	16	99.87	0.07	0	0	0	0.01	0	0	0	0	0.05
BO17	17	99.74	0.11	0.02	0	0.02	0	0.01	0	0.01	0.04	0.05
BO17	18	99.58	0.25	0.05	0.03	0.01	0.02	0	0.01	0	0.01	0.04
BO17	19	99.5	0.49	0	0	0.01	0	0	0	0	0	0
BO17	20	99.4	0.48	0.07	0	0	0.02	0	0	0	0	0.03
BO17-1	25	99.9	0.05	0.01	0	0.01	0.02	0	0.01	0	0	0
BO17-1	26	99.81	0.09	0.01	0	0	0.01	0.02	0.01	0	0	0.05
BO17-1	27	99.92	0.02	0.03	0	0	0	0	0	0.03	0	0
BO17-1	28	99.88	0	0.05	0.03	0	0	0	0	0	0	0.04
BO17-1	31	99.84	0.06	0	0	0	0.01	0	0	0	0.02	0.07
BO17-1	32	99.67	0.17	0	0	0.01	0	0.02	0.01	0	0	0.12
BO17-1	33	99.7	0.14	0	0	0	0	0	0	0.01	0	0.15
BO17-1	34	99.82	0.09	0.04	0.01	0	0.01	0.01	0	0.02	0	0
BO17-1	35	99.87	0.08	0	0	0.03	0	0	0.01	0.01	0	0
BO15	36	99.78	0.07	0.06	0	0	0.01	0.01	0	0	0	0.06
BO15	37	99.55	0.15	0.06	0	0.16	0.03	0	0.01	0.01	0	0.02
BO15	38	99.85	0.04	0.07	0.01	0	0	0	0.02	0	0	0
BO15	39	99.69	0.02	0.14	0	0	0.01	0.01	0.02	0.03	0.02	0.05
BO15	40	99.84	0.10	0.01	0	0.01	0	0.01	0.02	0	0	0.01
BO15	41	99.89	0	0	0	0	0	0	0.02	0	0.02	0.06
BO15	42	99.77	0.11	0.04	0	0	0.01	0	0.01	0	0	0.06
BO15	43	99.71	0.08	0.02	0	0.01	0.02	0.01	0	0	0	0.15
BO15	44	99.72	0.09	0.04	0.02	0	0.01	0	0	0	0	0.11
BO15	45	99.68	0.15	0.06	0	0.01	0.02	0	0.01	0	0.03	0.03
BO15	46	99.82	0.03	0.11	0.02	0	0	0	0.01	0	0.01	0
BO18	47	99.7	0.04	0.07	0	0	0	0	0	0.10	0	0.09	0
BO18	48	99.81	0.05	0.06	0	0.03	0	0	0.02	0.02	0.01	0
BO18	49	99.85	0.07	0.03	0	0	0	0	0.01	0.04	0	0
BO18	50	99.77	0.02	0.07	0	0	0	0.01	0	0.01	0.02	0.10
BO18	51	99.8	0.07	0	0.02	0.02	0	0	0.01	0.04	0.04	0
BO18	52	99.68	0.16	0.12	0	0	0.02	0.02	0	0	0	0
BO18	53	99.73	0.16	0.05	0	0	0.02	0.02	0.01	0	0.01	0
Mean		99.77	0.1	0.04	0.01	0.01	0.01	0	0.01	0.01	0.01	0.04
Std. Dev.		0.12	0.11	0.04	0.01	0.02	0.01	0.01	0.02	0.01	0.02	0.04

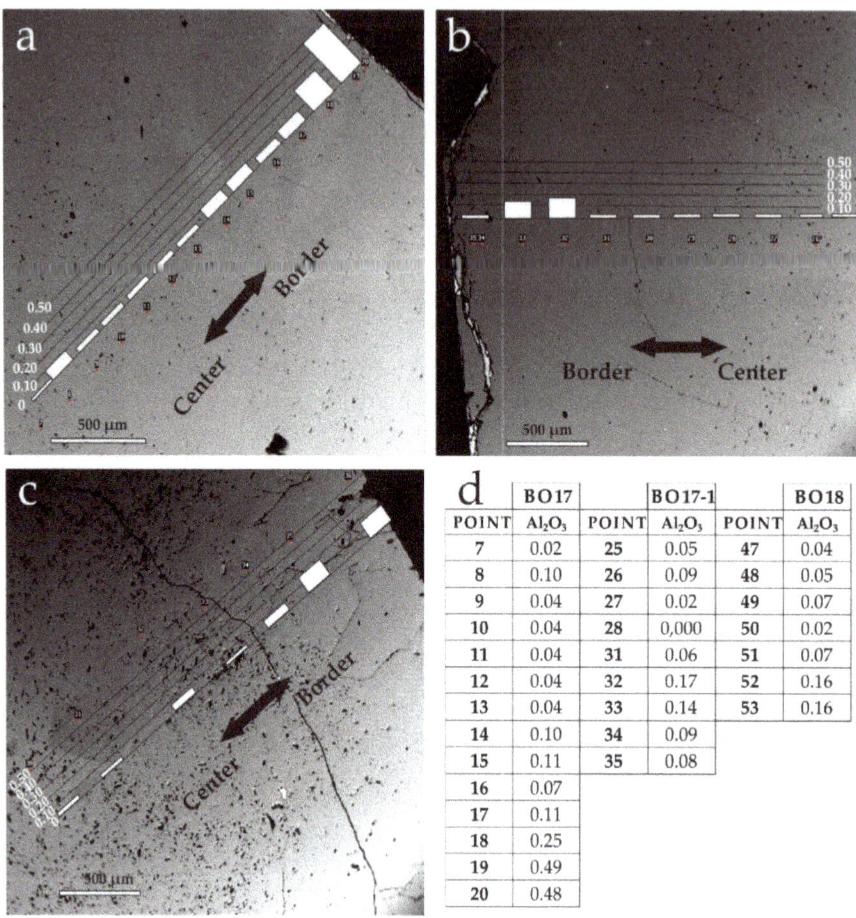

d **BO17**		**BO17-1**		**BO18**	
POINT	**Al$_2$O$_3$**	**POINT**	**Al$_2$O$_3$**	**POINT**	**Al$_2$O$_3$**
7	0.02	25	0.05	47	0.04
8	0.10	26	0.09	48	0.05
9	0.04	27	0.02	49	0.07
10	0.04	28	0,000	50	0.02
11	0.04	31	0.06	51	0.07
12	0.04	32	0.17	52	0.16
13	0.04	33	0.14	53	0.16
14	0.10	34	0.09		
15	0.11	35	0.08		
16	0.07				
17	0.11				
18	0.25				
19	0.49				
20	0.48				

Figure 4. Electron backscatter images of the quartz samples: they were mostly composed of silica (88%). (**a**) Sample BO17; (**b**) Sample BO17-1; and (**c**) Sample BO18. Al$_2$O$_3$ appeared as *ca* 0.11%, but slight variations were detected from lower values in the center towards higher values to the edges of the crystals; (**d**) Al$_2$O$_3$ values represented in the three a,b,c lines of the figure (for microprobe data see Table 1).

4.3. Scanning Electron Microscope Analyses

SEM analyses of the quartz crystals showed clearly a lack of abrasion caused by transportation (Figure 5a). No sign of dissolution processes was indicated by sharp crystal terminations and edges. There were anhydrite inclusions (Figure 5b) and lath patterns in the center of the quartz crystals (Figure 5c).

Figure 5. Scanning electron photomicrographs of Quartz (Q) crystals: (**a**) Idiomorphic edge of a bipyramidal quartz crystal with no signals of transport or dissolution; (**b**) High-resolution secondary electron images of quartz surface with abundant anhydrite laths (Anl); (**c**) Detail of the quartz central area of the quartz crystal with anhydrite laths, coated by quartz spherules (Sp); (**d**) Detail of the silica spherules (composed of Si, Fe, and Al) and microbial filaments (Mf) indicating the formation of the quartz within a saline environment rich in organic matter.

SEM observations led as well to the identification of a layer of micron-sized quartz spherules (Sp in Figure 5c). These were coating surfaces of the siliceous nodule (Figure 5d) and were associated with organic filaments (Mf).

4.4. Cathodoluminescence and Raman Spectra Analyses

Cathodoluminiscence analyses highlighted the presence of concentric layers or bands around a nucleus (Figure 6a). The inner part of the crystal (silica nodule) had abundant inclusions and smooth shape. The center of the crystal under CL appeared as evaporitic minerals that had been replaced by silica (Figure 6b,c) [44]. Other pseudomorphs appeared as needle, laths (Figure. 6b,c), or nodule form.

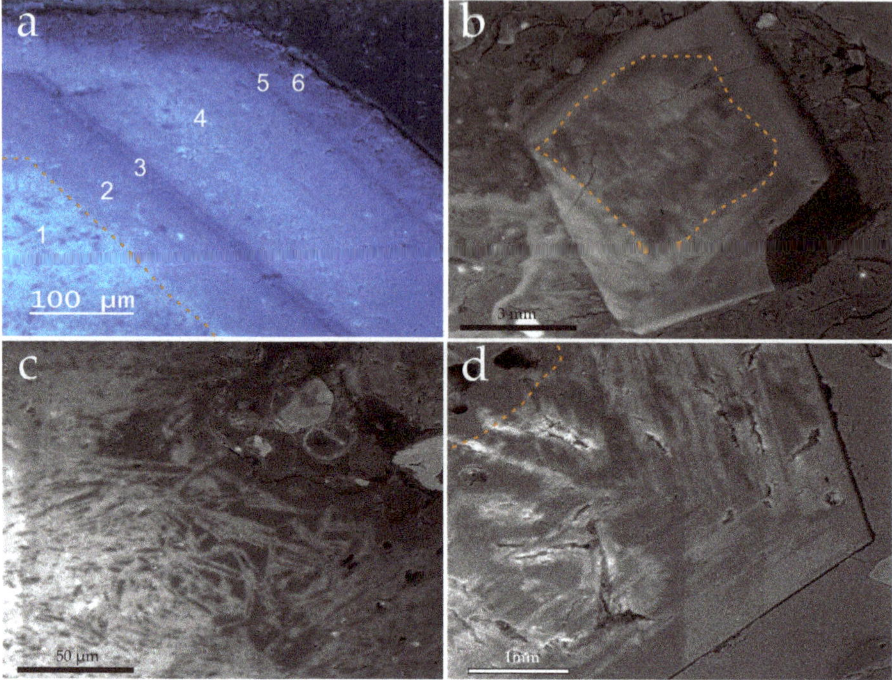

Figure 6. SEM-Cathodoluminescence (SEM-CL) analysis: (**a**) Light blue characterizes the luminescence response of the quartz crystals. Variation in the texture and color reveal changes between the center and the edges, separating the center, named silica nucleus (1) and five bands of quartz overgrowths (2–6); (**b**) The texture of the crystal reveal an irregular pattern in the center and a second phase of formation characterized by a regular overgrowth; (**c**) Detail of the center of the crystal (silica nucleus) showing quartz pseudomorphs after evaporite minerals (determined by SEM and microprobe analyses); the pseudomorphs appear either randomly oriented or subparallel to bedding; (**d**) The edges of the crystal appear as various bands, with variations in luminescence suggesting quartz overgrowth. To better distinguish textures, SEM-CL images are included in grey-scale colors.

Several thinner dark bands alternated with thicker lighter bands of quartz overgrowth over the silica nodule. (Figure 6b,d). Each lighter band followed the same pattern and direction as the previous darker band (Figure 6a,b,d). Laser Raman spectra of the center and outer part of bipyramidal quartz were obtained from different spots (Figure 7). The Raman spectra from three points of the central part of the crystal (Figure 7a,b) revealed the existence of quartz (Q), moganite (Mog), anhydrite (Anh), and polyhalite (Pol). Quartz and moganite both had similar Raman spectra which reflect they shared many structural elements [45]. The most intense quartz band appeared at 465 cm^{-1} while that of moganite was at 501 cm^{-1}. Rodgers and Cressey [46] discriminated the two phases by the use of broad, low frequency, vibrational band present in the spectrum of quartz at 205 cm^{-1} and at a slightly higher frequency of moganite at 220 cm^{-1}. The Raman spectra of four points from the external part of the crystal (Figure 7c,d) showed that there were only quartz and no moganite. The crystallinity in this part was higher, indicated by the sharper character of the Raman spectra picks towards the edges of the crystal.

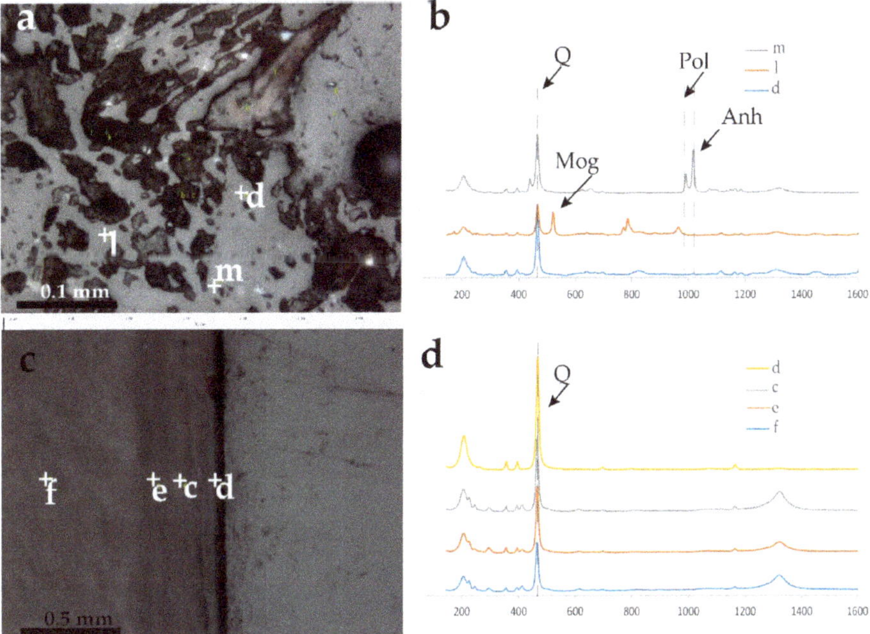

Figure 7. (**a**) Microphotograph showing the spots of measurement of Raman Spectrometry in the central part of euhedral quartz; (**b**) Image of Raman Spectra of micro-sampled spots d, l and m. Quartz (Q) appears together with moganite (Mog), anhydrite (Anh) and polyhalite (Pol); (**c**) Microphotograph showing the spots of measurement of Raman Spectrometry in the external part of a euhedral quartz; (**d**) Image of Raman Spectra of micro-sampled spots f, e, c and d. Quartz, in this case, was pure and more crystalline (X-axis represents Raman Shift /cm^{-1}).

5. Discussion

5.1. Relationship between Doubly Terminated Quartz Crystals, Evaporites, Organic Matter, and Silica Sources

The K4 facies of the Upper Triassic succession of the Iberian Basin [33] represents a saline mudflat in a clayey coastal plain environment with abundant shallow restricted Cl-rich brine salt ponds [47]. Salt pond deposits were composed of interlayered halite and mudstone beds with associated nodular anhydrite that formed a laterally extensive facies mosaic of sabkha and salinas across an evaporitic mudflat.

In the Iberian Basin, the authigenic bipyramidal quartz appears in scattered locations throughout the K4 facies of the Keuper succession (Figure 1) and does not show any concentration in a single layer or stratigraphic horizons indicating that they were not transported into the site of accumulation by aqueous currents. Euhedral quartz generally appears as crystals red or transparent, hyalines, doubly terminated crystals, and embedded in organic matter-rich crystals of white and black gypsum with abundant Fe-oxide and -hydroxides contents. Likewise, these quartz crystals also appear within mudstones and marls rich in chlorite, chlorite-smectite mixed layers, illite and scarce kaolinite [12]. The lithofacies where they appear are Mg-rich and poor in Ca, interpreted as caused by the algal community present in the environment of sedimentation [48].

Euhedral quartz crystals, which were not corroded (Figures 3 and 5), had a central part composed of various crystals with undulose extinction and sutured boundaries. This central part had a large number of anhydrite and prismatic inclusions filled with organic matter, homogeneously spread in the crystals, although they disappeared towards the edges (Figures 3, 5 and 6). The ubiquitous presence of anhydrite inclusions (Figure 5b) and laths patterns, together with the organic matter inclusion in the center of the quartz crystals (Figure 5c), indicated a replacement of an anhydrite precursor with abundant organic matter in the environment. Some of the original mineral matter had been incorporated during this process. Moganite clearly showed up in the Raman spectra analysis (Figure 7a,b). This mineral was a metastable silica polymorph that was structurally similar to quartz [49], and commonly appeared associated with it in chert deposits formed in evaporitic environments [17]. Petrographic analyses also revealed the existence of silica phases filling voids in the center of the quartz crystal. This silica crystal must have grown in the dissolution cavities of the primary evaporite. The anhydrite nodules were probably partially dissolved by the interaction with meteoric water during a sea-level low-stand that occurred during the Upper Triassic [50]. As a result of sea level fluctuations, the waters in contact with the sediments changed from marine-derived evaporative/sabkha waters to meteoric in composition, and at some point, they became undersaturated with respect to anhydrite. This led to the evaporites' dissolution, which typically starts in the shallow subsurface, as salt beds are flushed by meteoric or marine waters and then continue deeper in the subsurface [15]. The most saline salts (typically halite and carnallite) undergo partial dissolution and are flushed, leaving behind residues of the less saline salts (typically gypsum–anhydrite). Pyrite and chalcopyrite crystals appeared associated with the replacive quartz and their presence was related to the metabolic activity of sulfate-reducing bacteria [51]. The environment of deposition was organic-rich and, in contact with the organic carbon, the sulfates (gypsum and anhydrite) would have been reduced to sulfides and, subsequently, oxidized in the near-surface environment of the sabkha. Oxidation of sulfides to sulfates would have decreased the pH of the waters and could have contributed to the precipitation of the available silica. The main factor controlling silica precipitation is the pH variations around 9 [52]. The solubility of silica dramatically increases when pH exceeds 9, and silica precipitates when pH falls to a lower value [17] when silica is available. Sources of silica could include the release of silica by clay transformations [53] and even silica introduced by springs and seeps [54,55] which are common in continental rift basins and volcanic regions, such as the Iberian Basin during the Upper Triassic [1,56].

The dissolution of sulfate crystals and precipitation of quartz occurred at a very shallow burial and low pH conditions during early diagenesis [57]. The quartz crystals started growing in void spaces created by the dissolution of evaporite minerals. Probably, the original quartz phase was the amorphous silica phase opal-A [58]. Miliken [11] determined that the anhydrite replacement by quartz in sabkhas takes place at temperatures that were <40 °C. During burial diagenesis, opaline phases age by undergoing successive dissolution–precipitation–recrystallization reactions, including the opal A-opal CT to quartz transition [58,59]. Quartz and lutecite (chalcedony) are common silica phases that typically replace previous anhydrite/gypsum nodules [60–62]. SEM observations led to the identification of micron-sized silica spherules (Figure 5c,d) partially coating the surfaces of the internal quartz nodule. Amorphous silica coated the surfaces of many detrital grains and appeared related to organic matter content [63]. They have been described as resulting from the volume for volume replacement of anhydrite nodules by microcrystalline quartz and quartz spherulites [11].

5.2. Second Phase of Quartz Formation: Quartz Overgrowth

Subsequent to silicification, these deposits were affected by diagenesis due to the downward percolation of meteoric waters and burial, generating dissolution of the nucleus and a phase of silica overgrowth.

Cathodoluminescence and petrographic analysis revealed that the silica nodule (that resulted from the replacement of an anhydrite nodule) appeared to have an irregular contour and an overgrowth composed of several layers, which appeared in the Raman spectrum as a more crystalline quartz phase with no moganite or evaporite mineral relicts (Figure 7c,d). The luminescence of the quartz crystals was blue (Figure 5a) and showed variations of texture: 1) the silica nucleus which showed pseudomorphs of the anhydrite precursor (the center); 2) a first band on top of the replacing silica (80 μm in thickness), which was bright in CL; 3) a thin CL dark band (*ca* 20 μm thick); 4) another bright CL band, (150 μm thick), 5) a second dark band with *ca* 2 5 μm thickness; 6) the last phase of overgrowth which was light in CL color and had a thickness of less than 60 μm. CL emissions in quartz indicated the specific physicochemical conditions of crystal growth and can be used as a signature of genetic conditions of mineral formation. The most common CL emission bands in natural quartz were the 450 and 650 nm [64,65], which resulted in bluish-violet CL colors and were detectable in quartz crystals from igneous, volcanic, and metamorphic rocks as well as authigenic quartz from sedimentary environments. The resulting CL analysis (Figure 5a) appeared similar to the authigenic quartz found in the Zechstein salt deposit of the Triassic of Germany [66].

These bands appeared as quartz overgrowths (syntaxial quartz cements). During the formation of quartz overgrowths in sandstones, relatively large quartz crystals (50–100 mm) grew into pore spaces after burial, inheriting the exact crystallographic orientation of the detrital host grain upon which they have grown [67]. The silica needed to form the idiomorphic external part of the bipyramidal quartz in the Keuper facies could have resulted from the dissolution during the burial of part of the silicified anhydrite nodule. Organic acids, resulting from burial transformation of the organic matter present, could produce an increase in the solubility of quartz by bonding with silicic acids [68]. This organically mobilized silica precipitated as overgrowth around the siliceous nodule. Because silica is more soluble at high temperatures, the movement of silica-saturated solutions might be expected to be accompanied by growth variations. The temperature of formation of quartz overgrowths rarely takes place below 80 °C [69] and commonly appears in sandstones buried to depths where temperatures exceed 100 °C [68,70]. Therefore, the initial temperature of the formation of the phase of quartz as overgrowth would be at a minimum of *ca* 80 °C. Microprobe measurements of Al_2O_3 showed lower contents in the inner part of the crystals (Figure 4a–d) and a progressive rising towards the outer parts (Figure 4b). In the present study, the slight change in Al_2O_3 content in the quartz is considered to be indicative of variation in the temperature of formation, rising towards the edges [71]. Among temperature, other factors controlling silica formation include pH higher than 5 and changes in the chemistry of the pore-water, which could lead to variations of composition reflected in the CL bands [70].

5.3. Relation of the Bipyramidal Quartz with the Aragonite Presence in the Keuper Facies

In this work, the origin of bipyramidal quartz in the K4 unit of the Keuper facies in the Iberian Basin is described [33]. In these same facies, pseudohexagonal crystals of aragonite appear at some locations as well. The presence of aragonite seems to be related to saline pond environments where Mg-rich clay minerals originated [72]. Around 80% of the aragonite crystals outcrops are concentrated in three areas and their occurrence defines an NW to SE lineation [1] from north-central Spain to the Eastern Iberian Peninsula (Figure 10 in Reference [1]). These three areas coincide with the location of volcanic materials that correspond to a pre-Hettangian magmatic episode [56,73] coeval to the second period of intense synrift magmatic activity in the Iberian Basin [56].

The occurrence of aragonite is as well coincident with a set of normal faults of Middle-to-Upper Jurassic age (154–144 Ma) [74]. These faults resulted from a tectonic extensional episode which occurred at the same time interval as an important hydrothermal event (150 Ma) defined in the Eastern Iberian Central System, just west of the Iberian Basin. This hydrothermal event was caused by fluids that circulated within regional-scale convective cells. The fluids flowed along the Middle-to-Upper Jurassic faults, and they mixed with downwards percolating brines [75]. The mixing of these fluids through the K4 unit of the Keuper produced the transformation of the Mg-rich smectite into corrensite and lead to the precipitation of the aragonite [1]. Fluid inclusions analyses led to determine the temperatures of formation ranging between 160 and 260 °C [1].

The aragonite crystals were commonly engulfing corrensite but also euhedral quartz crystals, suggesting a temporal relationship [1]. Therefore, the bipyramidal quartz resulted from an early diagenetic evaporitic mineral replacement by silica, at temperatures of less than 40 °C and lately, mesodiagenetic quartz overgrowths formed at temperatures between 80 and 100 °C. Both quartz phases were formed during the rifting period of the Upper Triassic. On the other hand, aragonite was precipitated 40 Ma later during Callovian times, resulting from the mixing of hydrothermal fluids with infiltrated marine waters at temperatures ranging between 160 and 260 °C [1].

6. Conclusions

Euhedral quartz, in the form of individual doubly terminated crystals, appear throughout Upper Triassic evaporite-bearing sabkha deposits (K4) in the Iberian Basin.

Optical microscopy, SEM, microprobe, SEM-CL, XRD, and Raman spectra analyses performed in the bipyramidal quartz crystals revealed differences in formation between the inner and outer part of the crystals.

The inner part of the quartz crystals resulted from a replacement of anhydrite nodules that formed in a sabkha environment. Meteoric waters repeatedly flushed the sabkha deposits and dissolved some of the evaporitic minerals, producing relatively large pore spaces into which silica precipitated. Organic rich contents and the action of sulfate-reducing bacteria that transformed part of the sulfates into sulfides produced variations of pH that promoted silica precipitation. The replacement of evaporites by quartz occurred at a very shallow burial, low pH, and temperatures of less than 40 °C, soon after deposition.

The organic matter left in the inclusions was transformed with burial and generated organic acids that produced the dissolution of the siliceous nodule. This organically mobilized silica eventually precipitated as overgrowths, at temperatures between 80 and 100 °C, indicating that quartz overgrowth was generated during the mesodiagenesis. The slight change in Al_2O_3 content from the center to the outer part of the crystal also indicates the rising of the temperature of formation.

The Upper Triassic succession was affected by the reactivation of alpine NW–SE trending fractures during the Callovian, 40 Ma later. These fractures served as conduits for hydrothermal fluid that mixed with infiltrated marine waters and produced important transformations of Mg-rich smectite into corrensite and lead to the precipitation of the aragonite at temperatures between 160 and 260 °C.

Hence, crystals of compositions as different as quartz and aragonite, occurring within the same sedimentary succession and depositional environment, indicate different formation fluids, temperature regimes, and diagenetic, tectonic, and thermal stages within the geological history of the Iberian Basin.

This study demonstrates a great example of how a detailed establishment of the origin of different mineral phases within a sedimentary succession permits unraveling an accurate reconstruction of a basin's evolution.

Author Contributions: Conceptualization, M.J.H. and M.S.; Data curation, M.S.; Formal analysis, R.M. and J.I.E.; Funding acquisition, M.J.H. and I.A.-A.; Investigation, M.J.H., R.M., J.I.E., I.A.-A., and M.S.; Methodology, M.J.H., R.M., and J.I.E.; Project administration, M.J.H.; Software, J.I.E. and M.S.; Supervision, M.J.H.; Validation, I.A.-A.; Writing—original draft, M.J.H., R.M., J.I.E., and M.S.; Writing—review and editing, M.J.H. and R.M. All authors have read and agreed to the published version of the manuscript.

Funding: This research was funded by Projects FEI16/52-4159070 and 4157047 of the Fundación UCM.

Acknowledgments: We are grateful to A. Rodriguez from the CNME of the UCM for his assistance with the SEM studies and to M. Barajas and P. Lozano for the thin sections' preparation. SEM-Catodoluminescence was performed at the CAI de Ciencias Geológicas, and we would like to acknowledge X. Arroyo for assistance. We would like to thank the reviewers and the editorial team for their comments and suggestions that have led to improvements in the manuscript. Ihsan Al-Aasm acknowledges the continuous support by NSERC.

Conflicts of Interest: The authors declare no conflict of interest.

References

1. Herrero, M.J.; Marfil, R.; Escavy, J.I.; Scherer, M.; Arroyo, X.; Martín-Crespo, T.; López de Andrés, S. Hydrothermal activity within a sedimentary succession: Aragonites as indicators of Mesozoic Rifting (Iberian Basin, Spain). *Int. Geol. Rev.* **2020**, *62*, 94–112. [CrossRef]
2. Chafetz, H.S.; Zhang, J. Authigenic Euhedral Megaquartz Crystals in a Quaternary Dolomite. *J. Sediment. Pet.* **1998**, *68*, 994–1000. [CrossRef]

3. Winsborough, B.M.; Seeler, J.S.; Golubic, S.; Folk, R.L. Recent fresh-water lacustrine stromatolites, stromatolitic mats and oncoids from northeastern Mexico. In *Phanerozoic Stromatolites II*; Bertrand-Sarfati, J., Monty, C., Eds.; Kluwer Academic Publishers: Dordrecht, The Netherlands, 1994; pp. 71–100.

4. Folk, R.L. Petrography and petrology of the Lower Ordovician Beekmantown carbonate rocks in the vicinity of State College. Ph.D. Thesis, Pennsylvania State University, Pennsylvania, PA, USA, 1952.

5. Friedman, G.M.; Shukla, V. Significance of authigenic quartz euhedra after sulfates: Example from the Lockport Formation (Middle Silurian) of New York. *J. Sediment. Res.* **1980**, *50*, 1299–1304.

6. Wilson, R.C.L. Silica diagenesis in Upper Jurassic limestones of Southern England. *J. Sediment. Pet.* **1966**, *36*, 1036–1049.

7. Zenger, D.H. Definition of type Little Falls Dolostone (Late Cambrian), east-central New York. *Am. Assoc. Pet. Geol. Bull.* **1976**, *60*, 1570–1575.

8. Scholle, P.A. A Color Illustrated Guide to: Constituents, Textures, Cements, and Porosities of Sandstones and Associated Rocks. *Am. Assoc. Pet. Geol.* **1979**, *27*, 201.

9. Folk, R.L.; Pittman, J.S. Length-slow chalcedony: A new testament for vanished evaporites. *J. Sediment. Pet.* **1971**, *41*, 1045–1058.

10. Friedman, G.M. Dissolution-collapse breccias and paleokarst resulting from dissolution of evaporite rocks, especially sulfates. *Carbonates Evaporites* **1997**, *12*, 53–63. [CrossRef]

11. Milliken, K.L. The silicified evaporite syndrome—Two aspects of silicification history of former evaporite nodules from southern Kentucky and northern Tennessee. *J. Sediment. Pet.* **1979**, *49*, 245–256.

12. Marfil, R. Estudio Petrogenético del Keuper en el sector meridional de la Cordillera Ibérica. *Estudios Geológicos* **1970**, *24*, 113–161.

13. Calvo Rebollar, M. *Minerales Y Minas De España. Volumen VIII: Cuarzo Y Otros Minerales De La Sílice*; Fundación Gómez-Pardo: Madrid, Spain, 2016; p. 399.

14. Knauth, L.P. Petrogenesis of chert. In *Silica, Physical Behavior, Geochemistry and Materials Application*; Heaney, P.J., Prewitt, C.T., Gibbs, G.V., Eds.; Mineralogical Society of America: Chantilly, VA, USA, 1994; Volume 29, pp. 233–258.

15. Warren, J.K. *Evaporites: A Geological Compendium*; Springer: Berlin/Heidelberg, Germany, 2016; p. 1816.

16. Scotford, D.M. A Test of Aluminum in Quartz as a Geothermometer. *Am. Mineral.* **1975**, *60*, 139–142.

17. Bustillo, M.A. Silicification of Continental Carbonates. In *Developments in Sedimentology*; Alonso-Zarza, A.M., Tanner, L.H., Eds.; Elsevier: Amsterdam, The Netherlands, 2010; Volume 62, pp. 153–178.

18. Goudie, A.; Calcrete, S.; Goudie, A.S. *Chemical Sediments and Geomorphology*; Goudie, A.S., Pye, K., Eds.; Academic Press: London, UK, 1983; pp. 93–131.

19. Arche, A.; López-Gómez, J. Origin of the Permian-Triassic Iberian Basin, central-eastern Spain. *Tectonophysics* **1996**, *266*, 443–464. [CrossRef]

20. Sopeña, A.; López-Gómez, J.; Arche, A.; Pérez-Arlucea, M.; Ramos, A.; Virgili, C.; Hernándo, S. Permian and Triassic rift basins of the Iberian Peninsula. In *Triassic–Jurassic Rifting—Continental Breakup and the Origin of the Atlantic Ocean and Passive Margins*; Manspeizer, W., Ed.; Elsevier: Amsterdam, The Netherlands, 1988; Volume 22B, pp. 757–786.

21. Van Wees, J.D.; Arche, A.; Beijdorff, C.G.; López-Gómez, J.; Cloetingh, S. Temporal and spatial variations in tectonic subsidence in the Iberian Basin (eastern Spain): Inferences from automated forward modelling of high-resolution stratigraphy (Permian–Mesozoic). *Tectonophysics* **1998**, *300*, 285–310. [CrossRef]

22. Vargas, H.; Gaspar-Escribano, J.; López-Gómez, J.; Van Wees, J.D.; Cloetingh, S.; De la Horra, R.; Arche, A. A comparison of the Iberian and Ebro Basins during the Permian and Triassic, eastern Spain: A quantitative subsidence modelling approach. *Tectonophysics* **2009**, *474*, 160–183. [CrossRef]

23. Salas, R.; Guimerà, J.; Mas, R.; Martín-Closas, C.; Meléndez, A.; Alonso, A. Evolution of the Mesozoic Central Iberian Rift System and its Cainozoic inversion (Iberian Chain). In *Peri-Tethys Memoir 6: Peri-Tethyan Rift/Wrench Basins and Passive Margins*; Ziegler, P.A., Cavazza, W., Robertson, A.H., Crasquin-Soleau, F., Eds.; Muséum National Histoire Naturelle: Paris, France, 2001; pp. 145–185.

24. Sopeña, A.; Sánchez-Moya, Y. Tectonic systems tract and depositional architecture of the western border of the Triassic Iberian Trough (central Spain). *Sediment. Geol.* **1997**, *113*, 245–267. [CrossRef]

25. Sánchez-Moya, Y.; Sopeña, A. El rift Mesozoico Ibérico. In *Geología de España*; Vera, J.A., Ed.; SGE-IGME: Madrid, Spain, 2004; pp. 484–522.

26. Rosembaum, G.; Lister, G.S.; Duboz, C. Relative motions of Africa, Iberia and Europe during Alpine orogeny. *Tectonophysics* **2002**, *359*, 117–129. [CrossRef]

27. López-Gómez, J.; Alonso-Azcárate, J.; Arche, A.; Arribas, J.; Fernández Barrenechea, J.; Borruel-Abadía, V.; Bourquin, S.; Cadenas, P.; Cuevas, J.; De la Horra, R.; et al. Permian-Triassic Rifting Stage. In *The Geology of Iberia: A Geodynamic Approach*; Quesada, C., Oliveira, T., Eds.; Springer: Berlin/Heidelberg, Germany, 2019; Volume 3, pp. 29–112.

28. De Vicente, G.; Vegas, R.; Muñoz-Martín, A.; Van Wees, J.D.; Casas-Sainz, A.M.; Sopeña, A.; Sánchez-Moya, Y.; Arche, A.; López-Gómez, J.; Olaiz, A.; et al. Oblique strain partitioning and transpression on an inverted rift: The Castilian Branch of the Iberian Chain. *Tectonophysics* **2009**, *470*, 224–242. [CrossRef]

29. Murphy, J.B.; Nance, R.D.; Cawood, P.A. Contrasting modes of supercontinent formation and the conundrum of Pangea. *Gondwana Res.* **2009**, *15*, 408–420. [CrossRef]

30. Ziegler, P.A.; Stamplfli, G.M. Late Paleozoic-early Mesozoic plate boundary reorganization, collapse of the variscan orogeny and opening of neotethys. In *Permian Continental Deposits of Europe and other Areas. Regional Reports and Correlations*; Cassinis, G., Ed.; Museo Civico Science Naturali: Brescia, Italy, 2001; Volume 25, pp. 17–34.

31. López-Gómez, J.; Arche, A.; Pérez-López, A. Permian and Triassic. In *The Geology of Spain*; Gibbons, W., Moreno, T., Eds.; The Geological Society: London, UK, 2002; pp. 185–212.

32. Sopeña, A.; Gutiérrez-Marco, J.C.; Sánchez-Moya, Y.; Gómez, J.J.; Mas, J.R.; García, A.; Lago, M. Cordillera Ibérica y Costero Catalana. In *Geología de España*; Vera, J.A., Ed.; SGE-IGME: Madrid, Spain, 2004; pp. 465–527.

33. Ortí, F.; Pérez-López, A.; Salvany, J.M. Triassic evaporites of Iberia: Sedimentological and palaeogeographical implications for the western Neotethys evolution during the Middle Triassic-Earliest Jurasic. *Palaeogeogr. Palaeoclimatol. Palaeoecol.* **2017**, *471*, 157–180. [CrossRef]

34. Suarez Alba, J. La Mancha Triassic and Lower Lias stratigraphy, a well log interpretation. *J. Iber. Geol.* **2007**, *33*, 55–78.

35. Arche, A.; López-Gómez, J.; García-Hidalgo, J.F. Control climático, tectónico y eustático en depósitos del Carniense (Triásico Superior) del SE de la Península Ibérica. *J. Iber. Geol.* **2002**, *28*, 13–30.

36. Ortí, F.; García-Veigas, J.; Rosell Ortiz, L.; Jurado, M.J.; Utrilla, R. Formaciones salinas de las cuencas triásicas en la Península Ibérica: Caracterización petrológica y geoquímica. *Cuadernos De Geología Ibérica* **1996**, *20*, 13–35.

37. Escavy, J.I.; Herrero, M.J. The use of location-allocation techniques for exploration targeting of high place-value industrial minerals: A market-based prospectivity study of the Spanish gypsum resources. *Ore Geol. Rev.* **2013**, *53*, 504–516. [CrossRef]

38. Escavy, J.I.; Herrero, M.J.; Arribas, M.E. Gypsum resources of Spain: Temporal and spatial distribution. *Ore Geol. Rev.* **2012**, *49*, 72–84. [CrossRef]

39. Goy, A.; Martínez, G.; Pérez-Valera, F.; Pérez-Valera, J.A.; Trigueros-Ramos, L.M. Nuevos hallazgos de Cefalópodos (Ammonoideos y Nautiloideos) del Ladiniense Inferior en el sector oriental de las Cordillera Béticas. *Boletín de la Real Sociedad Española de Historia Natural* **1996**, *125*, 311–314.

40. Lago, M.; Dumitrescu, R.; Bastida, J.; Arranz, E.; Gil-Imaz, A.; Pocovi, A.; Lapuente, M.P.; Vaquer, R. Características de los magmatismos alcalino y subalcalino, pre-Hettangiense, del borde SE de la Cordillera Ibérica. *Cuadernos De Geología Ibérica* **1994**, *20*, 159–181.

41. Chun, F.H. Quantitative interpretation of X-ray diffraction patterns of mixtures. III. simultaneous determination of a set of reference intensities. *J. Appl. Chrystallogr.* **1975**, *8*, 17–19. [CrossRef]

42. Zinkernagel, U. Cathodoluminescence of quartz and its application to sandstone petrology. *Contrib. Sedimentol.* **1978**, *8*, 1–69.

43. Matter, A.; Ramseyer, K. Cathodoluminescence Microscopy as A Tool for Provenance Studies of Sandstones. In *Provenance of Arenites*; Zuffa, G.G., Ed.; Reidel: Boston, MA, USA, 1985; pp. 191–211.

44. El Tabakh, M.; Grey, K.; Pirajno, F.; Schreiber, B.C. Pseudomorphs after evaporitic minerals interbedded with 2.2 Ga stromatolites of the Yerrida basin, Western Australia: Origin and significance. *Geology* **1999**, *27*, 871–874. [CrossRef]

45. Kingma, K.J.; Hemley, R.J. Raman spectroscopic study of microcrystalline silica. *Am. Miner.* **1994**, *79*, 269–273.

46. Rodgers, K.A.; Cressey, G. The occurrence, detection and significance of moganite (SiO_2) among some silica sinters. *Miner. Mag.* **2001**, *65*, 157–167. [CrossRef]

47. Del Olmo, P.; Olivé, A.; Gutiérrez, M.; Aguilar, M.J.; Leal, M.C., Portero, J.M. Hoja geológica 438 (Paniza). In *Mapa Geológico de España, E. 1:50.000*; Segunda Serie I.G.M.E.: Madrid, Spain, 1983.

48. Castaño, R.; Doval, M.; Marfil, R. Naturaleza, origen y distribución de los minerales de la arcilla en la cuenca Triásica (Keuper) del área de Valencia. *Cuadernos de Geología Ibérica* **1987**, *11*, 339–361.

49. Miehe, G.; Graetsch, H.; Flörke, O.W. Crystal structure and growth fabric of length-fast chalcedony. *Phys. Chem. Miner.* **1984**, *10*, 197–199. [CrossRef]

50. Sánchez-Moya, Y.; Herrero, M.J.; Sopeña, A. Strontium isotopes and sedimentology of a marine Triassic succesion (upper ladinian) of the westernmost Tethys, Spain. *J. Iber. Geol.* **2016**, *42*, 171–186.

51. Palmer, S.E. Effect of Biodegradation and Water Washing On Crude Oil Composition. In *Organic Geochemistry: Principles and applications*, Engel, M.H., Macko, S.A., Eds.; Springer: Berlin/Heidelberg, Germany, 1993; Volume 11, pp. 511–533.

52. Siever, R. Silica solubility 0–200 °C and the diagenesis of siliceous sediments. *J. Geol.* **1962**, *70*, 127–150. [CrossRef]

53. Mahran, T.M. Late Oligocene lacustrine deposition of the Sodmin Formation, Abu Hammad Basin, Red Sea, Egypt; sedimentology and factors controlling palustrine carbonates. *J. Afr. Earth Sci.* **1999**, *29*, 567–592. [CrossRef]

54. De Wet, C.C.; J. F.; H. The Scots Bay Formation, Nova Scotia, a Jurassic carbonate lake with silica-rich hydrothermal springs. *Sedimentology* **1989**, *36*, 857–875. [CrossRef]

55. Smith, A.M.; Mason, T.R. Pleistocene, multiple-growth, lacustrine oncoids from the Poacher's Point Formation, Etosha Pan, northern Namibia. *Sedimentology* **1991**, *38*, 591–599. [CrossRef]

56. Lago, M.; Arranz, E.; Gil, A.; Pocovi, A. *Cordilleras Ibérica y Costero-Catalana*; Magmatismo Asociado. In Geología de, España, Vera, J.A., Eds.; SGE-IGME: Madrid, Spain, 2004; p. 890.

57. Maliva, R.G.; Siever, R. Nodular chert formation in carbonate rocks. *J. Geol.* **1989**, *97*, 421–433. [CrossRef]

58. Williams, L.A.; Parks, G.; Crerar, D.A. Silica diagenesis, I. Solubility controls. *J. Sediment. Pet.* **1985**, *55*, 301–311.

59. Williams, L.A.; Crerar, D.A. Silica diagenesis II. General mechanisms. *J. Sediment. Pet.* **1985**, *55*, 312–321.

60. Alonso-Zarza, A.M.; Sánchez-Moya, Y.; Bustillo, M.A.; Sopeña, A.; Delgado, A. Silicification and dolomitization of anhydrite nodules in argillaceous terrestrial deposits: An example of meteoric-dominated diagenesis from the Triassic of central Spain. *Sedimentology* **2002**, *49*, 303–317. [CrossRef]

61. Arbey, F. Les formes de la silice et l'identification des évaporites dans les formations silicifiées. *Bulletin des Centres de Recherche, Exploration-Production Elf Aquitaine* **1980**, *4*, 309–365.

62. Hesse, R. Silica diagenesis: Origin of inorganic and replacement cherts. *Earth-Sci. Rev.* **1989**, *26*, 253–284. [CrossRef]

63. Rimstidt, J.D.; Barnes, H.L. The kinetics of silica-water reactions. *Geochimica et Cosmochimica Acta* **1980**, *44*, 1683–1699. [CrossRef]

64. Götze, J.; Plötze, M.; Tichomirowa, M.; Fuchs, H.; Pilot, J. Aluminium in quartz as an indicator of the temperature of formation of agate. *Miner. Mag.* **2001**, *65*, 407–413. [CrossRef]

65. Ramseyer, K.; Baumann, J.; Matter, A.; Mullis, J. Cathodoluminescence colours of alpha-quartz. *Miner. Mag.* **1988**, *52*, 669–677. [CrossRef]

66. Götze, J. Mineralogy, Geochemistry and cathodoluminescence of authigenic quartz from different sedimentary rocks. In *Quartz: Deposits, Mineralogy and Analytics*; Götze, J., Möckel, R., Eds.; Springer: Berlin/Heidelberg, Germany, 2012; pp. 287–306.

67. Waugh, B. Formation of quartz overgrowths in Penrith Sandstone (Lower Permian) of northwest England as revealed by scanning electron microscopy. *Sedimentology* **1970**, *14*, 309–320. [CrossRef]

68. French, M.W.; Worden, R.H.; Mariani, E.; Larese, R.E.; Mueller, R.R.; Kliewer, C.E. Microcrystalline Quartz Generation and the Preservation of Porosity in Sandstones: Evidence from the Upper Cretaceous of the Subhercynian Basin, Germany. *J. Sediment. Res.* **2012**, *82*, 422–434. [CrossRef]

69. Wilkinson, M.; Haszeldine, R.S. Fibropus illite in oilfield sandstones—A nucleation kinetic theory of growth. *Terra Nova* **2002**, *14*, 56–60. [CrossRef]

70. Worden, R.H.; Morad, S. Quartz cementation in oil field sandstones: A review of the key controversies. *Spec. Publ. Int. Assoc. Sedimentol.* **2000**, *29*, 1–20.

71. Dennen, W.H.; Blackburn, W.H.; Quesada, A. Aluminium in quartz as a geothermometer. *Contrib. Miner. Pet.* **1970**, *27*, 32–42. [CrossRef]

72. García Palacios, M.C.; Lucas, J.; De la Peña, J.A.; Marfil, R. La Cuenca Triásica de la Rama Castellana de la Cordillera Ibérica. I Petrografía y Mineralogía. *Cuadernos de Geología Ibérica* **1977**, *4*, 341–354.

73. Lago, M.; Pocovi, A.; Bastida, J.; Arranz, E.; Vaquer, R.; Dumitrescu, R.; Gil-Imaz, A.; Lapuente, M.P. El magmatismo alcalino, hettangiense, en el dominio nor-oriental de la Placa Ibérica. *Cuadernos de Geología Ibérica* **1996**, *20*, 109–138.

74. Dercourt, J.; Gaetani, M.; Vrielynck, B.; Barrier, E.; Biju-Duval, B.; Brunet, M.F.; Cadet, J.P.; Crasquin, S.; Sandulescu, M. *Atlas or Peri-Tethys. Paleeogeographical Maps*; Commission for the Geological Map of the World: Paris, France, 2000; p. 268.

75. Tornos, F.; Spiro, B.F. The Geology and Isotope Geochemistry of the Talc Deposits of Puebla de Lillo (Cantabrian Zone, Northern Spain). *Econ. Geol.* **2000**, *95*, 1277–1296. [CrossRef]

Article

Clay Minerals and Element Geochemistry of Clastic Reservoirs in the Xiaganchaigou Formation of the Lenghuqi Area, Northern Qaidam Basin, China

Guoqiang Sun [1,2,*], Yetong Wang [1,3], Jiajia Guo [4], Meng Wang [1,3], Yun Jiang [1,3] and Shile Pan [1,3]

1   Northwest Institute of Eco-Environment and Resources, Chinese Academy of Sciences, Lanzhou 730000, China; wangyetong16@mails.ucas.edu.cn (Y.W.); wangmeng17@mails.ucas.ac.cn (M.W.); jiangyun1234554321@163.com (Y.J.); shilepan@163.com (S.P.)
2   Key Laboratory of Petroleum Resources, Lanzhou 730000, China
3   University of Chinese Academy of Sciences, Beijing 100049, China
4   Research Institute of Petroleum Exploration and Development-Northwest, Petrochina, Lanzhou 730020, China; guojiajia123@petrochina.com.cn
*   Correspondence: sguoqiang@lzb.ac.cn; Tel.: +86-0931-496-0858

Received: 17 October 2019; Accepted: 31 October 2019; Published: 3 November 2019

Abstract: We performed mineralogical and geochemical analyses of core samples from the Lenghuqi area in the northern marginal tectonic belt of the Qaidam Basin. The clay mineralogy of the Xiaganchaigou Formation sandstone is dominated by I + I/S + C types and characterized by high illite, a higher mixed-layer illite/smectite and chlorite, lesser smectite, and an absence of kaolinite. The clay minerals reflect that the Oligocene sedimentary basin formed in an arid-semi-arid climate with weak leaching and chemical weathering, and that diagenesis occurred in a K^+- and Mg^{2+}-rich alkaline environment. Measured major oxide concentrations show clear correlations. The lower Xiaganchaigou Formation is representative of a dry and cold freshwater sedimentary environment, whereas the upper Xiaganchaigou Formation is warmer and more humid. Trace element and rare earth element variations indicate that the paleoclimate conditions of the lower Xiaganchaigou Formation sedimentary period were relatively cold and dry, while the upper Xiaganchaigou Formation formed under warmer and more humid climate conditions. These findings reflect a global climate of a cold and dry period from the late Eocene to early Oligocene, and a short warming period in the late Oligocene.

Keywords: clay minerals; major elements; trace elements; sedimentary environment; diagenetic Environment; Qaidam Basin

1. Introduction

Clay minerals are widely distributed in sediments and are the products of sedimentation and diagenesis under certain conditions related to climate, water media, and provenance, which have important implications for interpreting the paleoenvironment, climate, and diagenesis processes. The study of clay mineralogy therefore provides important information regarding the characteristics of provenance, paleoclimate, and sedimentary and diagenetic environments, which shed insight on the sedimentary-diagenesis process [1–4]. The volatile elemental contents within sediments are commonly studied to extract information regarding environmental evolution. Particular environmental conditions influence the behavior of various elements differently during decomposition and migration, which affects their enrichment in different environments. Changes in element contents within sediments therefore often reflect a change in the environmental conditions during deposition [5–8]. Common major oxides (CaO, MgO, K_2O, Na_2O, SiO_2, Al_2O_3, Fe_2O_3, and TiO_2) [9,10] and trace elements (Sr, Ba, Ti, Fe, P, Mn, U, V, and Ni) [11,12] are sensitive to paleoenvironmental conditions with clear and observable indications [13–15]. The sedimentary-diagenetic environment and paleoclimate conditions of sedimentary rock formations can therefore be analyzed on the basis of major and trace elements in sedimentary rocks and changes in their ratios.

The increasing difficulty of shallow oil and gas extraction in China has led to a shift towards deep exploration and development [16,17]. Primary pores in reservoirs with a depth >3000 m are nearly absent owing to strong compaction and cementation [18], and deposits are dominated by secondary pores. However, recent studies [19–22] have shown that primary pores in a clastic reservoir can be preserved under specific geological conditions at about 3500–4500 m and form a high-quality reservoir. The Lenghuqi area is located in the center of the northern margin of the Qaidam Basin. Previous studies have summarized the characteristics and controlling factors of deep reservoirs in detail [23,24] and suggested that the sedimentary environment and deep ultra-high pressure layer are the main controlling factors of deep high-quality clastic reservoirs in the region. Carbonate cement also has a substantial impact on reservoir properties. However, few studies have addressed the paleoclimate conditions and diagenetic environments that existed during the formation of these reservoir rocks. The diagenesis process and diagenetic environment are critical factors required to clarify the evolution process of the reservoir rocks [25]. Further study on the sedimentary-diagenetic environment of the high-quality clastic reservoirs in the Xiaganchaigou Formation of the Oligocene can therefore provide a geological basis for the prediction of high-quality reservoir distribution laws.

2. Geological Setting

The Qaidam Basin located in northwest China is the largest continental basin in the northeastern Tibetan Plateau with a total area of 120,000 km^2 [26]. The basin has undergone compressional tectonic settings throughout the Cenozoic [26] and is bound by the Eastern Kunlun fault zone to the southwest, the South Qilian fault zone to the northeast, and the Altyn Tagh fault zone to the northwest [27] (Figure 1). The basin developed an exceptionally thick Tertiary sedimentary succession, of which the average thickness is up to 6 km and the maximum thickness is >10 km [28]. The basin is divided into four first-order structural units: the Western Qaidam Basin Uplift, the Yilingping Sag, the Sanhu Sag, and the Northern Qaidam Basin Uplift [29]. The Northern Qaidam Basin Uplift lies in front of the South Qilian Mountains and the Lenghuqi area is located in the center of the uplift.

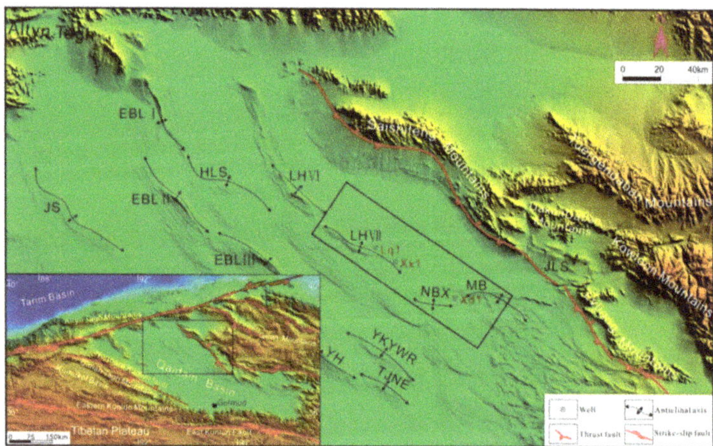

Figure 1. Structural location of the Lenghuqi area in the northern margin of the Qaidam Basin.

The ancient lake basin of the Qaidam Basin underwent three evolution stages from Paleogene to Neogene: the lake basin first appeared in the Paleocene and early Eocene, the paleolake then continuously expanded from late Eocene to Oligocene, and reached its maximum extent in the subsidence stage of the Oligocene. The paleolake then gradually shrank during the Early Miocene to

Pliocene [30,31]. The Paleogene and Neogene strata can be divided into six units on the basis of previous stratigraphic results from the northern Qaidam Basin. These six stratigraphic units (in ascending order) are the Lulehe Formation (E_{1+2}, Paleocene to early Eocene, ~65–~45 Ma), Xiaganchaigou Formation (E_3, middle to late Eocene, ~45–~35.5 Ma), ShangGanchaigou Formation (N_1, late Eocene to Oligocene, ~35.5–~22 Ma), XiaYoushashan Formation (N_2^1, early to middle Miocene, ~22–~15 Ma), ShangYoushashan Formation (N_2^2, middle to late Miocene, ~15–~8 Ma), and Shizigou Formation (N_2^3, late Miocene to Pliocene, ~8–2.8 Ma) [32–35]. In recent years, abundant oil and gas reservoirs in Paleogene strata have been found in the central area of the northern Qaidam Basin, which indicates abundant oil and gas resources in this area [36,37].

3. Materials and Methods

The Xiaganchaigou Formation (E_3) in the Lenghuqi area is deeply buried with an average depth of >4000 m. Samples were collected from drill core of the Xx1 well (Figure 1). The Xx1 well is located at 93°52′41″E, 38°7′34″N. It is 5500 m deep and its wellhead is 2829 m above sea level. One 5.64-m drill core was obtained between 4111.34 and 4117.09 m with a harvest rate of 98.1% (Figure 2a), and another drill core of 8.72 m was obtained between 4847.00 and 4855.82 m with a harvest rate of 98.9% (Figure 2b). The lithology and logging characteristics of the core section of the Xx1 well are shown in Figure 2. Whole rock and clay mineral compositions are listed in Table 1. X-ray diffraction (XRD) was performed following techniques outlined by the Qinghai Oilfield Exploration and Development Research Institute of PetroChina (SY/T5163-2010 X-ray Diffraction Analysis Method for Clay Minerals and Common Non-clay Minerals in Sedimentary Rocks). Twenty-five whole rock and clay mineral components were determined (Table 1). We use the XRD results in combination with microscopic observations, scanning electron microscopy, and electron probe analysis of the rock flakes to identify the clay mineral development characteristics in the Xiaganchaigou Formation from the Oligocene mudstones and sandstones in the Lenghuqi area of the Qaidam Basin.

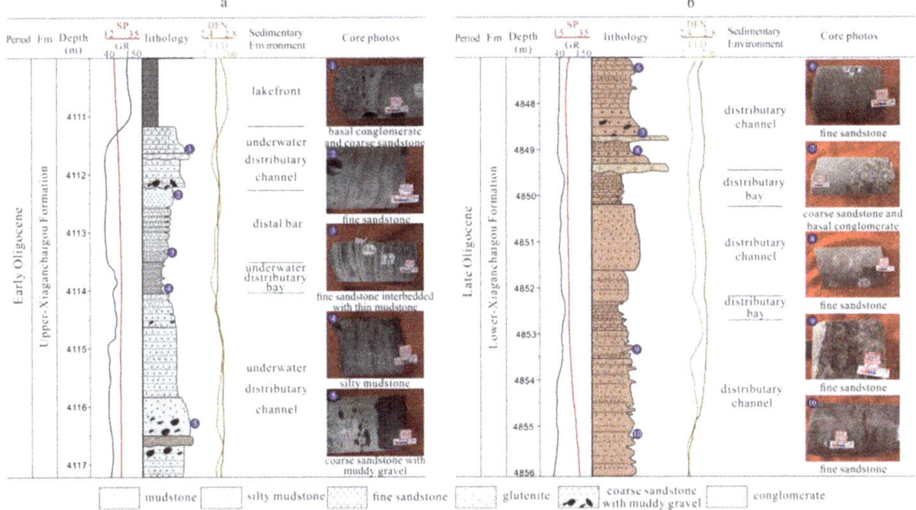

Figure 2. Sedimentological and petrophysical characteristics of the Xx1 well. (**a**): Upper-Xiaganchaigou Formation; (**b**): Lower-Xiaganchaigou Formation.

Table 1. X-ray Diffraction Results of Whole Rock and Clay Mineralogy of the Xiaganchaigou Formation.

Buried Depth	Formation	Mineral Type and Content (%)					Total Amount of Clay (%)	Relative Content of Clay Minerals (%)					Mixed Layer Ratio (%)
		Quartz	Potassium Feldspar	Sodium Feldspar	Calcite	Anhydrite		S	I/S	I	K	C	I/S
4111.51		68.2	6	9.4	3.8	/	12.6	/	32	44	/	24	6
4111.75		47.6	6	16.2	20.3	/	9.9	/	19	47	/	34	10
4112.11		47.4	9.5	10.3	11.4	/	21.3	/	11	72	/	17	12
4112.64	Upper Xia Ganchaigou Fm.	69.9	2.2	7.1	7.9	0.8	12.1	/	26	54	/	20	5
4112.99		45.1	3.9	12.1	16.6	/	22.4	/	7	80	/	13	9
4113.00		49.7	9.9	9.9	14.5	/	16.0	/	15	69	/	16	10
4114.16		46.7	9.9	9.9	14.3	1.1	18.1	/	30	61	/	9	10
4114.82		42.8	9.5	17.1	6.4	/	21.8	/	32	42	/	26	12
4115.27		70.2	3.6	8.4	5.2	/	12.7	/	25	58	/	17	8
4115.60		60.7	10.3	14.1	4.1	/	10.8	/	22	51	/	27	5
4115.72		64.7	5.2	11.9	2.9	/	15.4	/	11	62	/	27	7
4116.04		68.1	3.6	10	5.1	/	13.3	/	51	31	/	18	14
4116.34		53.9	14.9	12.6	6.4	/	12.3	/	38	41	/	21	15
Average		56.72	7.27	11.46	9.15	0.16	15.28	/	24.54	54.77	/	20.69	9.46
4847.08		49.1	11.9	20.9	11.5	/	6.6	/	9	84	/	7	11
4849.23		65.6	4.5	11.8	2.1	/	15.9	/	5	72	/	23	12
4850.46		57.7	10.5	23	1.4	/	7.4	/	4	78	/	18	13
4851.00	Lower Xia Ganchaigou Fm.	68.6	5	11.9	2.3	1.5	10.7	/	16	73	/	11	9
4851.39		59.9	17	13.7	1.8	/	7.7	/	24	53	/	23	7
4851.68		54.9	5.5	13	17.6	/	9.2	/	4	64	/	32	10
4852.40		42.6	21.7	7.3	24.2	/	4.1	/	18	69	/	13	13
4853.37		56.7	5.8	7.4	20.1	/	10.0	/	2	80	/	18	10
4853.91		62.5	5.7	12.9	6	/	13.0	/	3	79	/	18	11
4854.38		49	12.5	15.2	6.7	/	16.6	/	11	82	/	7	10
4854.62		59.7	5.5	10.6	7.5	1.5	15.2	/	5	69	/	12	14
4855.30		49	7.3	13.4	14.4	/	15.9	/	6	76	/	13	5
Average		56.28	9.41	13.43	9.63	0.27	11.03	/	8.92	73.25	/	16.25	10.42

Standard: SY/T 5163-1995. The reason that S + I/S + I +K +C + C/S = 101 or 99 is rounded numbers rather than data bias. The clay mineral with a mixed-layer ratio > 70% is smectite. S is smectite, I/S is mixed-layer illite/smectite, I is illite, K is kaolinite, and C is chlorite. The mixed layer ratio (e.g., 20%) indicates that the smectite content in the mixed layer illite/smectite is 20%.

Samples were observed microscopically prior to chemical analysis to monitor potential alteration, mineralization, or secondary weathering. All samples were ground with a non-contaminating crusher, screened in a 200-mesh sieve, and heated in an oven at 80 °C for 3 h to remove moisture. Major element measurements were performed using a fluorescence spectrometer *3080E3X* (Rishi Electric Corporation, Japan). We used HF + HNO_3 to seal and dissolve samples during trace element analysis using laser coupled plasma mass spectrometry (ICP-MS). Analyses were completed in the Lanzhou Oil and Gas Resources Research Key Laboratory in the Institute of Geology and Geophysics at the Chinese Academy of Sciences.

4. Results

4.1. Clay Minerals

The test results show that illite is the most abundant mineral in all samples, followed by chlorite and the mixed layer illite/smectite. Smectite and kaolinite were not detected. The quartz content in the lower Xiaganchaigou Formation ranges from 42.6% to 68.6% with an average of 56.28%. The clay mineral content ranges from 4.1% to 16.6% with an average of 11.03%. The illite content in the clay minerals ranges from 53% to 84% with an average of 73.25%. The chlorite content ranges from 7% to 23% with an average of 16.25%. The mixed layer illite/smectite content ranges from 6.6% to 16.6% with an average of 8.92% and average smectite content in the mixed layer illite/smectite of 10.42%. In the Upper Xiaganchaigou Formation, the quartz content ranges from 42.8% to 70.2% with an average of 56.72%. The clay mineral content ranges from 9.9% to 22.4% with an average of 15.28%. The illite content in the clay minerals is substantially lower than that in the lower Xiaganchaigou Formation; ranging from 31% to 80% with an average of 54.77%. The mixed layer illite/smectite content ranges from 7% and 51% with an average of 24.54% with an average smectite content in the mixed layer illite/smectite of 9.46%. The chlorite content is also somewhat higher, ranging from 9% to 34% with an average of 20.69%.

The illite in the upper member sandstone of the Xiaganchaigou Formation is mostly filamentous or bridging between particles (Figure 3a). The mixed layer illite/smectite is more developed, usually on particle surfaces or filled between particles (Figure 3a–d). The illite in the lower member sandstone of the Xiaganchaigou Formation is filled with curved leaves or scales between particles (Figure 3e). The mixed layer illite/smectite is less common, and flaky chlorite is visible on the particle surfaces (Figure 3f).

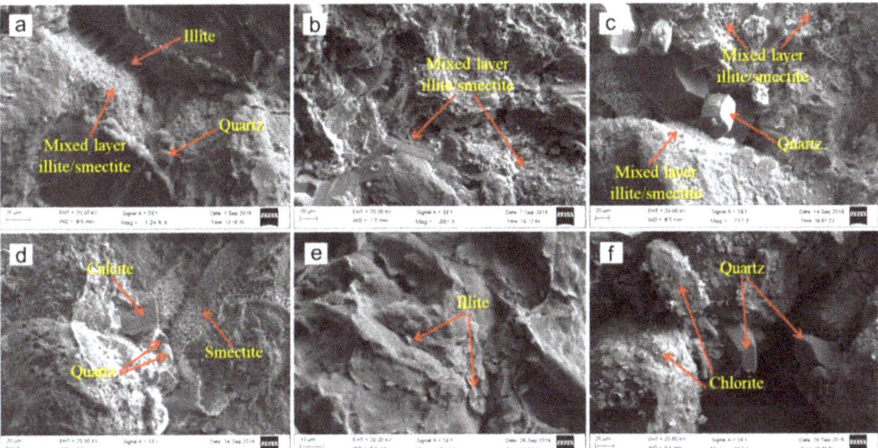

Figure 3. Scanning electron micrograph of sandstone from the Xiaganchaigou Formation (Xx 1 well). (**a**) 4111.51 m, upper Xiaganchaigou Formation, filamentous illite, mixed layer illite/smectite, and other clay minerals filled between particles and particle surfaces; (**b**) 4115.37 m, upper Xiaganchaigou Formation, mixed layer illite/smectite filled between particles and particles surfaces; (**c**) 4209.77 m, upper Xiaganchaigou Formation, interstitial calcite, self-generated and enlarged quartz. Particle surfaces develop a mixed layer illite/smectite clay mineral; (**d**) 4122.14 m, upper Xiaganchaigou Formation, interstitial calcite. Particle surfaces develop a mixed layer illite/smectite clay mineral; (**e**) 4850.00 m, lower Xiaganchaigou Formation, curved leaf-like illite clay completely filled in intergranular pores; (**f**) 4853.05 m, lower Xiaganchaigou Formation, intergranular pores are developed, self-generated and enlarged quartz. Chlorite is developed on the particle surfaces.

The illite content in the lower member is substantially higher than that in the upper member and the mixed layer illite/smectite content also reduces downwards. This indicates that the clay minerals in the sandstone are mainly self-generated. Their combined characteristics reflect their formation conditions, including the climate and sedimentary and diagenetic environments. Kaolinite is mostly distributed in moist tropical or subtropical regions [4,26] and mainly forms from feldspar leaching and weathering [27], as well as weathering and low-temperature hydrothermal metasomatism [28]. In contrast to the kaolinite formation environment, chlorite can only be preserved in areas with weak chemical weathering [1]. Smectite is associated with a warm–humid–cold climate, pervasive within sediments in temperate and semi-humid regions [26], and its contents tend to decrease with warmer temperatures [29]. Illite mainly forms in dry and cold climates [30,31] and the combination of clay minerals dominated by illite reflects an arid–semi-arid climate [26]. The clay mineral combination of the Xiaganchaigou Formation sandstone in the Lenghuqi area is dominated by the I + I/S + C type, which displays distribution characteristics of high illite content, higher mixed layer illite/smectite and chlorite content, low smectite, and missing kaolinite. This suggests that the sedimentary basin in the Oligocene (i.e., the Xiaganchaigou Formation sedimentary period) was generally in an arid–semi-arid climate with weak leaching and chemical weathering.

Authigenic kaolinite mainly forms by feldspar leaching in an acidic medium followed by the direct formation of pore water during diagenesis [26,32,33]. Kaolinite is therefore an indicator mineral for weakly acidic environments with leaching and strong chemical weathering [28,29]. Authigenic smectite usually forms in alkaline media that is depleted in K^+ but enriched in Na^+ and Ca^{2+} [4,34]. Illite mainly forms by the weathering of potassium feldspar, mica, and other aluminosilicate minerals in a hydrous K^+-rich alkaline medium under weak leaching [4,26,35,36]. Fe^{2+} and Mg^{2+} in smectite are replaced by Al^{3+} and K^+ during dehydration [32]. Smectite may also form from the precipitation of pore water in an alkaline medium [37]. Chlorite forms by the precipitation of pore water rich in Mg^{2+} in a relatively high-temperature and more strongly-alkaline environment [26]. Illite is more stable in a water environment with a high K^+/H^+ ratio. The presence of the mixed layer illite/smectite reflects the gradual transition of smectite to illite in a K^+-rich alkaline environment, and chlorite forms in a Mg^{2+}-rich alkaline environment [26,32]. The clay mineralogical characteristics of the Xiaganchaigou Formation sandstone in the Lenghuqi area reflect a diagenetic and primarily alkaline environment rich in K^+ and Mg^{2+}.

4.2. Major Element Analysis

Environmental conditions influence the behavioral characteristics of elements and compounds, including decomposition, migration, and the enrichment of various elements with different properties. A change of sediment chemistry can reflect a change in environmental conditions during deposition. Common major oxides (CaO, MgO, K_2O, Na_2O, SiO_2, Al_2O_3, Fe_2O_3, and TiO_2) are sensitive to paleoenvironmental conditions and have clear environmental indications [9,10]. The major oxide compositions measured from samples from the Lenghuqi area are listed in Table 2. The results (Figure 4) show a clear correlation between the oxides. The overall change of oxide content in the lower Xiaganchaigou Formation is small, and the oxide content in the upper member varies greatly. This indicates that climate change in the lower member was not substantial during the sedimentary period. On the other hand, paleoclimatic change in the upper member occurred relatively frequently during the sedimentary period, and the climate fluctuated considerably.

High $CaCO_3$ levels in a closed or semi-enclosed inland lake during the early stages of chemical deposition represent an arid climate, whereas low $CaCO_3$ levels represent a relatively humid climate [38–40]. Ca^{2+} and Mg^{2+} commonly exchange for one another but the large ionic radius of Ca^{2+} enhances its migration. Thus, an environment enriched in Ca^{2+} was likely drier than an environment enriched in Mg^{2+} [10]. The average CaO and MgO contents from the upper Xiaganchaigou Formation samples are 3.23% and 1.26%, respectively, whereas those of the lower Xiaganchaigou Formation are 6.14% and 0.78%. Compared with the upper Xiaganchaigou Formation, the lower Xiaganchaigou Formation is characterized by high CaO and Na_2O and low K_2O, MgO, and TiO_2 (Figure 5). This supports that the lower Xiaganchaigou Formation formed in a relatively dry climate with strong evaporation, while the upper member formed in a relatively warm and humid environment with less evaporation [41,42]. The change of Ti content and other elements reflects the extent of the addition of terrigenous materials. Higher Ti values are associated with richer terrigenous material contents, indicating a warm and humid climate background [43,44]. The paleoclimate characteristics of the sedimentary period of the Xiaganchaigou Formation therefore gradually became humid from early to late in the deposition sequence.

Table 2. Major oxide contents of the Xiaganchaigou Formation in the Lenghuqi structural belt.

Formation	Buried Depth	Lithology	Na₂O (%)	MgO (%)	Al₂O₃ (%)	SiO₂ (%)	K₂O (%)	CaO (%)	TiO (%)	Fe₂O₃ (%)	P₂O₅ (%)	MnO (%)
Upper Xia Ganchaigou Fm.	4111.49	Dark gray medium sandstone	2.15	0.51	7.90	79.75	1.66	1.97	0.09	2.26	0.05	0.12
	4111.64	Dark gray fine sandstone	2.00	0.77	8.88	76.77	1.80	2.47	0.13	3.13	0.05	0.11
	4112.24	Reddish-brown mud-bearing fine sandstone	1.52	2.98	16.69	64.37	4.05	0.54	0.32	5.90	0.13	0.07
	4113.54	Dark gray fine sandstone	2.04	1.05	10.07	68.64	2.03	5.18	0.24	2.96	0.09	0.09
	4114.04	Gray-white mud-bearing medium sandstone	1.93	0.98	9.68	68.78	1.94	6.87	0.26	2.76	0.09	0.13
	4114.50	Gray-white mud-bearing medium sandstone	1.86	1.39	11.99	71.18	2.53	3.15	0.26	3.49	0.10	0.10
	4115.44	Gray-white medium sandstone	2.19	0.44	7.91	80.38	1.76	2.01	0.08	2.05	0.04	0.12
	4116.24	Gray-white coarse sandstone	2.15	0.31	7.08	78.63	1.61	2.31	0.08	1.70	0.04	0.13
	4116.99	Gray-white mud-bearing medium sandstone	1.60	2.87	15.37	57.87	3.46	4.53	0.44	6.25	0.13	0.09
	Average		1.94	1.26	10.62	71.82	2.32	3.23	0.21	3.39	0.08	0.11
Lower Xia Ganchaigou Fm.	4847.45	Brown red fine sandstone	1.86	1.12	9.15	67.38	2.49	7.02	0.19	2.55	0.03	0.09
	4847.75	Brown red fine sandstone	2.07	1.22	9.90	73.81	2.54	2.98	0.20	3.30	0.02	0.15
	4848.30	Brown red fine sandstone	1.89	0.84	8.34	68.70	2.29	7.23	0.18	2.47	0.01	0.11
	4849.00	Gray-green medium sandstone	2.15	1.88	10.73	70.86	2.65	2.87	0.20	3.63	0.05	0.13
	4849.50	Brown red fine sandstone	1.88	0.62	7.45	69.66	2.10	7.76	0.13	2.11	0.02	0.12
	4850.00	Brown red fine sandstone	2.40	0.67	8.20	77.81	2.08	2.70	0.14	2.49	0.02	0.14
	4850.30	Brown red fine sandstone	2.63	0.49	8.03	83.31	1.91	0.74	0.12	2.16	0.02	0.12
	4851.00	Brown red fine sandstone	2.04	0.68	7.80	69.97	2.07	7.22	0.16	2.23	0.02	0.10
	4852.00	Gray-brown fine sandstone	3.14	0.29	7.22	83.19	1.81	0.69	0.08	1.90	0.01	0.13
	4852.70	Gray-brown fine sandstone	2.92	0.46	7.37	79.06	1.86	2.48	0.09	1.99	0.01	0.12
	4853.50	Gray-brown fine sandstone	2.02	0.30	6.09	68.05	1.73	9.94	0.10	1.76	0.01	0.13
	4853.90	Brown red fine sandstone	1.20	0.27	5.16	58.51	1.82	15.94	0.09	1.58	0.01	0.10
	4854.30	Brown red fine-silt sandstone	1.64	0.33	6.08	62.06	1.82	12.91	0.13	1.75	0.01	0.10
	4855.00	Brown red fine-silt sandstone	2.12	1.71	9.88	68.59	2.31	5.50	0.21	2.45	0.02	0.09
	Average		2.14	0.78	7.96	71.50	2.11	6.14	0.15	2.31	0.02	0.12

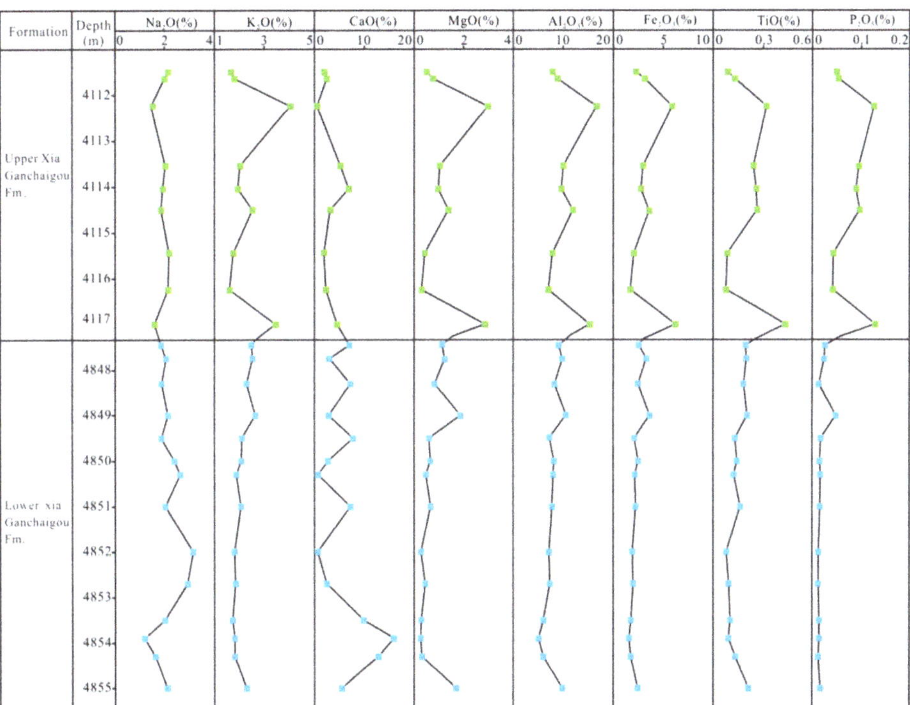

Figure 4. Major oxide content of the Xiaganchaigou Formation in the Lenghuqi structural belt.

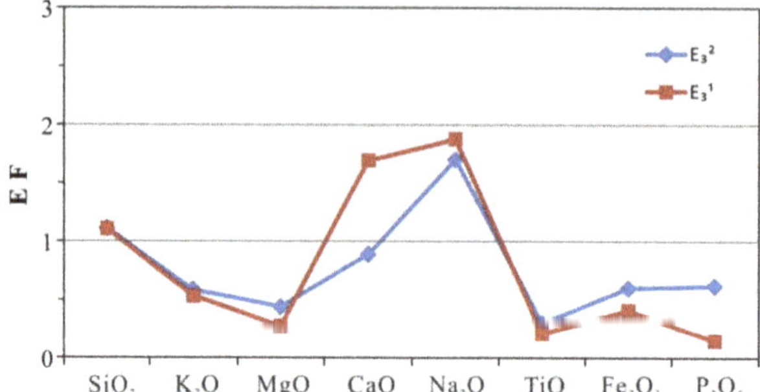

Figure 5. Comparison of major oxides measured in samples and those of the average shale [40]. (EF (enrichment factor) = Csediment/Cstandard rock). E_3^2 = Upper Xiaganchaigou Formation. E_3^1 = Lower Xiaganchaigou Formation.

4.3. Trace Element Analysis

The chemical distribution in a rock layer depends on the physical and chemical properties of the elements themselves, which are also substantially affected by paleoclimate and paleoenvironment. The distribution of trace elements and ratio changes are therefore also indicative of the paleoclimate evolution [11,12,14,45,46]. For example, the compositions of Sr, Ba, Ti, Fe, P, Mn, and the Mg/Ca and Sr/Cu ratios are sensitive geochemical indicators and parameters for identifying changes in sedimentary environment [15].

In lacustrine sediments without seawater intrusion, Sr/Ba > 1 typically indicates the onset of lake alkalization, whereas Sr/Ba < 1 typically indicates freshwater deposition [9,46]. Sr/Cu ratios between 1 and 10 indicate a warm and humid climate, and values > 10 represent arid climatic conditions [15]. The Sr/Ba values of all the samples in the Xiaganchaigou Formation are less than 1 (Table 3) and the average values of the upper and lower members are not substantially different: 0.29 and 0.22, respectively. All Sr/Cu values in the upper Xiaganchaigou Formation aside from one data point are <10 with an average value of 8.58. All Sr/Ba values in the lower member aside for two data points are >10 with an average of 14.39. The Ba content in the Xiaganchaigou Formation is significantly higher than that in the average shale [47] (Figure 6). The distribution of Sr/Ba and Sr/Cu demonstrates that the depositional environment of the Xiaganchaigou Formation was mainly dominated by a freshwater sedimentary environment during the sedimentary period. The climate was relatively dry in the early stage and humid in the later stage.

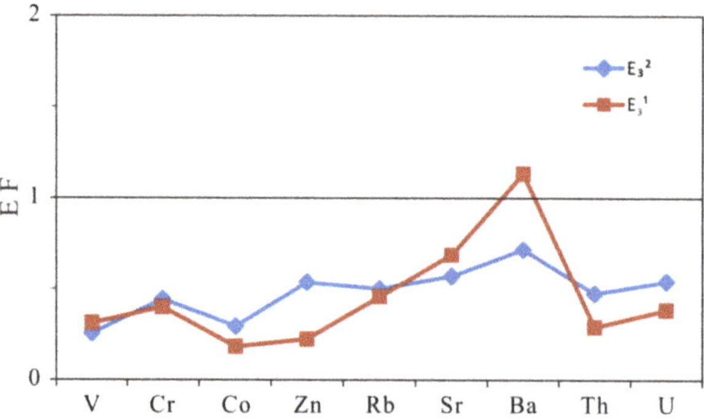

Figure 6. Comparison of trace elements between samples and the average shale. EF (enrichment factor) = Csediment/Cstandard rock. E_3^2 = Upper Xiaganchaigou Formation. E_3^1 = Lower Xiaganchaigou Formation.

Table 3. Partial trace element content of the Xiaganchaigou Formation in the Lenghuqi Area

Formation	Buried Depth	Element Content (mg/kg)										Ratio	
		V	Cr	Co	Ni	Cu	Zn	Sr	Ba	Th	U	Sr/Cu	Sr/Ba
Upper Xia Ganchaigou Fm.	4111.49	23.05	53.46	4.85	8.86	11.54	32.87	132.00	572.10	4.28	0.95	11.43	0.23
	4111.64	40.03	44.72	5.71	11.44	12.94	40.84	108.07	391.96	5.19	1.29	8.35	0.28
	4112.24	86.64	63.00	11.20	29.03	12.66	80.66	117.53	782.20	13.15	3.38	9.28	0.15
	4114.04	38.99	59.30	9.57	15.66	20.08	50.51	131.46	280.50	7.66	1.92	6.55	0.47
	4114.50	48.46	63.18	9.10	16.14	11.60	60.21	114.03	438.07	9.56	2.23	9.83	0.26
	4115.44	18.94	31.56	4.04	8.14	11.00	32.47	97.58	436.59	4.13	0.94	8.87	0.22
	4115.89	36.91	44.32	6.59	11.54	12.64	42.43	115.49	313.19	7.64	1.84	9.14	0.37
	4116.24	14.63	30.37	3.16	7.55	15.26	24.38	95.79	509.00	3.80	0.84	6.28	0.19
	4116.99	97.24	65.08	11.47	27.90	18.56	87.80	139.52	347.24	12.86	3.70	7.52	0.40
Lower Xia Ganchaigou Fm.	4847.45	38.40	43.68	4.88	10.33	9.11	20.60	121.31	469.56	5.27	1.41	13.31	0.26
	4847.75	46.57	61.62	6.52	11.22	12.26	27.12	118.02	718.14	5.66	1.63	9.63	0.16
	4848.30	35.11	45.67	4.34	9.34	9.14	16.69	119.87	427.95	5.32	1.34	13.12	0.28
	4849.00	53.52	58.99	8.97	11.09	15.72	49.91	110.48	400.82	8.60	1.60	7.03	0.28
	4849.50	34.68	32.86	3.73	7.75	9.04	14.26	137.74	819.29	3.64	1.02	15.24	0.17
	4850.00	55.20	47.37	4.21	6.65	10.74	18.27	121.99	611.88	4.16	1.22	11.35	0.20
	4850.30	59.62	48.62	3.48	6.17	8.23	18.23	141.71	859.52	3.58	1.10	17.22	0.16
	4851.00	53.16	46.37	3.69	6.39	8.36	16.91	121.62	429.92	4.34	1.27	14.55	0.28
	4852.00	41.39	41.39	2.83	5.01	8.77	10.13	175.04	1518.69	2.20	0.79	19.96	0.12
	4852.70	51.01	38.82	3.54	5.30	12.06	14.76	206.03	1835.48	2.65	0.88	17.08	0.11
	4853.50	42.24	31.89	2.63	4.28	8.03	14.64	142.31	716.11	2.61	0.91	17.71	0.20
	4853.90	32.35	37.23	2.23	5.03	7.06	12.02	120.66	547.27	3.00	0.82	17.10	0.22
	4854.30	40.76	29.14	2.56	3.44	8.33	8.88	132.45	500.18	3.20	1.02	15.90	0.26
	4855.00	70.70	50.93	5.43	10.67	12.36	24.68	152.30	459.19	5.29	1.65	12.32	0.33

4.4. Rare Earth Elements

Rare earth elements (REE) are a special group of elements that play an important role in geochemistry. Because of the similar chemical nature of the REE, they are typically found together in nature and exhibit only minor differences in their atomic structure. The REE therefore display characteristic fractionation patterns during different geological processes [48].

The REE content of the Xiaganchaigou Formation in the Lenghuqi area is particularly high (Table 4). The distribution of REE contents in the upper member is 52.40–217.73 mg/kg with an average of 117.87 mg/kg. The average concentration of light REE (LREE) is 87.58 mg/kg, accounting for 74.50%, whereas the average heavy REE (HREE) is 30.29 mg/kg, accounting for 25.50%. The distribution of REE in the lower Xiaganchaigou Formation ranges from 33.68 to 98.59 mg/kg with a mean of 75.59 mg/kg. The average LREE is 54.71 mg/kg, accounting for 71.90%, and the average HREE is 20.88 mg/kg, accounting for 28.10%. The normalized REE distribution patterns of in the upper and lower Xiaganchaigou Formation (Figure 7) are similar to those of chondrites [49] and the North American shale [41], which indicates that the sediments have the same material source and formation process [48,50]. The distribution of the total amount of REE ($\sum$REE) with depth direction is very close to the trend of La, Ce, and Er (Figure 8). The change of $\sum$REE is closely related to the change of climate and environment, that is, REE content is higher in warm and humid climates and lower in cold and dry climates [51–53]. The REE content of the lower Xiaganchaigou Formation is significantly lower than that of the upper member, indicating that the paleoclimate conditions of sedimentary period in the lower member of the lower Ganchaigou Formation are relatively cold and dry, while the climate in the upper Xiaganchaigou Formation is warmer and humid. This is consistent with the climatic characteristics reflected by clay minerals.

Table 4. Rare earth element content of the Xiaganchaigou Formation in the Lenghuqi Area.

Formation	Buried Depth	Rare Earth Element Content (mg/kg)															
		Sc	Y	La	Ce	Pr	Nd	Sm	Eu	Tb	Gd	Dy	Ho	Er	Tm	Yb	Lu
Upper Xia Ganchaigou Fm.	4111.49	2.56	7.12	11.78	20.84	2.48	9.05	1.68	0.41	0.24	1.49	1.35	0.27	0.77	0.12	0.78	0.12
	4111.64	3.58	9.71	15.21	25.79	3.03	10.97	2.11	0.51	0.30	1.86	1.75	0.35	1.04	0.16	1.05	0.16
	4112.24	13.35	20.86	38.25	69.93	7.89	28.64	4.91	1.01	0.68	4.21	4.05	0.83	2.45	0.38	2.46	0.38
	4114.04	8.47	17.70	23.41	43.86	5.39	20.09	3.82	0.78	0.56	3.48	3.20	0.63	1.81	0.28	1.80	0.28
	4114.50	8.76	15.74	26.29	46.65	5.37	19.46	3.51	0.71	0.50	3.11	2.93	0.60	1.75	0.28	1.81	0.28
	4115.44	2.15	6.98	11.47	20.82	2.45	9.22	1.70	0.42	0.24	1.48	1.37	0.27	0.73	0.12	0.79	0.12
	4115.89	6.74	14.40	21.44	39.09	4.64	17.38	3.31	0.67	0.47	2.95	2.72	0.54	1.53	0.25	1.60	0.25
	4116.24	1.54	6.56	9.34	17.95	2.13	8.14	1.53	0.38	0.23	1.38	1.29	0.26	0.73	0.11	0.73	0.11
	4116.99	12.15	24.85	42.17	74.27	8.48	30.82	5.49	1.09	0.82	4.95	4.87	1.00	2.95	0.46	2.93	0.45
Lower Xia Ganchaigou Fm.	4847.45	4.86	13.11	17.70	31.64	3.94	14.72	2.92	0.69	0.42	2.62	2.39	0.47	1.34	0.21	1.35	0.21
	4847.75	6.46	12.07	15.44	29.87	3.46	13.11	2.57	0.61	0.39	2.36	2.28	0.46	1.32	0.21	1.33	0.21
	4848.30	4.67	11.91	16.02	27.19	3.49	12.95	2.58	0.62	0.37	2.34	2.15	0.43	1.20	0.19	1.24	0.19
	4849.00	6.75	16.24	29.41	56.97	7.24	24.39	4.26	0.89	0.55	3.32	3.18	0.63	1.80	0.30	1.91	0.29
	4849.50	2.85	11.22	12.48	26.60	3.10	12.58	2.66	0.69	0.36	2.45	1.95	0.38	1.03	0.15	0.95	0.14
	4850.00	3.54	10.39	11.60	24.91	2.85	11.66	2.55	0.64	0.36	2.31	2.01	0.39	1.02	0.17	1.04	0.16
	4850.30	2.89	6.25	8.17	17.33	1.86	7.17	1.43	0.40	0.21	1.25	1.22	0.25	0.72	0.11	0.75	0.12
	4851.00	4.14	11.13	13.36	23.17	3.12	11.91	2.31	0.57	0.34	2.08	1.97	0.40	1.17	0.19	1.20	0.18
	4852.00	0.87	4.37	5.13	12.88	1.24	4.65	0.96	0.32	0.15	0.84	0.88	0.18	0.51	0.08	0.52	0.08
	4852.70	2.27	5.35	6.39	13.65	1.57	5.88	1.15	0.40	0.17	1.00	1.02	0.21	0.61	0.10	0.66	0.10
	4853.50	2.51	11.28	10.20	21.02	2.79	11.91	2.67	0.72	0.38	2.47	1.98	0.37	0.99	0.15	0.91	0.13
	4853.90	2.73	8.20	9.05	13.48	2.06	7.54	1.44	0.41	0.22	1.38	1.35	0.28	0.83	0.13	0.84	0.13
	4854.30	2.60	8.16	9.81	15.49	2.34	8.65	1.61	0.43	0.24	1.47	1.43	0.29	0.87	0.14	0.93	0.14
	4855.00	4.44	12.27	15.24	28.99	3.57	13.23	2.59	0.62	0.39	2.32	2.32	0.47	1.40	0.22	1.41	0.22

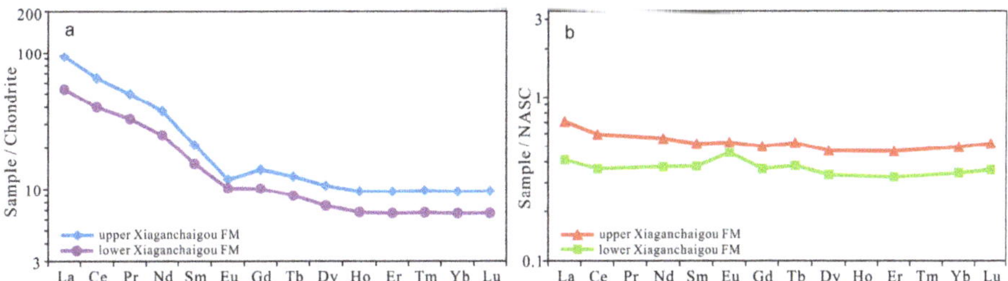

Figure 7. Chondrite-normalized rare earth element (REE) patterns (**a**) and NASC (North American shale composite)-normalized (REE) patterns (**b**) of sandstones from the Xiaganchaigou Formation in the Lenghuqi Area. Chondrite data are from [49]; NASC data are from [47].

Figure 8. Rare earth elements contents in the Xiaganchaigou Formation of the Lenghuqi area.

5. Discussion

Sedimentary processes, diagenesis, and weathering may all cause migration and elemental depletion. Sources should be therefore be analyzed when using elemental geochemical properties to restore sedimentary environments and paleoclimatic conditions. Only trace elements that are mainly self-generated and maintain their initial content can be used to accurately determine paleoenvironmental conditions [54]. The influence of terrigenous debris on trace element contents can be assessed using the content covariant diagram of Al, Ti, or Th. If the two exhibit a good linear relationship and the Al, Ti, and Th contents are similar to the average shale, the trace elements contained in the sediments or sedimentary rocks are mainly provided by terrigenous debris and therefore not suitable for environmental analysis. Otherwise, it can be used for environmental analysis as reported previously [55–58]. REEs have strong stability during weathering, transport, deposition, and diagenesis, and their solubility in water is very low. The ΣREE in sediments or sedimentary rocks is therefore an indicator of the amount of terrestrial material [59]. The ΣREE of the average shale (PAAS), which is often used as a standard, is 184.8×10^{-6}. If the ΣREE is close to or higher than this value, the trace elements contained in sediments or sedimentary rocks are mainly provided by terrigenous debris. If the ΣREE in the analyzed sediments or sedimentary rocks is much lower than this value, this shows that it is less affected by terrestrial materials and is mainly authigenic [60,61]. According to the covariant diagram (Figure 9), the correlation between Th and Al, Ti is not very good and the Th, Al, and Ti contents differ significantly from PAAS. The ΣREE content is also substantially lower than the PAAS value, indicating that the REE in the samples are likely mostly authigenic with only a small amount of added terrigenous clastic materials. These data can therefore be used to restore the sedimentary environment and paleoclimate conditions.

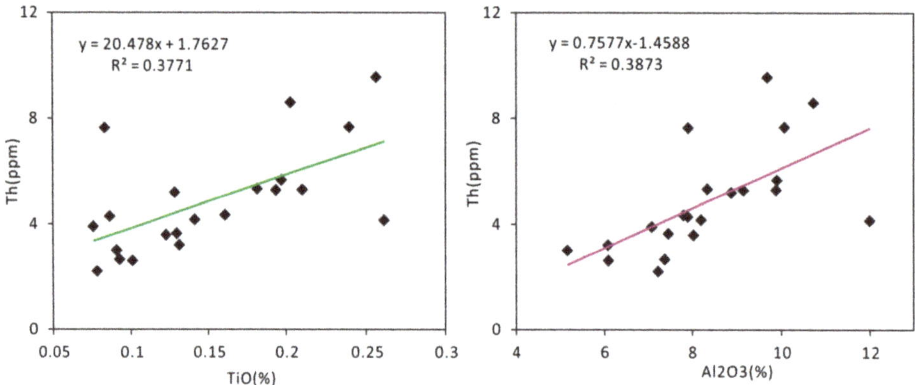

Figure 9. Covariation diagram of Th with Al and Ti from the Xiaganchaigou Formation of the Lenghuqi Area.

Cu and Zn are both transition metal elements. According to the Nernst equation, the pH of the medium environment affects the redox process, and Cu and Zn will partition into the environmental medium with decreasing oxygen fugacity. The redox environment during deposition can therefore be divided according to Cu/Zn. A ratio less than 0.21 represents a reducing environment, between 0.50 and 0.63 represents a weak oxidizing environment, and greater than 0.63 represents an oxidizing environment [62]. The geochemical properties of Th and U vary greatly in an oxidizing environment but are similar in the reducing environment; thus, this feature can be used to interpret the sedimentary environment, where $\&U = U/[0.5 \times (Th/3 + U)]$. A $\&U$ value greater than one represents a reducing environment and less than 1 represents an oxidizing environment [63]. A Th/U ratio between 0 and two indicates a reducing environment, and the ratio under strongly oxidizing conditions can reach eight [64].

The Cu/Zn values in the samples ranged from 0.32 to 0.94 with a mean value of 0.58, indicating a weakly oxidizing environment. The $\&U$ value is between 0.72 and 1.04 with a mean of 0.93, and the Th/U ratio is between 2.78 and 5.36 with a mean of 3.52, both of which indicate an oxidizing environment. Ce is the only element among the REE with redox properties: it is depleted under oxidizing conditions and enriched in an oxygen-deficient reducing environment. Wright et al. [65] proposed that the sign of $Ce_{anom} = \lg [3Ce_n/(2La_n + Nd_n)]$ can be used to assess Ce enrichment. A positive Ce_{anom} value indicates enrichment whereas a negative value indicates a deficit [65], where "n" signifies shale-normalized concentrations using the convention established by Gromet et al. (1984) [41]. The Ce_{anom} values of the samples from the Xiaganchaigou Formation of Lenghuqi ranged from -0.564 to -0.524, with an average of -0.536. The samples were therefore depleted in Ce, which mainly reflects an oxidizing environment.

The most important climate warming event since the Cenozoic era occurred from the middle Paleocene to early Eocene when global temperature and sea levels rose rapidly and reached a peak in the early Eocene. A long-term gradual cooling process occurred from the beginning of the middle Eocene to the early Oligocene with global sea levels dropping slowly and permanent ice sheets forming on the Antarctic continent. It was not until the late Oligocene when the climate warmed again that the permanent ice sheets began to melt [66–70]. During the Eocene-Oligocene transition, studies of sedimentation [71], vegetation changes [72], and animal evolution [73] on the North American continent all show that the climate changed significantly from the late Eocene to early Oligocene, from warm and humid to cold and dry. Studies on the European continent [74,75], Asia, and Oceania [76] also indicate a trend of the Eocene-Oligocene climate becoming colder and drier. This indicates that the global climate entered a dry and cold period during the late Eocene to early Oligocene. The measurements obtained from the Xiaganchaigou Formation (Oligocene) of Lenghuqi also reflect characteristics of a dry and cold global climate during this period. After entering the late Oligocene (as recorded by deposition of the upper Xiaganchaigou Formation), the global climate also entered a brief warming period. Our findings show that the sedimentary environment in the upper Xiaganchaigou Formation was warmer and more humid than the lower member, which reflects the response of this climatic phenomenon.

6. Conclusions

Oligocene strata (the Xiaganchaigou Formation) in the Lenghuqi area of the North Qaidam Basin are deeply buried with an average depth of >4000 m. The illite content in clay minerals is the highest among all phases (73.25%), followed by chlorite (16.25%), and a mixed layer illite/smectite (8.92%). Smectite and kaolinite are not observed in these samples. The average illite, chlorite, mixed layer illite/smectite contents in the upper Xiaganchaigou Formation are 54.77%, 20.69%, and 24.54%, respectively. The combination of clay minerals reflects that the sedimentary basins of the Oligocene (sedimentary period of the Xiaganchaigou Formation) originate from an arid-semi-arid climate with weak leaching and chemical weathering. The diagenetic environment is dominated by a K^+- and Mg^{2+}-rich alkaline environment. Clear characteristics of CaO and Na_2O enrichment with K_2O, MgO, and TiO_2 loss are observed, especially in the lower Xiaganchaigou Formation. The lower member of the Xiaganchaigou Formation shows characteristics of a dry and cold freshwater sedimentary environment with high evaporation, whereas the upper Xiaganchaigou Group is relatively warm and humid with a significantly increased recharge of terrigenous detrital materials. The paleoclimate conditions in the Oligocene period gradually became moister. The variation of Sr, Ba, Cu, Zn, U, Th, and Ce contents and ratios, as well as REE characteristics, suggest that the paleoclimate conditions of the lower Xiaganchaigou Formation during sedimentation were relatively cold and dry, while the climate of the upper Xiaganchaigou Formation became warmer and more humid.

Author Contributions: G.S. conceived the research plan of this study, analyzed the experimental data, and wrote the manuscript. Y.W. collated the experimental data and designed the figures. J.G. conducted the scanning electron microscope image analyses of the sandstones. M.W., Y.J., and S.P. collected data from the Xx 1 well and drew a single well map.

Funding: This work was financially supported by the CAS "Light of West China" Program (Y304RC1SGQ), Nature Science Foundation of Gansu Province (Grant No. 1308RJZA310), and Key Laboratory Project of Gansu Province (Grant No.1309RTSA041).

Acknowledgments: We would like to thank Junli Qiu. for their technical assistance with chemical analyses.

Conflicts of Interest: The authors declare no conflict of interest.

References

1. Xie, Y.; Wang, J.; Li, L.X.; Xie, Z.W.; Deng, G.H.; Li, M.H.; Jiang, X.S. Distribution of the Cretaceous clay minerals in Ordos basin, China and its implication to sedimentary and diagenetic environment. *Geol. Bull. China* **2010**, *29*, 93–104. [CrossRef]
2. Singer, A. The paleoclimatic interpretation of clay minerals in soils and weathering profiles. *Earth Sci. Rev.* **1980**, *15*, 303–326. [CrossRef]
3. Deconinck, J.F.; Blanc-Valleron, M.M.; Rouchy, J.M.; Camoin, G.; Badaut-Trauth, D. Palaeoenvironmental and diagenetic control of the mineralogy of Upper Cretaceous Lower Tertiary deposits of the Central Palaeo Andean basin of Bolivia (Potosi area). *Sediment. Geol.* **2000**, *132*, 263–278. [CrossRef]

4. Tang, Y.J.; Jia, J.Y.; Xie, X.D. Environment significance of clay minerals. *Earth Sci. Front.* **2002**, *9*, 337–341. [CrossRef]
5. Sun, L.M.; Yang, Y.B.; Wu, Y.X.; Fang, X.F.; Li, L.H.; Li, Y.C.; Zhao, Y.F.; Tian, L.F. Sedimentary environment of Neogene laterite records in the Xihe basin, Inner Mongolia. *Geol. China* **2008**, *35*, 683–690. [CrossRef]
6. Yu, S.H.; Zheng, H.H. REE of sediments of the Chang Liushui section at Zhong Wei County of Ning Xia and the environmental significance. *Acta Sedimentol. Sin.* **1999**, *17*, 149–155. [CrossRef]
7. Li, Y.C.; Wang, S.M.; Huang, Y.S. The lake sediments rediments to environmental and climatic change. *Adv. Earta Sci.* **1999**, *14*, 412–416. [CrossRef]
8. Fan, Y.H.; Qu, H.J.; Wang, H.; Yang, X.C.; Feng, Y.W. The application of trace elements analysis to identifying sedimentary media environment. A case study of Late Triassic strata in the middle part of western Ordos Basin. *Geol. China* **2012**, *39*, 382–389. [CrossRef]
9. Zhong, W.; Fang, X.M.; Li, J.J.; Zhu, J.J. The Geochemical Record of Paleoclimate during about 7.0 Ma–0.73 Ma in Linxia Basin, Gansu Province. *J. Arid Land Resour. Environ.* **1998**, *12*, 36–43. [CrossRef]
10. Kuang, S.P.; Xu, Z.; Zhang, S.S.; Ma, Z.D. Applying Geochemistry to Research into Meso-Cenozoic Climate: Discussion on Jurassic Climatic change in Sichuan Basin, China. *J. Qingdao Inst. Chem. Technol.* **2002**, *3*, 4–9. [CrossRef]
11. Miller, E.K.; Blum, J.D.; Friedland, A.J. Determination of soil exchangeable-cation loss and weathering rates using Sr isotopes. *Nature* **1993**, *362*, 438–441. [CrossRef]
12. Reinhardt, E.G.; Blenkinsop, J.; Patrerson, R.T. Assessment of a Sr isotope (^{87}Sr/^{86}Sr) vital effect in marine taxa from Lee Stocking Island, Bahamas. *Geo-Mar. Lett.* **1998**, *18*, 241–246. [CrossRef]
13. Dominik, J.; Stanley, D.J. Boron, beryllium and sulfur in holocene sediments and peats of the Nile delta, Egypt: Their use as indicators of salinity and climate. *Chem. Geol.* **1993**, *104*, 203–216. [CrossRef]
14. Bailey, S.W.; Hornbeck, J.W.; Priscoll, C.T.; Gaudette, H.E. Calcium inputs and transport in a base poor forest ecosystem as interpreted by Sr isotopes. *Water Resour. Res.* **1996**, *32*, 707–719. [CrossRef]
15. Liu, G.; Zhou, D.S. Application of microelements analysis in identifying sedimentary environment—Taking Qianjiang formation in the Jianghang Basin as an example. *Pet. Geol. Exp.* **2007**, *29*, 307–311. [CrossRef]
16. Zhong, D.K.; Zhu, X.M.; Wang, H.J. Analysis on the characteristics and formation mechanism of high quality clastic reservoirs in China. *Sci. China* **2008**, *38*, 11–18. [CrossRef]
17. Feng, J.R.; Gao, Z.Y.; Cui, J.G.; Zhou, C.M. The exploration status and research advances of deep and ultra-deep clastic reservoirs. *Adv. Earth Sci.* **2016**, *31*, 718–736. [CrossRef]
18. Schmidt, V.; Mcdonald, D.A. The role of secondary porosity in the course of sandstone diagenesis. *SEPM Spec. Publ.* **1979**, *26*, 178–207.
19. Taylor, T.R.; Giles, M.R.; Hathon, L.; Diggs, T.N.; Braunsdorf, N.R.; Birbiglia, G.V.; Kittridge, M.G.; Macaulay, C.I.; Espejo, I.S. Sandstone diagenesis and reservoir quality prediction: Models, myths and reality. *AAPG Bull.* **2010**, *94*, 1093–1132. [CrossRef]
20. Bjrlykke, K.; Jahren, J. Open and closed geochemical systems during diagenesis in sedimentary basins: Constrains on mass transfer during diagenesis and the prediction of porosity in sandstone and carbonate reservoir. *AAPG Bull.* **2013**, *96*, 2193–2214. [CrossRef]
21. Jin, Z.K.; Su, K.; Su, N.N. Origin of Jurassic deep burial-quality reservoirs in the central Junggar Basin. *Acta Pet. Sin.* **2011**, *32*, 26–31. [CrossRef]
22. Gao, C.L.; Ji, Y.L.; Gao, Z.Y.; Wang, J.; Ren, Y.; Liu, D.W.; Duan, X.B.; Huan, Z.J.; Cheng, T.R. Multi-factor coupling analysis on property preservation process of deep buried favorable reservoir in hinterland of Junggar Basin. *Acta Sedimentol. Sin.* **2017**, *35*, 577–591. [CrossRef]
23. Guo, J.J.; Sun, G.Q.; Men, H.J.; Zhu, W.J.; Ma, J.Y.; Zhu, J.; Guan, B.; Shi, J.A. Genetic analysis of anomalously high porosity zones in deeply buried reservoirs in the west part of northern edge of Qaidam Basin, NW China. *Acta Sedimentol. Sin.* **2018**, *36*, 777–786. [CrossRef]
24. Wang, Y.T.; Sun, G.Q.; Yang, Y.H.; Wang, M.; Ma, F.Q.; Zhang, C.J.; Zhao, J.; Shi, J.A. U-Pb geochronology analysis of detrital zircons in the Paleogene lower Xiaganchaigou formation in the northern Mahai area, northern margin of Qaidam Basin. *J. Lanzhou Univ. Nat. Sci.* **2019**, *55*, 141–149. [CrossRef]
25. Guo, R.T.; Ma, D.D.; Zhang, Y.S.; Liu, B.; Chen, X.D.; Cui, J.; Wang, P.; Zhang, Q.H.; Jiang, Y.H.; Li, Y.F. Characteristics and formation mechanism of overpressure pore-fracture reservoirs for Upper Member of Xiaganchaigou Formation in the west of Yingxiong ridge, Qaidam Basin. *Acta Pet. Sin.* **2019**, *40*, 411–422. [CrossRef]

26. Long, H.; Wang, C.H.; Liu, Y.P.; Ma, H.Z. Application of clay minerals in paleoenviroment research. *J. Salt Lake Res.* **2007**, *15*, 21–25. [CrossRef]

27. Jin, N.; Li, A.C.; Liu, H.Z.; Meng, Q.Y.; Wan, S.M.; Xu, Z.K. Clay minerals in surface sediment of the northwest Parece Vela basin: Distribution and provenance. *Oceanol. Limnol. Sin.* **2007**, *38*, 504–511. [CrossRef]

28. Wu, B.H.; Yang, H.N.; Li, S.J. *The Mineral Composition of Sediment and Sedimentation of Central Pacific*; Geological Publishing House: Beijing, China, 1993; pp. 44–45.

29. Lu, C.X. Clay minerals as indicators of paleoenvironment. *J. Desert Res.* **1997**, *17*, 456–460.

30. He, L.B. The relation between clay mineral variation of marine sediment cores and paleoclimate. *Chin. Sci. Bull.* **1982**, *27*, 809–812.

31. Yang, Z.S. Mineralogical assemblages and chemical characteristics of clays from sediments of the Huanghe, Changjiang, Zhujiang rivers and their relationship to the climate environment in their sediment source areas. *Oceanol. Limnol. Sin.* **1988**, *19*, 336–346.

32. Fu, W.J. Influence of clay minerals onsandstone reservoir properties. *J. Palaeogeogr.* **2000**, *2*, 59–67. [CrossRef]

33. Dong, W.H.; Su, X.S.; Hou, G.C.; Lin, X.Y.; Liu, F.T. Study on distribution law of TDS and main ion concentration in groundwater in the Ordos Cretaceous Artesian Basin. *Hydrogeol. Eng. Geol.* **2008**, *35*, 11–16. [CrossRef]

34. Liu, Y. The clay mineral characteristics and the analysis for sedimentary environment of the Late Cretaceous in Songliao Basin. *Acta Sedimentol. Sin.* **1985**, *3*, 131–137.

35. Yuan, H.R.; Nie, Z.; Liu, J.Y.; Wang, M. Paleogene sedimentary characteristics and their paleoclimatic implications in the Baise Basin, Guangxi. *Acta Geol. Sin.* **2007**, *81*, 1692–1697. [CrossRef]

36. Hower, J.; Eslinger, E.V.; Hower, M.E.; Perry, E.A. Mechanism of burial metamorphism of argillaceous sediments: I. Mineralogical and chemical evidence. *Geol. Soc. Am. Bull.* **1976**, *87*, 725–737. [CrossRef]

37. Boles, J.R.; Surdam, R.C. Diagenesis of volcanogenic sediments in a Tertiary saline lake; Wagon Bed Formation, Wyoming. *Am. J. Sci.* **1979**, *279*, 832–853. [CrossRef]

38. Wang, Y.F. Lacustrine carbonate chemical sedimentation and climatic-environmental evolution—A case study of Qinghai Lake and Daihai Lake. *Oceanol. Limnol. Sin.* **1993**, *24*, 31–35.

39. Xi, X.X.; Mu, D.F.; Fang, X.M.; Li, J.J. Climatic change since the Late Miocene in west China. *Acta Sedimentol. Sin.* **1998**, *16*, 155–160. [CrossRef]

40. Sun, G.Q.; Yin, J.G.; Zhang, S.C.; Lu, X.C.; Zhang, S.Y.; Shi, J.A. Diagenesis and Sedimentary Environment of Miocene Series in Eboliang III Area. *Environ. Earth Sci.* **2015**, *74*, 5169–5179. [CrossRef]

41. Gromet, L.P.; Haskin, L.A.; Korotev, R.L.; Dymek, R.F. The "North American shale composite": Its compilation, major and trace element characteristics. *Geochim. Cosmochim. Acta* **1984**, *48*, 2469–2482. [CrossRef]

42. Guo, J.J.; Sun, G.Q.; Liu, W.M. Formation phases of carbonate cements and sedimentary environments in lower jurassic sandstones of the lenghu v tectonic belt, north Qaidam basin, China. *Carbonates Evaporites* **2018**. [CrossRef]

43. Ma, B.L.; Wang, Q. Distribution characteristics of elements in modern sediments of Qinghai lake. *Acta Sedimentol. Sin.* **1997**, *15*, 120–125. [CrossRef]

44. Yuan, B.Y.; Chen, K.Z.; Bowler, J.M.; Ye, S.J. The formation and evolution of the Qinghai Lake. *Quat. Sci.* **1990**, *10*, 233–243.

45. Wang, S.J.; Huang, X.Z.; Tuo, J.C.; Shao, H.S.; Yan, C.F.; Wang, Q.S.; He, Z.R. Evolutional characteristics and their Paleoclimate significance of trace elements in the Hetaoyuan Formation, Biyang Depression. *Acta Sedimentol. Sin.* **1997**, *15*, 65–70. [CrossRef]

46. Song, M.S. Sedimentary Environment Geochemistry in the Shasi Section of Southern Rama, Dongying Depression. *J. Mineral. Petrol.* **2005**, *5*, 67–73. [CrossRef]

47. Taylor, S.R.; McLennan, S.M. *The Continental Crust: Its Composition and Evolution*; Blackwell Scientific Pub.: Palo Alto, CA, USA, 1985; p. 312.

48. Shi, J.A.; Guo, X.L.; Wang, Q.; Yan, N.Z.; Wang, J.X. Geochemistry of REE in QH1 sediments of Qinghai Lake since Late Holocene and its Paleoclimatic significance. *J. Lake Sci.* **2003**, *15*, 28–34. [CrossRef]

49. McDonough, W.F.; Sun, S.S. The composition of the earth. *Chem. Geol.* **1995**, *120*, 223–253. [CrossRef]

50. Cao, J.J.; Zhang, X.Y.; Wang, D.; Zhou, J. REE geochemistry of late Cenozoic eolian sediments and the paleoclimate significance. *Mar. Geol. Quat. Geol.* **2001**, *21*, 97–101. [CrossRef]

51. Kashiwaya, K.; Yamanoto, A.; Fukuyama, K. Time variations of erosional force and grain size in Pleistocene lake sediments. *Quat. Res.* **1985**, *28*, 61–68. [CrossRef]

52. Zhao, Y.Y.; Wang, J.T.; Qin, C.Y.; Chen, Y.W.; Wang, X.J.; Wu, M.Q. Rare-earth elements in continental shelf sediments of the China Seas. *Acta Sedimentol. Sin.* **1990**, *8*, 37–43. [CrossRef]
53. Chen, J.A.; Wan, G.J.; Xu, J.Y. Sediment particle sizes and the dry-humid transformation of the regional climate in Erhai Lake. *Acta Sedimentol. Sin.* **2000**, *18*, 341–345. [CrossRef]
54. Tribovillard, N.; Algeo, T.J.; Lyons, T.; Riboulleau, A. Trace metals as paleoredox and paleoproductivity proxies: An update. *Chem. Geol.* **2006**, *232*, 32. [CrossRef]
55. Böning, P.; Brumsack, H.J.; Böttcher, M.E.; Schnetger, B.; Kriete, C.; Kallmeyer, J.; Borchers, S.L. Geochemistry of Peruvian near-surface sediments. *Geochim. Cosmochimica. Acta* **2004**, *68*, 4429–4451. [CrossRef]
56. Calvert, S.E.; Pedersen, T.F. Geochemistry of recent oxic and anoxic sediments: Implications for the geological record. *Mar. Geol.* **1993**, *113*, 67–88. [CrossRef]
57. Hild, E.; Brumsack, H.J. Major and minor element geochemistry of Lower Aptian sediments from the NW German Basin (core Hoheneggelsen KB 40). *Cretacous Res.* **1998**, *19*, 615–633. [CrossRef]
58. Tribovillard, N.; Desprairies, A.; Lallier-Vergès, E.; Bertrand, P.; Moureau, N.; Ramdani, A.; Ramanampisoa, L. Geochemical study of organic-matter rich cycles from the Kimmeridge Clay Formation of Yorkshire (UK): Productivity vs. anoxia. *Palaeogeogr. Palaeoclimatol. Palaeoecol.* **1994**, *108*, 165–181. [CrossRef]
59. Murray, R.W.; Brink, M.R.B.T.; Gerlach, D.C.; Russ, G.P.; Jones, D.L. Rare earth, major, and trace element composition of Monterey and DSDP chert and associated host sediment: Assessing the influence of chemical fractionation during diagenesis. *Geochim. Cosmochim. Acta* **1992**, *56*, 2657–2671. [CrossRef]
60. Chang, H.J.; Chu, X.L.; Feng, L.J.; Huan, J.; Zhang, Q.R. Redox sensitive trace elements as paleoenvironments proxies. *Geol. Rev.* **2009**, *55*, 91–99. [CrossRef]
61. Nothdurft, L.D.; Webb, G.E.; Kamber, B.S. Rare earth element geochemistry of Late Devonian reefal carbonates, Canning Basin, Western Australia: Confirmation of a seawater REE proxy in ancient limestones. *Geochim. Cosmochim. Acta* **2004**, *68*, 263–283. [CrossRef]
62. Mei, S.Q. Application of rock chemistry in the study of presinian sedimentary environment and the source of uranium mineralization in Hunan province. *Hunan Geol.* **1988**, *7*, 25–32.
63. Steiner, M.; Wallis, E.; Erdtmann, B.D.; Zhao, Y.L.; Yang, R.D. Submarine-hydrothermal exhalative ore layers in black shales from South China and associated fossils—Insights into a Lower Cambrian facies and bio-evolution. *Palaeogeogr. Palaeoclimatol. Palaeoecol.* **2001**, *169*, 165–191. [CrossRef]
64. Wignall, P.B.; Twitchett, R.J. Oceanic Anoxia and the End Permian Mass Extinction. *Science* **1996**, *272*, 1155–1158. [CrossRef] [PubMed]
65. Wright, J.; Schrader, H.; Holser, W.T. Paleoredox variations in ancient oceans recorded by rare earth elements in fossil apatite. *Geochim. Cosmochim. Acta* **1987**, *51*, 631–644. [CrossRef]
66. Chang, Y.S.; Zhao, H.; Qin, J.; Li, S.; Zhang, J.P. Sedimentary Response to Paleoclimate Change in the East China Sea Shelf Basin. *Acta Sedimentol. Sin.* **2019**, *37*, 320–329. [CrossRef]
67. Tuo, S.T.; Liu, Z.F. Global climate event at the Eocene-oligocene transition: From greenhouse to icehouse. *Adv. Earth Sci.* **2003**, *18*, 691–696. [CrossRef]
68. Zachos, J.; Pagani, M.; Sloan, L.; Thomas, E.; Billups, K. Trends, rhythms, and aberrations in global climate 65 Ma to present. *Science* **2001**, *292*, 686–693. [CrossRef] [PubMed]
69. Haq, B.U.; Hardenbol, J.; Vail, P.R. Mesozoic and Cenozoic Chronostratigraphy and Cycles of Sea-Level Change: An Integrated Approach. In *Sea-Level Changes: An Integrated Approach*; Wilgus, C.K., Hastings, B.S., Kendall, C.G.S.C., Posamentier, H.W., Ross, C.A., Van Wagoner, J.C., Eds.; SEPM Special Publication: Broken Arrow, OK, USA, 1988; pp. 71–108. [CrossRef]
70. Miller, K.G.; Kominz, M.A.; Browning, J.V.; Wright, J.D.; Mountain, G.S.; Katz, M.E.; Sugarman, P.J.; Cramer, B.S.; Christie-Blick, N.; Pekar, S.F. The Phanerozoic record of global sea-level change. *Science* **2005**, *310*, 1293–1298. [CrossRef]
71. Prothero, D.R. Does climatic change drive mammalian evolution? *GSA Today* **1999**, *9*, 1–7.
72. Wolfe, J.A. Climatic, floristic, and vegetational changes near the Eocene/Oligocene boundary in North America. In *Eocene-Oligocene Climatic and Biotic Evolution*; Prothero, D.R., Berggren, W.A., Eds.; Princeton University Press: Princeton, NJ, USA, 1992; pp. 421–436.
73. Ivany, L.C.; Patterson, W.P.; Lohmann, K.C. Cooler winters as a possible cause of mass extinctions at the Eocene/Oligocene boundary. *Nature* **2000**, *407*, 887–890. [CrossRef]

74. Cavagnetto, C.; Anddon, P. Preliminary palynological data on floristic and climatic changes during the Middle Eocene-Early Oligocene of the eastern Ebro Basin, northeast Spain. *Rev. Palaeobot. Palynol.* **1996**, *92*, 281–305. [CrossRef]
75. Blondel, C. The Eocene-Oligocene ungulates from western Europe and their environment. *Palaeogeogr. Palaeoclimatol. Palaeocol.* **2001**, *168*, 125–139. [CrossRef]
76. Buenhy, N.; Carlson, S.J.; Spero, H.J.; Lee, D.E. Evidence for the Early Oligocene formation of a proto-Subtropical Convergence from oxygen isotope records of New Zealand Paleogence brachiopods. *Palaeogeogr. Palaeoclimatol. Palaeocol.* **1998**, *138*, 43–68. [CrossRef]

Article

Origin and Sources of Minerals and Their Impact on the Hydrocarbon Reservoir Quality of the PaleogeneLulehe Formation in the Eboliang Area, Northern Qaidam Basin, China

Bo Chen [1,2,*], Feng Wang [1,*], Jian Shi [2], Fenjun Chen [3] and Haixin Shi [1,*]

1   Guangxi Colleges and Universities Key Laboratory of Beibu Gulf Oil and Natural Gas Resource Effective Utilization, Beibu Gulf University, Qinzhou 515000, China

2   Key Laboratory of Petroleum Resources, Gansu Province/Key Laboratory of Petroleum Resources Research, Institute of Geology and Geophysics, Chinese Academy of Sciences, Lanzhou 730000, China

3   Exploration & Production Research Institute of Qinghai Oilfield Company, petroChina, Dunhuang 736202, China

*   Correspondence: cbo-11@163.com (B.C.); fefe12138@163.com (F.W.); shihaixin2006@163.com (H.S.)

Received: 13 May 2019; Accepted: 12 July 2019; Published: 15 July 2019

Abstract: The Lulehe sandstone in the Eboliang area is a major target for hydrocarbon exploration in the northern Qaidam Basin. Based on an integrated analysis including thin section analysis, scanning electron microscopy, X-ray diffraction, cathodoluminescence investigation, backscattered electron images, carbon and oxygen stable isotope analysis and fluid inclusion analysis, the diagenetic processes mainly include compaction, cementation by carbonate and quartz, formation of authigenic clay minerals (i.e., chlorite, kaolinite, illite-smectite and illite) and dissolution of unstable materials. Compaction is the main factor for the deterioration of reservoir quality; in addition, calcite cement and clay minerals are present, including kaolinite, pore-filling chlorite, illite-smectite and illite, which also account for reservoir quality reduction. Integration of petrographic studies and isotope geochemistry reveals the carbonate cements might have originated from mixed sources of bioclast- and organic-derived CO_2 during burial. The quartz cement probably formed by feldspar dissolution, illitization of smectite and kaolinite, as well as pressure solution of quartz grains. Smectite, commonly derived from alteration of volcanic rock fragments, may have been the primary clay mineral precursor of chlorite. In addition, authigenic kaolinite is closely associated with feldspar dissolution, suggesting that alteration of detrital feldspar grains was the most probable source for authigenic kaolinite. With the increase in temperature and consumption of organic acids, the ratio of K^+/H^+ increases and the stability field of kaolinite is greatly reduced, thereby transforming kaolinite into mixed layer illite/smectite and illite. Within the study area, porosity increases with chlorite content up to approximately 3% volume and then decreases slightly, indicating that chlorite coatings are beneficial at an optimum volume of 3%. A benefit of the dissolution of unstable minerals and feldspar grains is the occurrence of secondary porosity, which may enhance porosity to some extent. However, the solutes cannot be transported over a large scale in the deep burial environment, and simultaneous precipitation of byproducts of feldspar dissolution such as authigenic kaolinite and quartz cement will occur in situ or in adjacent pores, resulting in heterogeneity of the reservoirs.

Keywords: diagenesis; authigenic minerals; reservoir quality; Eboliang; Qaidam Basin

---

1. Introduction

Reservoir quality is considered a key control on hydrocarbon migration and accumulation and is ultimately a decisive factor for effective hydrocarbon exploration and production projects [1–4].

Consequently, quantitative evaluation of reservoir quality is important to obtain a detailed understanding of reserve estimations and productivity improvement [5–7]. However, the reservoir quality and heterogeneity have been strongly affected by various parameters, such as depositional parameters (e.g., primary depositional textures, detrital composition, grain size and sorting) and different types and sequences of diagenesis [8–10]. Depositional parameters have a significant effect on primary reservoir quality (i.e., porosity and permeability) and consequently on diagenetic alteration, which, in turn, results in complex spatial and temporal patterns of reservoir quality and heterogeneity [11–14]. Diagenesis, including physical processes of mechanical compaction and chemical action of mineral dissolution and precipitation, has a great influence on subsequent evolution paths of porosity and permeability [15–17]. According to burial history and depositional composition, the porosity of shallow sandstone with little cementation can be accurately predicted [18,19]. However, under a deeper burial environment, porosity is much more difficult to predict because of the import and export of materials related to dissolution and precipitation of minerals involved in chemical diagenesis [18,20]. Therefore, a comprehensive understanding of the diagenesis process of dissolution and precipitation of cements and the origin and sources of minerals are the key to predicting the quality of deeply buried sandstone.

Minerals, including carbonate cements, quartz cements and clay cements, are widely distributed in the clastic reservoirs and have been extensively concerned in the hydrocarbon exploration [21,22]. Formation and transformation of minerals are sensitive to the change of diagenetic fluid properties, which can record the information of diagenetic environment and reveal the mechanism of fluid-rock interaction [23]. In addition, fluid-rock interaction, especially the precipitation and dissolution of carbonate cement, the origin and sources of quartz cement and clay minerals, is of great significance for evaluating and predicting reservoir performance and porosity evolution [24,25]. Qaidam Basin, one of the most petroliferous basins in China, is located in western China (Figure 1a). The Eboliang area in the north of the basin, which comprises a significant structural belt and is a major target for hydrocarbon exploration and development [26]. However, to date, exploration in Eboliang has been limited because the main controlling factors of the reservoir quality have not been thoroughly studied, and only a few studies have been conducted on the Lulehe sandstone mainly focusing on different aspects, including sedimentary facies [27], heavy mineral characteristics and source analysis of the depositional environment [28–32]. These previous studies have shown that the Lenghu area is adjacent to source rocks, with sufficient oil sources in contact with sandbodies, and has become a key area for extended exploration at the north margin of the Qaidam Basin. Moreover, previous studies have also confirmed that the variations in porosity and permeability are mainly due to the development of carbonate cement, authigenic quartz cementation and clay minerals, which indicates that diagenetic alteration is the primary control on reservoir quality. Various types and generations of minerals are present in the Eboliang area [28–32], providing an excellent example to study the fluid-rock interaction and porosity evolution. However, little attention has been paid to the diagenesis and origin and source of the major cements and authigenic minerals in this area, which results in an increased risk of improving the accuracy of reservoir quality evaluation and prediction. Therefore, it is important to have a detailed understanding of origin and sources of minerals and their impact on the hydrocarbon reservoir quality.

Given a recent increase in petroleum exploration in the Eboliang region, factors that affect oil and gas exploration (e.g., the diagenetic alteration characteristics and reservoir quality) have become important areas of focus. In this regard, by integrating sandstone diagenesis with detrital composition, burial history, thermal history, and fluid inclusion and stable isotope analyses of cements, a multidisciplinary approach is developed to carry out a systematic investigation into origin and sources of minerals and their impact on reservoir quality of the Lulehe sandstone in the Eboliang area. Such an approach will provide important insights into the evolution of diagenetic patterns of sandstone, and ultimately improve our ability to understand and predict the evolution of reservoir quality. Thus, the main objectives of this study are to (1) investigate the main diagenetic minerals and identify the origin and source of cements and clay minerals; (2) analyze the evolution of diagenesis and reconstruct the history and paragenetic sequences of diagenesis; and (3) identify how difference in diagenetic alteration impact reservoir quality.

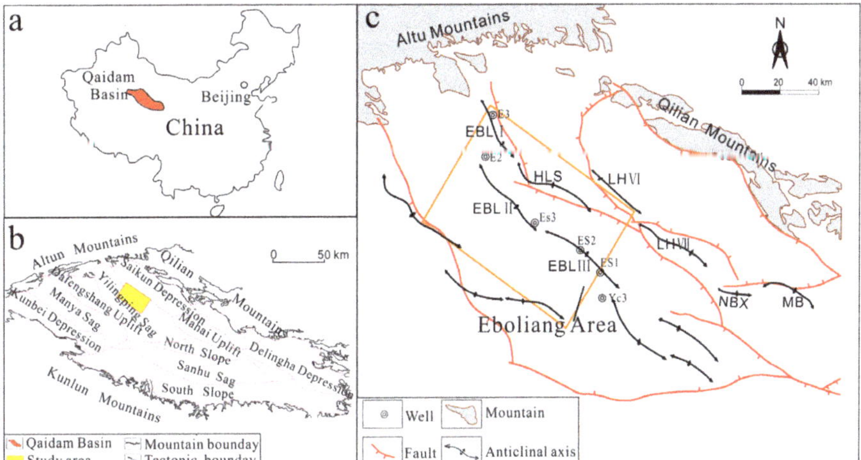

Figure 1. Sketch maps showing (**a**) location of the Qaidam Basin in NW China; (**b**) major structural units and location of the study area within the Qaidam Basin; and (**c**) tectonic structures and wells in the study area (modified after Sun [27]). EBL: Eboliang structure; LH: Lenghu structure; MB: Mabei structure; NBX: Nanbaxian structures.

2. Geological Setting

The Qaidam Basinis is a large intra-continental sedimentary basin, with a total area of approximately 120,000 km^2, located on the northeastern margin of the Tibetan Plateau [31–34]. The basin is a Mesozoic-Cenozoic continental hydrocarbon basin [27] with an irregular rhombic shape and is bounded by three mountain systems: the Altun Mountains to the northwest, the Qilian Mountains to the northeast and the Kunlun Mountains to the south (Figure 1b). With respect to administrative division, the basin is divided into ten first-order structural units: the Saikun Depression, the Yilingping Sag, the Dafengshan Uplift, the Manya Sag, the Kunbei Depression, the Mahai Uplift, the North Slope, the Sanhu Sag, the South Slope, and the Delingha Depression (Figure 1b). The northwest basin fault block belt is located in the southern region of the Qilian Mountains, mainly in the Yilingping Sag, and contains the Eboliang (Figure 1c) and Lenghu structural belts, along with the Mahai and Nanbaxian structures. Due to tectonic evolution, I, II, and III Eboliang structural belts developed from northwest to southeast (Figure 1c) and interbedded main source rocks contain abundant oil and gas source [27]. There is a strong relationship between generation-migration and the Eboliang structure, and the region is a target for commercial hydrocarbon flow.

Stratigraphic development extends from the Paleogene–Neogene–Quaternary, and in ascending order [21], consists of the Paleogene Lulehe Formation (E$_{1+2}$), the Lower (E$_3{}^1$) and Upper (E$_3{}^2$) Xia Ganchaigou Formation, the NeogeneShang Ganchaigou Formation (N$_1$), the Lower (N$_2{}^1$) and Upper (N$_2{}^2$) Youshashan Formation, the Shizigou Formation (N$_2{}^3$) and the Quaternary Qigequan Formation (Q$_{1+2}$), as shown in Figure 2. The Paleogene-Neogene ancient lake basin of Qaidam Basin underwent three evolutions stages [26]: the ancient lake basin first appeared in E$_{1+2}$ period; subsequently, the paleolake expanded continuously from E$_{1+2}$ to the N$_1$ period, and reached the maximum surface in the subsidence stage of the N$_1$ period; then, the paleolake gradually shrankduring the N$_2{}^1$ to N$_2{}^3$ period. The development center was located in the western part of the Qaidam Basin in the Paleogene and migrated to the Yiliping area during the Miocene to Pliocene [35]. Core analysis statistics show that the Paleogene Lulehe Formation (E_{1+2}) mainly consists of dark-gray mudstones and interbedded gray fine silt and argillaceous siltstone deposited in alluvial fan and fluvial-deltaic sedimentary environments [26,27].

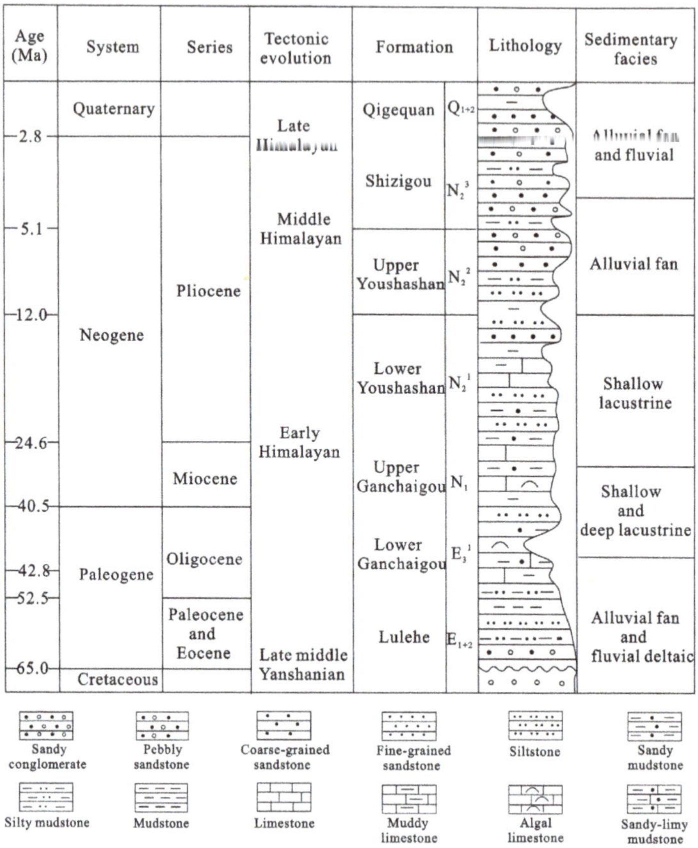

Age (Ma)	System	Series	Tectonic evolution	Formation		Lithology	Sedimentary facies
	Quaternary		Late Himalayan	Qigequan	Q_{1+2}		Alluvial fan and fluvial
-2.8				Shizigou	N_2^3		
-5.1			Middle Himalayan	Upper Youshashan	N_2^2		Alluvial fan
	Neogene	Pliocene					
-12.0				Lower Youshashan	N_2^1		Shallow lacustrine
-24.6			Early Himalayan	Upper Ganchaigou	N_1		Shallow and deep lacustrine
		Miocene					
-40.5				Lower Ganchaigou	E_3^1		
-42.8	Paleogene	Oligocene					
-52.5		Paleocene and Eocene		Lulehe	E_{1+2}		Alluvial fan and fluvial deltaic
-65.0	Cretaceous		Late middle Yanshanian				

Legend: Sandy conglomerate; Pebbly sandstone; Coarse-grained sandstone; Fine-grained sandstone; Siltstone; Sandy mudstone; Silty mudstone; Mudstone; Limestone; Muddy limestone; Algal limestone; Sandy-limy mudstone.

Figure 2. Generalized stratigraphic column for the northwestern Qaidam Basin. Figure after Feng [26].

Burial and thermal history of the Lulehe Formation have been analyzed in detail using data from wells, the history of regional erosion and heat flow was reconstructed with the BasinMod software by previous studies [31–34]. During Lulehe period, the formation shows a progressive subsidence, and the maximum burial depth occurs at approximately 4500 m (Figure 3). Previous studies indicated that the geothermal gradient of the Lulehe Formation ranges from 25 °C/km to 27.5 °C/km [34], and the present-day average surface temperature is 10 °C; thus, the highest temperature is approximately 110–125 °C with a maximumburied depth of 4500 m.

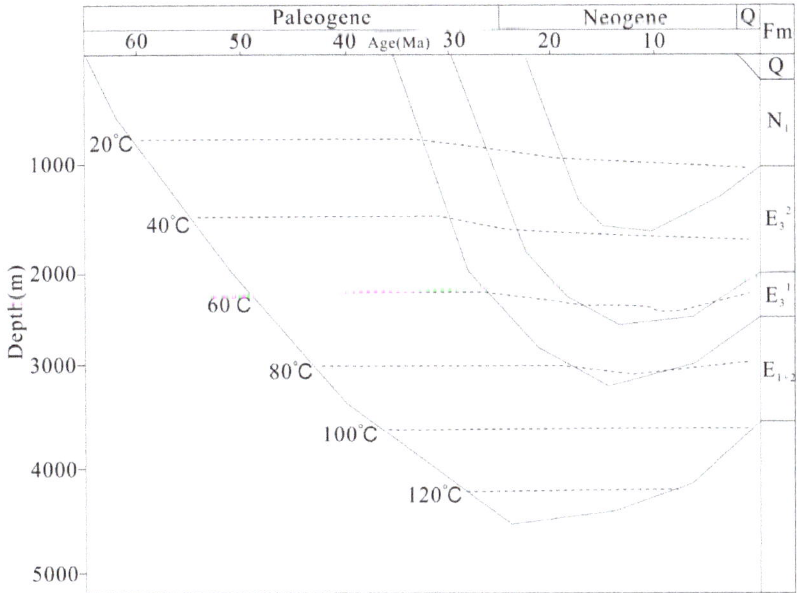

Figure 3. Burial and thermal historiesof the Paleogene Lulehe Formation (E_{1+2}). Figure after Li [34].

3. Samples and Methods

More than 160 core samples were collected from sections of the Lulehe Formation in the Eboliang area and were derived from more than 30 wells. Quantitative determination of porosity and permeability were made by a CMS^{tm}-300 Core Measurement system. The mineralogy and structure of selected samples were described based on optical microscope observations of thin sections. The thin sections were prepared by vacuum impregnation with blue epoxy and stained with Alizarin Red S and potassium ferricyanide to facilitate petrographic recognition of visual porosity and identification of rock mineralogy such as dolomites, ferroan calcite, and nonferroan calcite. The samples for compositional analysis were based on 300 points, and petrographic image modal analysis was used to determine the percentage of framework, matrix, cement and porosity in the samples.

A Philips XL 20 scanning electron microscope (SEM, Philips, Amsterdam, The Netherland) equipped with a digital imaging system (Version I) was used to image pore structures, authigenic mineralogy, cements, clay occurrence and paragenetic relationships. The SEM samples were prepared from freshly broken rock fragments coated with gold and analyzed under an acceleration voltage of 20 kV using a beam current of 33 mA. Back scattered electron (BSE) images were also used to determine the relative timing of mineral growth and textural relationships between diagenetic phases. A cathodo-luminescence (CL) detector containing an Olympus microscope equipped with a CL8200-MKS CL, with a beam voltage of 17 kV and a current of 600 mA, was used to help differentiate carbonate and quartz cement. X-ray diffraction (XRD) analyses of more than 30 samples were performed using a Dmax 12 kW powder diffractometer for clay mineralogy identification. Clay mineral fractions <2 μm were obtained by centrifugation of clay slurries, and each sample was air-dried, glycol-saturated, and heated to 550 °C.

The carbonate cements in 26 representative organic matter free sandstones samples were selected for analysis of the carbon and oxygen stable isotope compositions. These samples were dissolved in phosphoric acid (H_3PO_4), and the evolved CO_2 gas was analyzed for carbon and oxygen isotopes in a Finnigan MAT 253 isotope ratio mass spectrometer (Thermo Fisher Scientific, Waltham, MA, USA). The carbonate cements samples that reacted with 100% phosphoric acid (H_3PO_4) at 25 °C for one hour (calcite), at 50° C for 24 h (Fe-dolomite/ankerite), respectively. The precision of the carbon and oxygen isotope ratios were ±0.04‰ and ±0.06‰, respectively. The carbon and oxygen isotope data are presented in the δ notation relative to the Vienna Pee Dee Belemnite (V-PDB) standards. Six core samples were prepared by doubly polished fluid inclusion wafers for observation of fluid inclusion and microthermometric measurements. Microthermometry was conducted using calibrated Linkam A TH-600 heating and cooling stages (Linkam, Epsom, UK) of fluid inclusions, and the phase transition temperature range can be measured from 180 °C to 500 °C with an accuracy of ±0.1 to ±1.0. The chemical composition of the carbon-coated, polished thin sections of carbonate cements was analyzed by a JEOLJXA-8100 electron microprobe analyzer (EMPA) (JEOL, Tokyo, Japan) under an accelerating voltage of 20 kV with a beam current of 15 nA

All the core samples and statistical data from wells were collected from the Petroleum Exploration and Production Research Institute of Qinghai Oilfield Company, PetroChina. Point counting, fluid inclusion, and optical and scanning electron microscopy analyses were performed at the laboratory of PetroChina Qinghai Oilfield Company (Qinghai, China). XRD was conducted in the Micro Structure Analytical Laboratory, Beijing, China. The carbon and oxygen isotopic and chemical composition analyses of carbonate cements were conducted at the key laboratory of Petroleum Resources Research, Institute of Geology and Geophysics, Chinese Academy of Sciences (Lanzhou, China).

4. Results

4.1. Sandstone Petrography

4.1.1. Detrital Composition and Texture

The Lulehe Formation sandstones are primarily composed of brown mudstones, gray fine siltstone, argillaceous siltstone and fine sandstone (Figure 4a). The sandstones are predominantly classified as feldspathic litharenites and lithic arkose according to the classification by Folk [36] (Figure 4b). Overall, the petrographic framework compositions of the studied sandstones are immature. Quartz is the most common detrital constituent, with monocrystalline quartz ranging from 13% to 79%, with an average value of 47.9%, and polycrystalline quartz ranging from 0.3% to 12.6, with an average value of 5.7%. Feldspar consisting of both plagioclase and K-feldspar varies from 12% to 36% and averages 18.6%, while the K-feldspar is the dominant type of the feldspar. In addition, the rock fragment volume fraction ranges from 3% to 52%, with an average of 22.7%, primarily consisting of metamorphic rock debris and volcanic rock grains. Other detrital grains include micas, biotite, muscovite, heavy minerals (garnet, zircon, rutile) and mud intraclasts. On average, the sandstones are mainly composed of subangular to subrounded grains, most of which are in linear to concave-convex contacts, thus indicating moderate to poor sorting (Figure 4c).

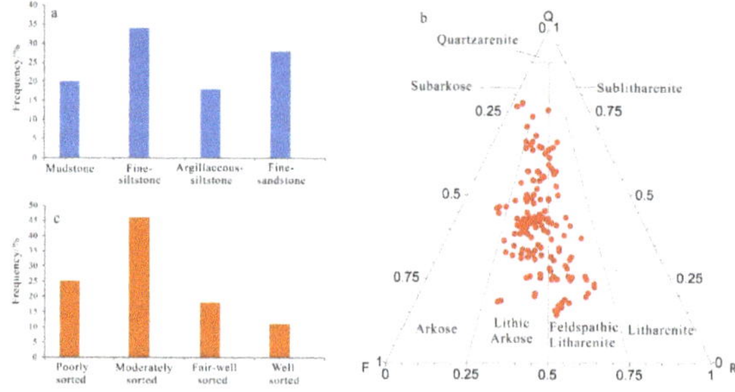

Figure 4. Rock composition lithology and sorting distribution of the Lulehe Formation sandstone reservoirs: (**a**) Lithology distribution; (**b**) Ternary diagram illustrating the classification of sandstone on base of Folk's [36] classification; (**c**) Sorting distribution.

4.1.2. Reservoir Physical Characteristics

According to conventional core analysis of more than 160 plugs sampled from the Lulehe Formation, the porosity and permeability of sandstone vary widely. A statistical analysis shows that the porosity ranges from 1.69% to 31.33%, with an average of 9.65% (Figure 5a), whereas the permeability ranges from 0.03 to 366.52 mD, with an average of 13.75 mD (Figure 5b). In general, the porosity and permeability of the Lulehe sandstones decrease with progressive burial but slightly increase in some depth intervals (300–3200 m, 3400–3600 m and 4000–4100 m respectively) (Figure 5a,b). The petrographic microscopy and SEM analyses revealed that depositional primary and diagenetic secondary pores are both present. Depositional primary pores mainly include residual intergranular pores (Figure 6a–c), while diagenetic secondary pores can be subdivided into dissolution pores (e.g., intergranular and intragranular dissolution pores), with minor amounts of microfractures (Figure 6d). In addition, point count thin section porosity is generally consistent with core plug porosity, but in most cases, thin section porosity is lower than core plug porosity, indicating the presence of micro-pores within the clay minerals (Figure 6a,c). The statistical analysis data show that diagenetic secondary pores (average 36% in the total porosity) are more abundant than depositional primary pores (up to 62%), suggesting that diagenesis plays an important role in the modification of the reservoir pores and properties.

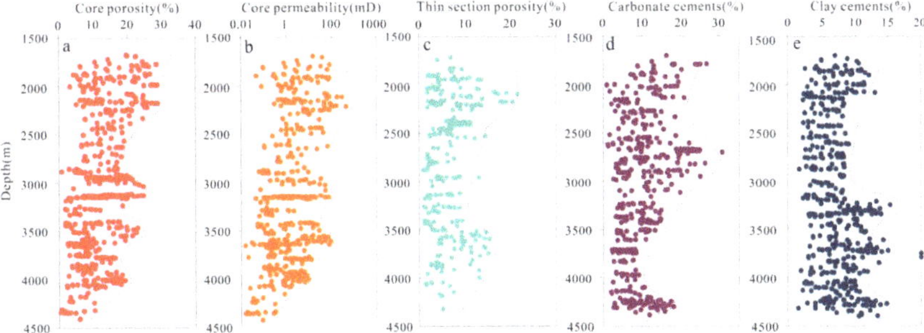

Figure 5. Plot showing variations of (**a**) core porosity; (**b**)core permeability; (**c**) thin section porosity; (**d**) carbonate cement; (**e**) and clay cementswith respect to depth.

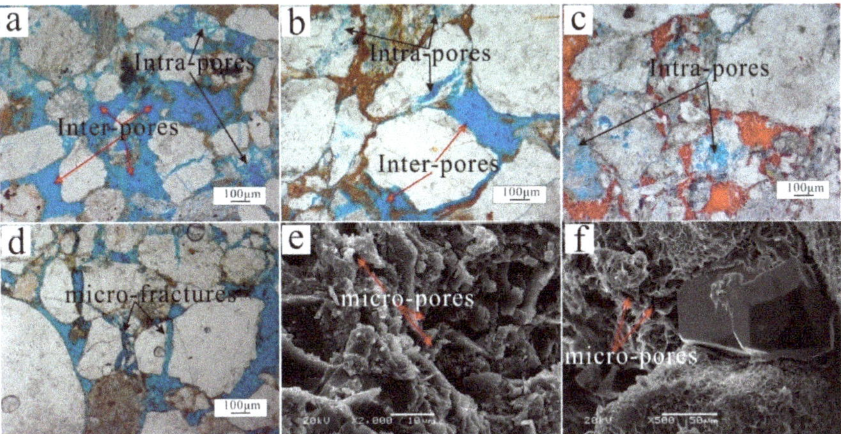

Figure 6. Pore types (**a–f**) in Lulehe sandstones reservoir. Inter-pores: intergranular pores; Intra-pore: intragranular pores.

4.2. Diagenetic Alteration Features

The petrographic and XRD as well as SEM analyses of the samples revealed complicated diagenetic alteration. The major diagenetic processes that controlled the reservoir properties include compaction, cementation by carbonate and quartz, authigenic clay minerals and dissolution of unstable materials.

4.2.1. Compaction

Compaction is one of the most important types of diagenesis in reservoirs, and it can be subdivided into mechanical and chemical compaction, which occurs widely in the Lulehe sandstones. Mechanical compaction is identified by grain rearrangement (Figure 7a) and deformation of argillaceous detritus, mica, and other ductile grains (Figure 7b), sometimes resulting in the formation of a pseudomatrix (Figure 7c). However, there is also evidence for deep burial depth, during which chemical compaction plays a major role. In this case, grain contacts are dominated by long contacts and concavo-convex contacts (Figure 7d). Furthermore, sutured contacts were occasionally observed in the thin sections (Figure 7e,f), implying the occurrence of chemical compaction in the Lulehe sandstone. In addition, the compaction effect was more intense in the samples containing more ductile fragments (e.g., clay matrix, mica and volcanic fragments) but was relatively weak in the samples with high quartz content and good sorting.

Figure 7. Thin-section photomicrographs showing compaction and calcite cement features. (**a**) Grain contacts present dominated by linear types (yellow arrow) due to the influence of compaction, cross-polarized light (CPL). (**b**) Mechanical compaction is identified by deformation of mica, plane-polarized light (PPL). (**c**) Deformation of mica and ductile grains resulting in the formation of a pseudomatrix (yellow arrow), CPL. (**d**) Grain contacts characterized by long contacts (yellow arrow), PPL. (**e**) Grain contacts characterized by long and concavo-convex contacts (yellow arrow), PPL. (**f**) Chemical compaction evidenced by grain rearrangements with concavo-convex contacts (yellow arrow), PPL. (**g**) Calcite cements filling between loosely packed grains, PPL. (**h**) As a result of the amount of calcite cements filling in the pores, the clastic grains appear to be floating, PPL. (**i**) Calcite cements represented the partial replacement of grains and filling in the intergranular pores, PPL. Cal: calcite.

4.2.2. Carbonate Cements

Carbonate cements (calcite, dolomite, ferroan calcite, ankerite) are the most abundant authigenic minerals in the studied sandstone and range from 0.86 to 32% of the total rock volume composition (avg. 13.6%). The EMPA data show that the carbonate cements are dominated by calcite, with the average amounts of $CaCO_3$, $FeCO_3$, $MgCO_3$ and $MnCO_3$ being 94.62 mol %, 1.66 mol %, 3.08 mol % and 0.65 mol %, respectively (Table 1), followed by dolomite and a minor amount of ferroan calcite and ankerite. In addition, at the depth of 2700–2800 m and 4400–4500 m, there are abnormally high values of carbonate cement content, corresponding to the low values of porosity and permeability (Figure 5d).

Table 1. Element (Ca, Fe, Mg and Mn) content of carbonate cements in the Lulehe sandstones.

Sample	Depth (m)	Analyzed Spot	Content (in wt. %)					Content (in mol%, Normalized with Calcium to 100 mol%)			
			CaO	FeO	MgO	MnO	Total	Ca	Fe	Mg	Mn
ES-I2	4216.82	1	53.46	0.68	1.56	0.06	55.76	94.13	1.36	4.35	0.16
ES-I2	4313.35	1	54.48	0.17	0.62	0.03	55.30	97.98	0.26	1.63	0.13
ES-I2		1	54.38	1.21	2.11	0.46	58.16	91.96	1.89	5.46	0.69
ES-I2		2	52.36	0.89	0.47	0.19	53.91	97.68	0.86	0.93	0.53
ES-I2	4323.58	1	53.67	0.16	1.76	0.08	55.67	97.31	0.49	2.16	0.04
ES-3	3622.05	1	54.76	3.21	3.86	0.96	62.97	83.32	5.63	9.87	1.18
ES-3		1	54.46	1.18	0.68	0.67	56.99	94.41	2.02	2.45	1.12
ES-3	3623.30	2	53.68	0.97	1.37	0.14	56.16	94.71	1.31	3.72	0.26
ES-3		1	54.37	1.28	1.35	0.62	57.62	93.25	1.76	3.76	1.23
ES-3	3627.63	2	53.67	0.65	0.49	0.03	54.84	97.44	0.98	1.06	0.52
ES-3	3631.18	1	54.35	1.02	1.16	0.76	57.29	96.10	1.36	1.68	0.86
ES-3	3632.57	1	53.69	1.32	1.25	0.67	56.93	95.27	1.98	1.86	0.89
LS-1	2810.23	3	53.96	1.98	1.16	0.36	57.46	95.13	1.73	2.46	0.68
LS-1	2813.65	1	54.39	0.96	0.68	0.32	56.35	96.60	1.46	1.49	0.45
LS-1	2820.34	2	54.36	1.16	1.53	0.68	57.73	93.97	1.76	3.25	1.02

There are two types of calcite in the Lulehe reservoir sandstone, calcite (Ca-I) is dominated bycoarse-crystalline calcite occurs as coarse crystalline poikilotopic, pore-filling blocky, and aggregate pore-filling forms (10–400 μm). Conversely, another type of calcite (Ca-II) occurs as micritic to microcrystalline and was in close contact with detrital grains, the different textures of calcite ments suggest they may precipitation in various diagenetic regimes [15]. Calcite (Ca-I) have presumably formed near the sediment–water interfaceduring early diagenesis (shallow burial realm), as evidenced by fills relatively large pores between loosely packed framework grains or represented the partial replacement of grains (Figure 7g–i), suggests that formed pre-dates significant mechanical compaction in shallow depth. However, dense micritic texture calcite (Ca-II) filled the gaps between tightly packed grains and covers and engulfs quartz overgrowths (Figure 8a), and is interpreted to have precipitated post-dates quartz overgrowths which was commonly occur in mesogenetic stage (deep burial realm). Moreover, based on the BSE and CL image, calcite cements also occur as partial grain replacements, locally replacing feldspar, lithic grains or mica (Figure 8b,c), and some of the calcite cements appear in a circular band (Figure 8d). In addition, dolomite cements generally fill in intergranular pores between floating grains (Figure 8e), indicating that the cements formed before intense compaction during early diagenesis. Ferrocalcite and ankerite typically occur as scattered patchy crystals or euhedral rhombs (20–150 μm) and are minor cements, they usually fill in feldspar dissolution pores and replace detrital grains or calcite and dolomite (Figure 8f), indicating that the ferrocalcite and ankerite formed post-datescalcite and dolomite cements and feldspar dissolution, and is can be inferred that it have precipitated in mesogenetic stage during deep burial depth.

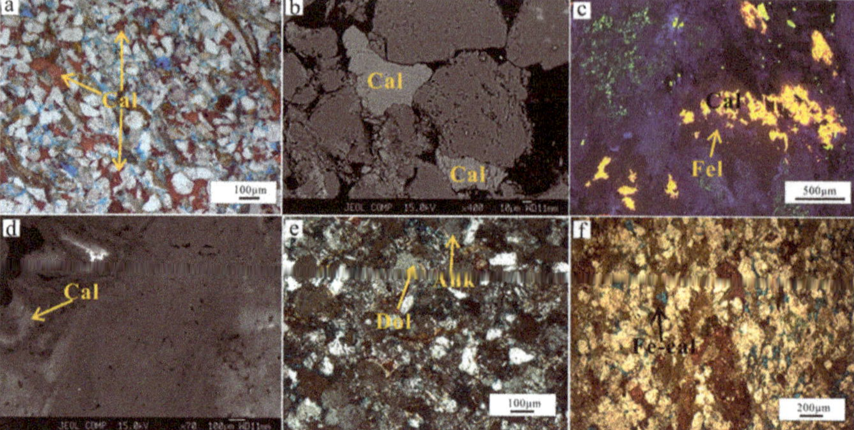

Figure 8. Photomicrographs showing the characteristics of carbonate cementation (**a**) Dense micritic texture calcite (Ca-II) filled the gaps between tightly packed grains and covers and engulfs quartz overgrowths (yellow arrow), PPL. (**b**) Backscattered electron (BSE) image of isolated pore-filling carbonate cement. (**c**) CL image showing calcite cements locally replacing feldspar. (**d**) BSE image showing pore-filling calcite cements appear in a circular band. (**e**) Dolomite and ankerite cement occur mainly as isolated rhombic and euhedral rhombs crystals, respectively. (**f**) Ferrocalcite filling in feldspar dissolution pores and replace detrital grains (black arrow), PPL. Cal: calcite; Fel: Feldspar; Dol: Dolomite; Ank: Ankerite; Fe-cal: Ferrocalcite.

4.2.3. Quartz Cements

Thin section and SEM analyses revealed that quartz cements are common and observed constituting 0.3–8.36% (avg. 3.8%) of the total rock volume and mainly occurs as syntaxial overgrowths around detrital quartz grains (quartz overgrowths, thickness of 20–200 μm). Some examples of authigenic quartz occurring as euhedral and hexagonal crystals partially or completely infilling intergranular pores (Figure 9a) were observed but represent a relatively minor amount. In most cases, quartz overgrowths can be discriminated from the detrital grains due to the existence of dust rims or clay mineral coatings and fluid inclusions, with pores often engulfed by large syntaxial overgrowths (Figure 9b–d). However, quartz overgrowths were inhibited by localized chlorite grain coating. In contrast, Quartz overgrowths were also observed where chlorite rims and grain coating are absent or rare, with thicknesses of ~20–200 μm and regularly varying morphology (Figure 9e). In other words, only continuous chlorite rims or a certain thickness of chlorite grain coating can delay and inhibit the quartz overgrowths. Quartz overgrowths are usually associated with feldspar dissolution (Figure 9f) and engulfed by ferroan calcite and ankerite cements, indicating that the quartz cement postdates feldspar dissolution but predates the ferroan calcite and ankerite cements. Quartz cement is more abundant in the matrix-rich sandstones, which are mainly poorly sorted, fine-grained sandstones.

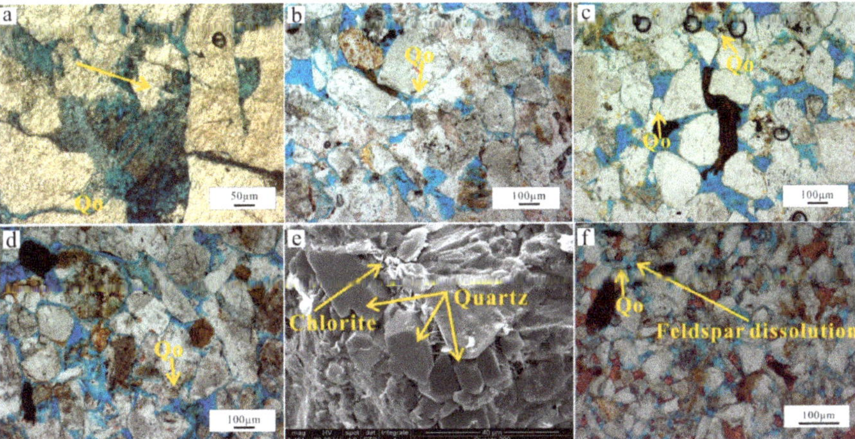

Figure 9. Photomicrographs showing characteristics of quartz cements. (**a**)Authigenic quartz occurring as euhedral crystals partially infilling intergranular pores (yellow arrow), PPL. (**b**) Quartz overgrowths can be discriminated from the detrital grains due to the existence of dust rims, PPL. (**c**) Quartz overgrowths occluding the pore space, PPL. (**d**) Pores space engulfed by quartz syntaxial overgrowths, PPL. (**e**) Quartz overgrowths due to discontinuous chlorite. (**f**) Feldspar dissolution was accompanied by quartz overgrowths. Qo: Quartz overgrowths.

4.2.4. Clay Cements

Four main clay mineral types (Figure 10) were observed via thin section petrography, SEM and XRD analyses ranging from 2% to 20% of the total rock volume (average 5.6%). Illite is the most abundant diagenetic constituent, accounting for 16% to 78% (average 62.87%) of the total clay content, followed by chlorite (average 15.16%), illite-smectite (average 13.69%) and kaolinite (average 8.27%). The content of clay minerals increased with the depth (Figure 5e), but the variation trend of various clay minerals was different. XRD data of clay minerals in the Lulehe sandstone show that illite dominates in sandstones with depth deeper than 3300m (Figure 9a), whereas the content of kaolinite increased significantly above 3330m (Figure 10b). The mixing layer of illite and smectite increased with the depth (Figure 10c), while the content of smectite in the mixing layer of illite and smectite decreased with the depth (Figure 10d). Chlorite tends to decrease with increasing depthand mainly concentrated above 3000m (Figure 10e).

In general, honeycomb-like crystals of illite-smectite and fibrous illite are the major clay cements and are widely distributed. In some cases, illite-smectite occurs as a replacement product of detrital micas and fills in pores as pore-filling clay. Furthermore, the thickness of the illite coating can be up to 50 μm, which usually encircles the grain at the place where the intergranular pore develops but rarely occurs at the grain contact. However, the fibrous morphology of illite typically appears as fibrous aggregates that bridge and filling in primary pores and secondary pores (Figure 11a,b). SEM observations also show that irregular flakes with lath-like illite usually attach to the quartz overgrowth surface and fill in the intergranular pores, indicating that quartz overgrowth precedes the precipitation of illite (Figure 11c).

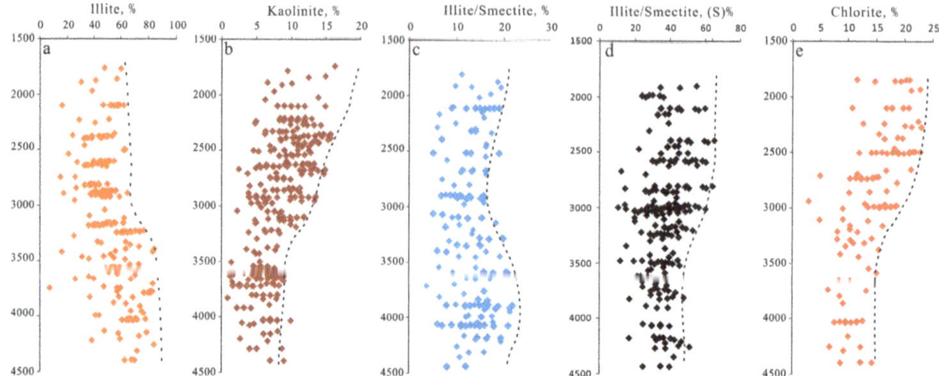

Figure 10. Vertical distribution characteristics of clay minerals in the Lulehe sandstones (From XRD data). (**a**) Cross plot of illite with depth; (**b**) Cross plot of kaolinite with depth; (**c**) Cross plot of mixing layer of illite and smectite with depth; (**d**) Cross plot of the content of smectite in the mixing layer of illite and smectite; (**e**) Cross plot of chlorite with depth.

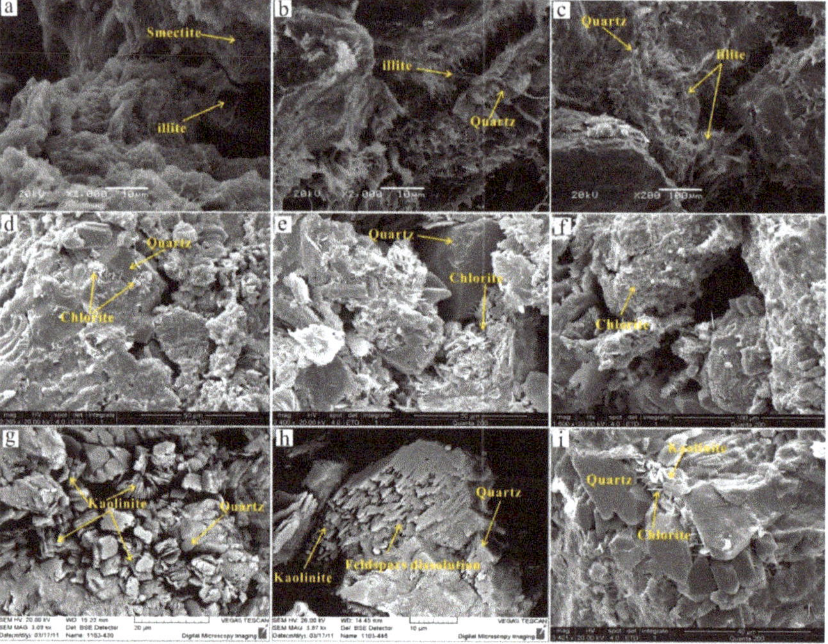

Figure 11. Scanning electron micrographs of sandstones showing characteristics of clay cements. (**a**) Pore filling illite and smectite. (**b**) Fibrous morphology of illite typically appears as fibrous aggregates that bridge and filling in pores. (**c**) Irregular flakes with lath-like illite usually attach to the quartz overgrowth. (**d**) The pore-lining chlorite occurs locally as irregular needle crystals perpendicular and parallel to the quartz grain surfaces. (**e**) Pore-filling aggregate chlorite minerals consist of rosette-shaped platelets oriented radially and occluding the intergranular pores. (**f**) Platelets of authigenic chlorite seem to inhibit quartz overgrowth. (**g**) Kaolinite occurs primarily as stacks of booklets aggregates. (**h**) Kaolinite is commonly associated with feldspar grain dissolution and silica precipitation. (**i**) Platelets of kaolinite that retards the growth of chlorite crystals.

Chlorite is slightly more abundant than kaolinite and mainly includes grain-coating (pore-lining) chlorite and pore-filling chlorite based on petrographic and SEM image observations. The pore-lining chlorite occurs locally as irregular needle crystals perpendicular and parallel to the detrital grain surfaces (Figure 11d), whereas a minor amount of pore-filling aggregate chlorite minerals consists of rosette-shaped platelets oriented radially and occluding the intergranular pores (Figure 10e). Generally, the quartz grains with discontinuous chlorite rims are locally enclosed by quartz overgrowths (Figure 10d,e), while continuous chlorite coating is generally characterized by a small amount of quartz cementation (Figure 11f). In addition, detrital biotite and igneous rock fragments were observed to be locally replaced by chlorite.

In contrast, kaolinite is generally a minor phase and occurs primarily as stacks of booklets, vermicular aggregates with individual crystals up to 6 μm long that sit in intergranular pores (Figure 11g). Kaolinite is commonly associated with feldspar grain dissolution (Figure 11h) and alteration of mica, and mica grains was partially replaced by kaolinite, suggesting that feldspar and mica grains may be a source for the precipitation of kaolinite. Less commonly, kaolinite also occurs as blocky aggregates composed of euhedral and pseudohexagonal plates filling in primary pores, and in the deeply buried areas, high-temperature kaolinite seems to be locally transformed into dickite. SEM images show that kaolinite seems to inhibit the growth of chlorite crystals (Figure 11i); thus, it can be inferred that kaolinite predates the precipitation of chlorite.

4.3. Mineral Chemistry

4.3.1. Stable Isotopes of Carbonate Cement

The type and contents of carbonate cements were determined on the basis of carbon and oxygen isotopic analysis of thin sections from 31 sandstone samples, which are plotted in Figure 12a. Both the carbon and oxygen isotopes (i.e., Vienna Pee Dee Belemnite normalized isotopes: $\delta^{13}C_{PDB}$ and $\delta^{18}O_{PDB}$) spanned a wide range of values (Table 2). The $\delta^{13}C_{PDB}$ values range from -15.21‰ to -3.21‰, and the $\delta^{18}O_{PDB}$ values range from -21.07‰ to -11.48‰. The $\delta^{13}C_{PDB}$ values in the Cal-I cements have a relatively wide range from -9.84‰ to -3.21‰, with an average value of -5.63‰, and the $\delta^{18}O_{PDB}$ values range from -14.28‰ to -11.74‰, with an average value of -13.75‰. Cal-II cements have only sporadic data, the $\delta^{13}C_{PDB}$ values range from -13.78‰ to -12.86‰, and the $\delta^{18}O_{PDB}$ values range from -19.56‰ to -18.75‰. Dolomite has a wider range of $\delta^{13}C_{PDB}$ values from -6.70‰ to -4.37‰, with an average value of -5.18‰, and $\delta^{18}O_{PDB}$ values from -13.74‰ to -11.48‰, with an average value of -12.21‰. The $\delta^{13}C_{PDB}$ values of ferroan-calcitecements range from -15.21‰ to -6.58‰, with an average value of -10.49‰, and the $\delta^{18}O_{PDB}$ values range from -20.20‰ to -16.58‰, with an average value of -18.41‰. While ankerite cements have a narrow range of $\delta^{13}C_{PDB}$ values that range from -12.84‰ to -9.38‰, with an average value of -10.81‰, and $\delta^{18}O_{PDB}$ values that range from -21.07‰ to -18.67‰, with an average value of -19.72‰.

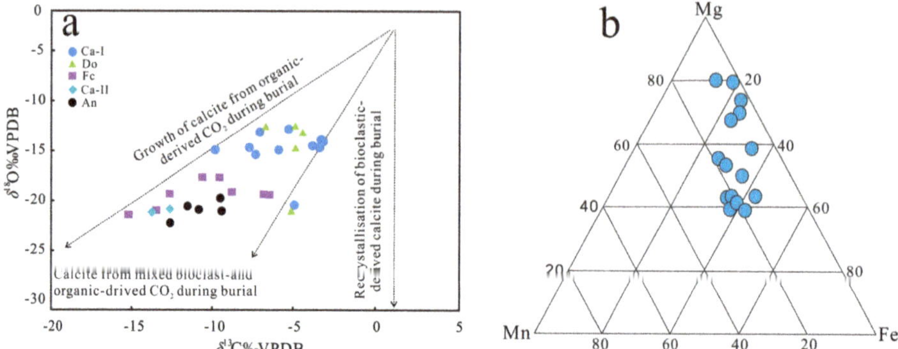

Figure 12. (**a**) Scatter diagram of $\delta^{13}C$ and $\delta^{18}O$ for calcite cements in the Lulehe sandstone reservoirs. The $\delta^{13}C$ values indicate that sparry carbonate cements were influenced by organic matter (the model modified after Rahman [37]). (**b**) Ternary plot of Fe–Mg–Mn element compositions of carbonate cements, data from Table 1.

Table 2. Mineralogical and carbon and oxygen isotopic composition of carbonate cements, and calculated formation temperature of carbonate cements in Lulehe sandstone.

Sample	Depth	Litho-logy	Carbonate Minerals	Carbonate Content (%)	$\delta^{13}C_{PDB}$ ‰	$\delta^{18}O_{PDB}$ ‰	T/°C
LS-1	2798.35	Bs	100% Ca-I	22	−3.33	−12.79	65.60
LS-1	2799.52	Bs	100% Ca-I	20	−3.85	−13.37	69.24
LS-1	2810.24	Gss	100% Ca-I	21	−3.21	−12.95	66.60
LS-1	2814.64	Bs	100% Ca-I	26	−7.36	−14.28	75.08
LS-1	2816.73	Gfs	100% Ca-I	12	−5.91	−13.82	72.11
LS-1	2817.67	Bss	100% Ca-I	24	−3.44	−13.57	70.51
LS-1	2819.36	Bs	100% Ca-I	18	−7.75	−13.57	70.51
LS-1	2820.45	Bs	100% Ca-I	19	−7.1	−12.02	60.87
ES-I2	2837.85	Gbps	100% Ca-I	16	−9.84	−13.82	72.11
ES-I2	2781.54	Bs	100% Ca-I	21	−5.28	−11.74	59.18
ES-I2	2976.39	Bps	70% Ca-I + 30% Do	11	−4.94	−19.34	-
ES-I2	2836.36	Bs	100% Do	14	−4.81	−13.74	71.60
ES-I2	2979.74	Gfs	100% Do	16	−6.7	−11.54	57.98
ES-I2	2959.67	Gfs	100% Do	8	−4.87	−11.48	57.62
ES-I2	3005.46	Bs	100% Do	15	−4.37	−12.08	61.23
ES-I2	4333.57	Bss	60% Do + 40% Fc	12	−5.12	−19.94	-
ES-I2	4334.56	Bs	100% Fc	14	−6.58	−18.36	103.33
ES-I2	4335.15	Bs	100% Fc	8	−10.62	−16.58	90.60
ES-3	3622.05	Bs	100% Fc	8	−8.81	−18.06	101.14
ES-3	3623.30	Gbs	100% Fc	8	−9.57	−16.61	90.81
ES-3	3626.30	Bs	60% Fc + 40% Do	10	−15.21	−20.2	-
ES-3	3627.56	Bs	90% Fc + 10% Ca-II	13	−13.56	−19.68	113.19
ES-3	3226.98	Gbps	100% Ca-II	9	−12.86	−18.75	106.21
ES-3	3356.87	Bs	100% Ca-II	12	−13.87	−19.56	112.28
ES-3	3598.48	Bs	80% Fc + 20% Ca-II	8	−12.89	−19.36	110.77
ES-3	3629.60	Gbps	90% Fc + 10% Ca-II	12	−12.84	−21.07	123.94
ES-3	3631.60	Bs	80% An + 20% Fc	8	−9.51	−18.67	105.62
ES-3	3633.29	Bs	70% An + 30% Do	6	−9.38	−19.73	-
ES-3	3635.10	Gbs	80% An + 20% Do	6	−11.51	−19.5	111.82
ES-3	3637.20	Bs	90% An + 10% Fc	5	−10.82	−19.64	112.88
ES-3	3639.10	Bs	70% Fc + 30% An	6	−6.69	−18.32	-

Note: Only isotopic data of samples with the content of one specific type of carbonate up to 80% were used to calculate the temperature. Bs: Brown sandstone; Gss: Gray silty sandstone; Gfs: Gray fine sandstone; Bss: Brown silty sandstone; Gbps: Gray-brown pelitic siltstone; Bps: Brown pelitic siltstone; Bss: Brown silty sandstone; Gbs: Gray-brown sandstone;Ca:calcite; Do:dolomite; Fc:ferrocalcite; An: ankerite.

4.3.2. Fluid Inclusions Analysis

Fluid inclusions record valuable information for diagenetic events and the precipitation temperature of authigenic minerals [38] and provide pressure-volume-temperature-composition (PVT) information. Therefore, to determine the temperature of the precipitation of the authigenic clay minerals and quartz cements, six core samples were prepared as doubly polished thin sections and were examined for aqueous inclusions, and the fluid inclusions occurring in quartz cements were selected for microthermometric measurements according to the thin section analysis. Most of the selected fluid inclusions had diameters of approximately 2–10 μm and two phases, with only a few single-phase gas inclusions at room temperature.

The homogenization temperatures (Th) were measured for the aqueous inclusions that were trapped during quartz cementation and range from 70–120 °C (Figure 13). Moreover, the temperatures calculated from $\delta^{18}O$ in carbonate cements (Ca-II, ferroan calcite and ankerite) range from −21.07‰ to −16.58‰, thus yield Th ranges from 90–123 °C (Table 2). The results suggest that precipitation of Ca-II, ferroan calcite, ankerite and quartz cementation mainly occurred in the mesodiagenetic stages, which is consistent with the thin section observations.

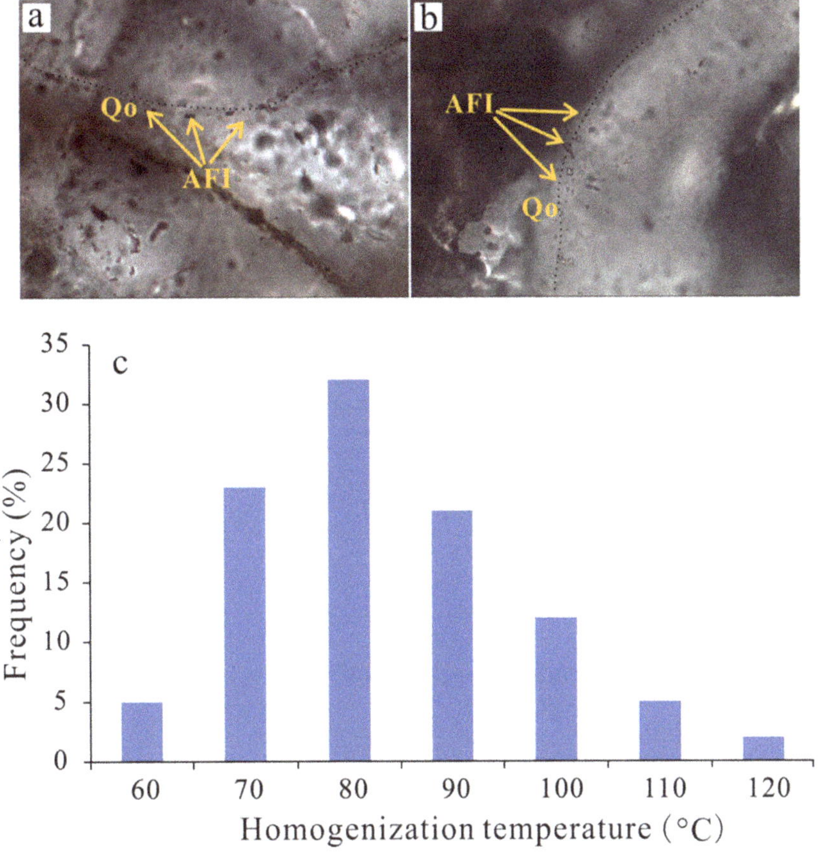

Figure 13. (a,b) Photomicrographs of aqueous fluid inclusions (AFI) in quartz overgrowth (Qo). (c) Histograms of homogenization temperatures (Th) of aqueous inclusions in quartz overgrowths in Lulehe sandstone reservoirs.

5. Discussion

5.1. Origin and Sources of Carbonate Cements

Carbonate minerals (calcite, ferroan-calcite, dolomiteand ankerite) are the most common authigenic minerals in the Lulehe sandstones and can be formed by various pathways and multiple diagenetic stages. The divalent cations Ca^{2+}, Mg^{2+}, Fe^{2+}, Mn^{2+} and carbon dioxide in carbonate cements may exist as internal, external or mixed sources [38–40]. Internal sources include dissolution of detrital anorthite grains, calcretes, carbonate rock fragments, calcite fossil fragments and reworked carbonate intraclasts [41–43]. External sources include compaction drainage and mutual conversion of clay minerals such as illitization of kaolinite and smectite, which will release the Ca^{2+}, Mg^{2+} and Fe^{2+} ions into the pore fluids, while CO_2 may be derived from the maturation of kerogen and decarboxylation of organic matter from the adjacent mudstone or source rocks [44–46]. The dissolution of detrital anorthite and carbonate grains is not obvious in the Lulehe sandstone, and the distribution pattern and occurrence form of the carbonate cements (Figure 8) suggests that external or mixed sources may be the main sources for these cements, which is consistent with the estimation scatter diagram of the material source based on the $\delta^{13}C_{PDB}$ and $\delta^{18}O_{PDB}$ isotopic values (Figure12a). The integration of the $\delta^{13}C$ and $\delta^{18}O$ isotopic values of carbonate cements and the burial and thermal history can effectively elucidate the origin of the potential source material and precipitation temperature involved in the cementation process. Generally, the $\delta^{13}C$ isotopic value distribution in carbonates is mainly related to carbon sources. For instance, in carbonates of the atmospheric source (meteoric water) are approximately −7‰ [47], whereas CO_2 associated with organic matter tends to have more negative $\delta^{13}C$ values of −18‰ to −33‰ because the organic matter is depleted in ^{13}C [48]. However, CO_2 in the carbonates that precipitate under methanogenic conditions has higher $\delta^{13}C_{PDB}$ values, ranging from −5‰ to +15‰ [49]. The $\delta^{13}C_{PDB}$ values of the carbonate cements in the Lulehe sandstones vary from −15.21‰ to −3.21‰, suggesting a mixed source origin [49]. As a consequence of the calcite (Ca-I) have presumably formed near the sediment–water interface, and there two points where the $\delta^{13}C_{PDB}$ values of the Ca-I are around −7‰ (Table 2), suggesting that the source of Ca-I may be affected by meteoric water to some extent. However, the Lulehe Formation being progressively buried and not suffering from erosion during uplift to the surface in the burial history (Figure 3), it can be concluded that meteoric water exerted a negligible influence on the source of carbonate cement (Ca-II, ferroan calcite and ankerite) which occurred in deep buried depth [49]. In addition, the ternary diagram of the relative Fe–Mn–Mg contents of diagenetic calcites shows that most of the points are close to the distribution of the end members of Fe–Mg (Figure 12b), also suggesting that the formation of cement was controlled by pore water and fluid formed during organic diagenesis [50,51], while the influence of meteoric water was relatively insignificant. Thus, the growth of carbonate cements might be derived from bioclast- and organic-derived CO_2 during burial (Figure 12a).

The $\delta^{18}O$ isotopic values of carbonate cements could reflect the precipitation temperature and composition of the diagenetic fluid [52]. The $\delta^{18}O$ isotopic values of meteoric water have a large range from −50‰ to 0‰ (SMOW, standard mean ocean water), while the $\delta^{18}O_{SMOW}$ values of ocean water are approximately 0‰, where as pore fluids in the reservoir usually deviate towards higher $\delta^{18}O$ isotope values due to water-rock interactions [52]. The precipitation temperatures for carbonate cements can be calculated in terms of the equation by Craig [53]:

$$T(°C) = 16.9 - 4.2(\delta^{18}O_{PDB} - \delta^{18}O_{SMOW}) + 0.13(\delta^{18}O_{PDB} - \delta^{18}O_{SMOW})^2$$

$\delta^{18}O_{PDB}$ represents the oxygen isotope value of carbonate cements, ‰ PDB, and $\delta^{18}O_{SMOW}$ means the oxygen isotopic value of pore fluid in equilibrium with calcite and dolomite,‰ SMOW. However, there is no direct research on the $\delta^{18}O_{SMOW}$ of pore fluids in the Lulehe Formation. Previous studies on the linear relationship between δD and $\delta^{18}O_{SMOW}$ defined the relationship as a global meteoric water line, and the relationship between $\delta^{18}O_{SMOW}$ and δD in China was established based on 30 GINP (Global Network of Isotopes in Precipitation) monitoring results [54]:

$$\delta D = 7.57 \times \delta^{18}O_{SMOW} + 6.02$$

Since the calculated precipitation temperature of carbonate cement is based on the estimation of a global meteoric water line, it may have an error of approximately 10 °C. The average δD of the Qaidam Basin is −21‰ [55]; thus, the $\delta^{18}O_{SMOW}$ value is −3.6‰ according to the calculation of the equation. With the pore fluid $\delta^{18}O_{SMOW}$ value and the carbonate cement precipitation temperatures, 57.62 °C to 75.08 °C (Table 2), which corresponds to the late phase of early diagenesis.

Differences in the carbon isotopic composition and the calculated precipitation temperatures among calcite, dolomite, ferran-calcite, and ankerite imply that they may have different precipitation mechanisms. The $\delta^{13}C_{PDB}$ of Ca-II, ferran-calcite and ankerite are relative light, ranging from −15.21‰ to −6.58‰, and the calculated precipitation temperature of the Ca-II, ferran-calcite and ankerite with an average value of 109.24 °C, 99.82 °C, 113.57 °C, respectively, indicating that the increasing input of organic acids and CO_2 may be induced by decarboxylation of organic matter at high temperatures. Changes in Mn^{2+} in carbonate cement are most likely related to alteration of volcanic materials (Table 1); moreover, alteration of the unstable volcanic rock fragments and illitization of smectite and kaolinite at the right temperatures also produce abundant Mg^{2+} and Fe^{2+} cations [37,40]. When the other cations in carbonates are gradually replaced by Fe^{2+} and Mg^{2+}, ferran-calcite and ankerite are formed.

5.2. Origin and Sources of Quartz and Clays

5.2.1. Source of Quartz Cements

Quartz cement may have multiple sources, including dissolution of biogenic amorphous silica and volcanic rock fragments, alteration of clay minerals, dissolved feldspar grains, and pressure solution of quartz grains, and all of them have internal and external sources [56,57]. However, unlike the solubility of Ca^{2+} and Mg^{2+} ions, that of SiO_2 (aq.) and Al^{3+} is extremely low in deeply buried geological systems, and together with constraints of heterogeneity in reservoir quality and fluid volume and without certain driving forces, massive transport of external SiO_2 (aq.) over long distances into sandstones is almost impossible [18,58]. Thus, with unrealistic external sources, the most likely silica sources for quartz cement were derived from internal sources.

Observations of thin sections reveal that there is no occurrence of abundant biogenic silica materials, so amorphous silica seems unlikely as a potential source for the quartz cements in the Lulehe sandstones. However, feldspar minerals are rich in Lulehe sandstones and petrographic evidence reveals that dissolution of feldspar is commonly accompanied by kaolinite and quartz cement (Figure 8f), which suggests the dissolution of feldspar is likely an important candidate source for quartz cements. Meanwhile, the maximum concentration of short chain carboxylic acid produced by organic maturation is between 75 °C and 90 °C, and the optimum temperature for the preservation of organic acids is 80–120 °C [59]. This temperature regime increased the rate of reaction of dissolution of feldspar grains [15]; thus, the growth of quartz cements due to an influx of SiO_2 has been associated with the concomitant alteration of feldspar to silica and kaolinite by feldspar-clay reactions due to acid buffering. In addition, fluid inclusion homogenization temperatures for quartz cements range from 70 °C to 120 °C (Figure 13c), which is consistent with optimum temperatures for the preservation of organic acids, also providing potential evidence for silica sources for quartz cements partly originating from feldspar dissolution.

Volcanic rock fragments are also common in the Lulehe sandstones, and they can be prone to transformation into smectite during eodiagenesis [60]; with progressive burial and increasing temperature, unstable smectite and kaolinite will transform to illite at temperatures of approximately 70 °C and 120 °C, respectively [61], and then SiO_2 will be released into the pore fluid. The XRD analysis statistical data reveal that illite comprises the most abundant content, ranging from 17% to 85% (avg.

6.8%), of the total clay minerals in the Lulehe sandstones, while smectite and kaolinite comprise only approximately 1.5% and 2% on average, respectively. This finding suggests that illitization of smectite and kaolinite is a possible source for quartz cements. In addition, the appearance of concavo-convex and sutured contact types suggesting the acquisition of silica through pressure dissolution can be another internal source for quartz cements.

5.2.2. Source of Clay Minerals

Authigenic clay minerals (e.g., chlorite, kaolinite, illite-smectite and illite) are ubiquitous types of cement in the Lulehe sandstones. Several previous studies have proposed that the origin of chlorite is generally associated with precursor clay mineral and high Fe and Mg ion content environments [61,62]. Chlorite forms in a variety of precursor clay minerals, but the main precursors include berthierine and smectite [63]. Berthierine is a Fe-rich clay mineral precursor developed from the interaction of the Al-Fe minerals with pore fluids under reducing conditions at the time of deposition [60]. Smectite reacts with iron oxide and saline pore waters to form chlorite, which can be prone to derived from alteration of volcanic rock fragments during the eodiagenesis stages [64]. Since berthierine is rarely detected and volcanic rock fragments are widely developed in the Lulehe sandstone, smectite, which is commonly derived from alteration of volcanic rock fragments, may be the primary clay mineral precursor of chlorite. In addition, the Lulehe sandstones were mainly deposited in fluvial deltaic environments, and proximal and distal delta fronts are particularly rich in Fe [65]. Moreover, the dissolution of volcanic rock fragments and feldspar grains also provides rich Fe and Mg ions as the main source for chlorite development.

Authigenic kaolinite is closely associated with feldspar dissolution (Figure 11h), which suggests that alteration of detrital feldspar grains is the most probable source for authigenic kaolinite [18,39]. Decomposed feldspar grains and the formation of authigenic kaolinite require a low K^+/H^+ activity ratio, and acidic (low K^+/H^+) pore fluids are presumably related to the decarboxylation of organic matter from the adjacent mudstone or source rocks [58]. However, the solubility of Al^{3+} ions derived from dissolved feldspar is extremely low [18,58]; as a result, kaolinite is likely precipitated in situ or in adjacent pores and is spatially synchronous with feldspar dissolution. With the increase in temperature and consumption of organic acids, the ratio of K^+/H^+ increases; consequently, the stability field of kaolinite is greatly reduced and can be easily converted to mixed layer illite/smectite and illite [6]. In addition, XRD data of clay minerals in the Lulehe sandstone show that illite dominates in sandstones with depth deeper than 3300m (Figure 9a), whereas the content of kaolinite increased significantly above 3330m (Figure 10b), also suggesting that kaolinite may be one of the material sources of illite. The mixing layer of illite and smectite increased with the depth (Figure 10c), while the content of smectite in the mixing layer of illite and smectite decreased with the depth (Figure 10d), indicating that some smectite may be transformed into illite.

5.3. Diagenetic Sequence

The diagenetic evolution regime generally includes eodiagenesis and mesodiagenesis stages sensu Morad et al. [66]. Eodiagenesis occurs mainly in an environment where the burial depth <2.7 km and the temperature <70 °C in the Lulehe sandstone reservoir, and the interstitial fluid chemistry is affected by depositional environment, whereas mesodiagenesis commonly occurs at >2.7 km of burial depth and temperatures >70 °C (Figure 3), which is mediated by evolved diagenetically altered fluids [67]. Although it is difficult to determine the precise timing and burial depth of diagenetic alterations [61], a general paragenetic sequence can be reconstructed and established by synthesizing the petrologic features, burial and thermal histories, various diagenetic minerals and relationships of mineral texture, clay mineral reactions associated with origin and source, and relative timing of the major diagenetic events of the Lulehe sandstones, as shown in Figure 14.

The sediments were subjected to mechanical compaction once deposited, as characterized by framework grain rearrangement and deformation of ductile grains (Figure 7a–c). In addition, Calcite

(Ca-I) have presumably formed near the sediment–water interface during early diagenesis (shallow burial realm), as evidenced by poikilotopic, pore-filling blocky, and aggregate pore-filling forms, it can be interpreted to have formed in the early stage of diagenesis (shallow burial realm) before significant compactional porosity-loss could occur. As the sediment is exposed to mechanical penetration near the surface, grains coated with clays (Figure 7a,b) may have developed in shallow depth. Authigenic chloritetypically develops at temperatures of >60 or 70 °C [12], corresponding to a depth of about 2400–2800 m (Figure 3) and belongs to eodiageneticstages. During the early diagenetic stage, volcanic rock fragments can be easily transformed into smectite, as burial depth and temperature increase, illitization of smectite will occurat a temperature around 70 °C [13], corresponds to a depth of about 2700 m in study area (Figure 3). In addition, the onset of authigenic quartz was interpreted at about 70–80 °C [45], corresponding to the burial depth from 2700~3200 m (Figure 3). Moreover, the $\delta^{13}C$ and $\delta^{18}O$ isotopic data of the carbonate cements and the homogenization temperatures (Th) of the aqueous inclusions in the diagenetic minerals may help to decipher the origin and source as well as the timing of precipitation of the minerals, integrated with the distribution pattern of the diagenetic minerals. It can be inferred that the significant eodiagenesis regime features mainly include (1) mechanical compaction, (2) precipitation of Ca-I and dolomite, (3) alteration of volcanic rock fragments and smectite development, (4) development of grain-coating chlorite, (5) precipitation of authigenic quartz, and (6) a minor amount of smectite transformed into illite.

As burial depth and temperature progressively increase during mesodiagenesis, porosity is destroyed predominantly by chemical compaction. Organic matter gradually becomes thermally mature, which generates organic acid and CO_2 above the critical onset temperature of 75 °C commonly occurs at >2.7 km of burial depth in Lulehe sandstone reservoir (Figure 3), and the optimum temperature for the preservation of organic acids is 80−120 °C [18], corresponding to the burial depth from 3200~4000 m (Figure 3). The pH values of the formation water decrease thereby dissolving early stage carbonate cements (Ca-I) and feldspar by low pH fluids, contributing to secondary porosity development. However, dissolution of feldspar releases aluminum and aqueous silica, which have low mobility and are considered to migrate over short distances, resulting in authigenic kaolinite and quartz cement precipitation in situ or in adjacent pores [58], indicate that kaolinite is closely associated with alteration of feldspar grains in the low pH (acidic) fluid which plausibly have been generated at depth burial depth from 3200~4000 m. Simultaneously, with the increase in K^+ released by feldspar dissolution, under high temperatures, the hair-like, spiny terminations and honeycomb-like habits of illite occur. During the late stage of mesodiagenesis, due to the consumption of organic acids and CO_2 during aluminosilicate dissolution, the pH value of the formation water increased, which is favorable for the precipitation of late carbonate cement (Ca-II, ferroan calcite and ankerite) (Figure 14). As a result, the mesogenetic process (deep burial realm) mainly includes: (1) chemical compaction, (2) dissolution of feldspar grains, (3) precipitation of kaolinite, (4) quartz overgrowth, (5) illitization of kaolinite, (6) development of pore-filling chlorite, and (7) precipitation of Ca-II, ferroan calcite and ankerite.

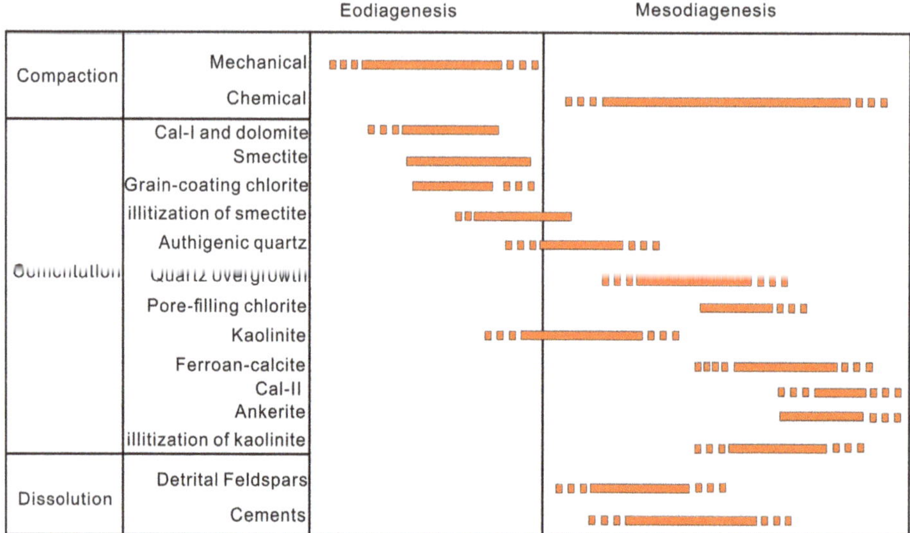

Figure 14. Paragenetic sequences of the Lulehe sandstone.

5.4. Impacts of Diagenesis on Reservoir Quality

Reservoir quality of the Lulehe Formation has been significantly controlled by various types of diagenetic alteration. The main diagenetic processes that controlled reservoir quality were compaction, precipitation of carbonate cements, formation of chlorite, quartz cements, development of illite, and dissolution of unstable grains and cements.

Compaction is the principal factor responsible for the deterioration of reservoir quality, followed by cementation, which is confirmed by using the methodology of Houseknecht [68] and Ehrenberg [69], as shown in Figure 15a. Porosity loss is due to rapid and progressive burial (Figure 3), and the petrographic analysis data revealed that mechanical compaction diminished porosity, as indicated by the presence of grain rotation, bending of micas, and deformation of mud intraclasts and ductile grains forming a pseudomatrix. The grain contacts gradually change from floating to point and line contact, indicating that mechanical compaction occurred progressively during shallow depth burial realm (eodiageneticstages). The primary intergranular porosity decreased by the recombination of grains. In addition, the appearance of concavo-convex and sutured contact types suggests that chemical compaction occur in deep burial realm (mesodiagenetic stages), is a possible silica source for quartz overgrowth, which may further occupy the pore space and harm the reservoir quality.

Carbonate minerals are the most common cements in the Lulehe sandstones, indicating that carbonate cement (e.g., calcite, dolomite, ferroan calcite and ankerite) also played a major role in drastically reducing reservoir quality. The sum of total calcite and dolomite along with ferroan calcite and ankerite cements exhibits a relatively good inverse correlation with porosity and permeability (Figure 16a,b), suggesting that reservoir quality was reduced by carbonate cement. The carbonate cement formed in the eodiagenetic stage (Ca-I) of shallow buried depth, due to its shallow burial, the particles have not been completely consolidated by the compaction stress. As a result, Ca-I exert a certain supporting effect on the particle skeleton against compaction, so it can enhance the compressive ability to some extent. The carbonate cements (Ca-II) including ferrocalcite and ankerite which were mainly formed in the mesodiagenetic stages, with the burial depth increasing, the reservoir has experienced strong compaction, and its supporting effect on the rock skeleton can be ignored. On the contrary, the late carbonate cements will occupy part of the dissolved pore, which resulted in the possibility of forming intergranular and intragranular pores are reduced.

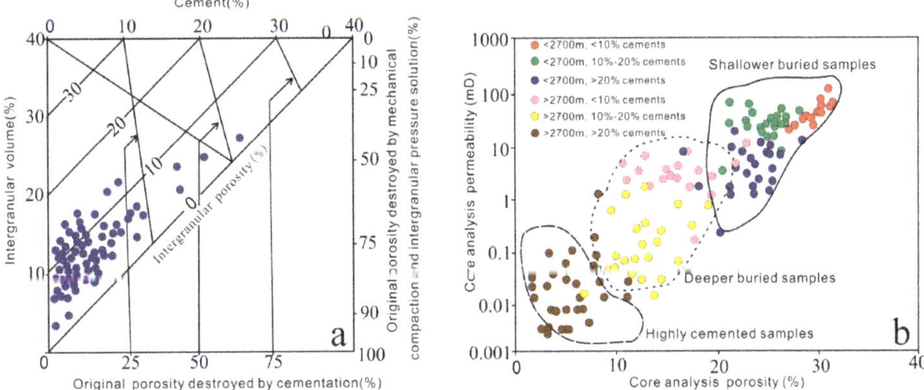

Figure 15. (**a**) Plot of intergranular volume (IGV) versus volume of cement, diagram after Houseknecht [68] and Ehrenberg [69]. (**b**) Core porosity versus core permeability of the Lulehe sandstones with different depth and cements content.

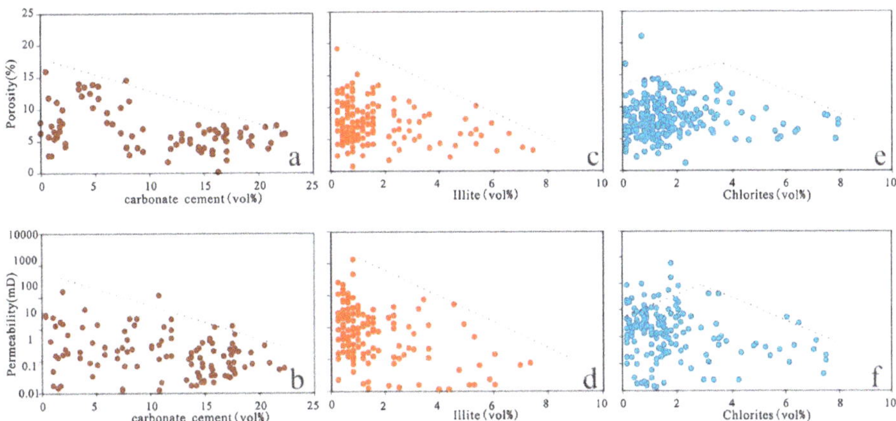

Figure 16. (**a**) Porosity versus carbonate cements. (**b**) Permeability versus carbonate cements. (**c**) Porosity versus illite cements. (**d**) Permeability versus illite cements. (**e**) Porosity versus chlorite cements. (**f**) Permeability versus chlorite cements.

Authigenic clays present in the Lulehe Formation consist mostly of chlorite, kaolinite and illite. As a byproduct of feldspar dissolution, kaolinite usually develops intercrystalline microporosity, which does not possess a significant influence on reservoir properties. Under high temperatures and deep burial depths, illitization of kaolinite and precipitation of quartz cement occur after feldspar dissolution. The presence of quartz cements mostly occurs as syntaxial overgrowths around detrital quartz grains and illite typically appear as fibrous aggregates that bridge and fill in primary pores and secondary pores. Thus, illite and quartz cement reduce reservoir quality by occluding pore spaces and decreasing pore radii (Figure 16c,d). However, grain-coating chlorites have been widely accepted as important mechanisms for reservoir quality preservation because chlorite rims delay or inhibit the precipitation of quartz overgrowth during burial [70–72]. Within the study area, porosity increases with chlorite content up to approximately 3% volume and then decreases slightly (Figure 16e,f); thus, chlorite coatings are beneficial at an optimum volume of 3%. In other words, if chlorite coatings are too thick, they will occur as pore-filling cements and become important causes of porosity loss.

Through dissolution, ductile fragments and calcite cements often formed irregular intergranular and intragranular pores (Figure 6d). In addition, feldspar grains commonly dissolved in Lulehe sandstone reservoir, and as a result, embayment-shaped intergranular pores were observed (Figure 9c). Furthermore, a small number of honeycomb intergranular pores resulted from severe feldspar dissolution, which was always accompanied by silica and kaolinite precipitation (Figure 9f, Figure 11h). A benefit of dissolution of unstable minerals (e.g., ductile fragments and carbonate cements) and feldspar grains is the occurrence of secondary porosity, which may enhance porosity to some extent. However, the deep burial environment seems to be a closed system, and the solutes cannot be transported over a large scale [6,58]. As a result, simultaneous precipitation of byproducts of feldspar dissolution such as authigenic kaolinite and quartz cement will occur in situ or in adjacent pores [58] Consequently, the net porosities created by dissolution of unstable minerals and feldspar grains may be insignificant. Moreover, reprecipitated cement material occupies the pore space and diminishes pore throat connectivity, resulting in reduced permeability and increasing reservoir heterogeneity.

In addition, the porosity and permeability of the reservoir are directly related to the burial depth and cement content as show in Figure 15b. The shallowest buried depth corresponds to the lowest compaction effective stress, so the corresponding samples have the highest porosity and permeability. Also, the sandstones with the higher cement contents tend to have lower permeability for a given porosity than those with lower cement contents. The difference of cement content and buried depth plays an important role in sandstone reservoirs in the study area.

6. Conclusions

1. The Lulehe Formation sandstones are very fine-grained, moderate to well sorted, feldspathic litharenites and lithic arkose mainly deposited in alluvial fan and fluvial deltaic environments. The sandstones have considerable heterogeneity with a wide range of porosity from 1.69% to 31.33%, with an average of 9.65%, and permeability from 0.03 to 366.52 mD, with an average of 13.75 mD.

2. The diagenetic regimes range from eodiagenesis to mesodiagenesis. Eodiagenesis includes mechanical compaction, precipitation of calcite and dolomite, alteration of volcanic rock fragments and smectite development, development of grain-coating chlorite, and a minor amount of smectite transformed into illite. Mesodiagenesis includes chemical compaction, development of pore-filling chlorite, formation of late carbonate cement and illite, quartz overgrowths, and dissolution of metastable cements and feldspar grains.

3. Carbonate cements might originate from mixed sources of bioclast- and organic-derived CO_2 during burial. Quartz cement is probably the result of feldspar dissolution, alteration of volcanic rock fragments, transformation of smectite into illite, and pressure solution of quartz grains. Smectite commonly derived from alteration of volcanic rock fragments may be the primary clay mineral precursor of chlorite. In addition, the dissolution of volcanic rock fragments and feldspar grains also provides abundant Fe^{2+} and Mg^{2+} as a potential source for chlorite development. Authigenic kaolinite is closely associated with feldspar dissolution, suggesting that alteration of detrital feldspar grains is the most probable source for authigenic kaolinite. With the increase in temperature and consumption of organic acids, the ratio of K^+/H^+ increases and the stability field of kaolinite is greatly reduced, thereby transforming kaolinite into mixed layer illite/smectite and illite.

4. Compaction is the principal factor for the deterioration of reservoir quality. In addition, carbonate cement, quartz overgrowth and the formation of mixed illite-smectite and illite are the subsequent main factors for porosity reduction, whereas grain-coating chlorites are mainly mechanisms for preservation of the reservoir quality, but immoderately thick chlorite coatings will also obstruct the pore throats and result in reservoir quality reduction.

5. Dissolution of metastable minerals and grains may lead to the occurrence of secondary porosity and enhance porosity to some extent. However, the solute cannot be transported over a large scale in the deep burial environment; as a consequence, simultaneous precipitation of byproducts of feldspar

dissolution, such as authigenic kaolinite and quartz cement, will occur in situ or in adjacent pores, which results in further heterogeneity of the reservoirs.

Author Contributions: Conceptualization, B.C. and F.W.; Methodology, B.C.; Validation, B.C. and H.S; Investigation, J.S; Resources, F.C.; Project Administration, J.S. and H.S.

Funding: This research was funded by the High level innovation teams and excellent scholars program in Guangxi Universities (Guijiao Ren [2016]42), the Basic Ability Improvement Project for Young and Middle-aged Teachers in Guangxi (2018KY0610) and Scientific research project of Beibu Gulf University (17KYQD57).

Acknowledgments: We thank the Qinghai Oilfield Company of petroChina for providing all related research core samples and some of the geological data.

Conflicts of Interest: The authors declare no conflict of interest.

References

1. Ajdukiewicz, J.M.; Lander, R.H. Sandstone reservoir quality prediction: The state of the art. *Am. Assoc. Pet. Geol. Bull.* **2010**, *94*, 1083–1091. [CrossRef]
2. Taylor, T.R.; Kittridge, M.G.; Winefield, P.; Bryndzia, L.T.; Bonnell, L.M. Reservoir quality and rock properties modeling—Triassic and Jurassic sandstones, greater Shearwater area, UK Central North Sea. *Mar. Pet. Geol.* **2015**, *65*, 1–21. [CrossRef]
3. Zhang, W.; Jian, X.; Fu, L.; Feng, F.; Guan, P. Reservoir characterization and hydrocarbon accumulation in late Cenozoic lacustrine mixed carbonate-siliciclastic fine-grained deposits of the northwestern Qaidam basin, NW China. *Mar. Pet. Geol.* **2018**, *98*, 675–686. [CrossRef]
4. Busch, B.; Hilgers, C.; Lander, R.H.; Bonnell, L.M.; Adelmann, D. Reservoir quality and burial model evaluation by kinetic quartz and illite cementation modeling: Case study of Rotliegendes, north Germany. *Am. Assoc. Pet. Geol. Bull.* **2018**, *102*, 293–307. [CrossRef]
5. Morad, S.; Ketzer, J.M.; De Ros, L.F. Linking Diagenesis to Sequence Stratigraphy: An Integrated Tool for Understanding and Predicting Reservoir Quality Distribution. *Link Diagenes Seq. Stratigr.* **2013**, *45*, 1–36. [CrossRef]
6. Bjørlykke, K. Relationships between depositional environments, burial history and rock properties. Some principal aspects of diagenetic process in sedimentary basins. *Sediment. Geol.* **2014**, *301*, 1–14. [CrossRef]
7. Taylor, T.R.; Giles, M.R.; Hathon, L.A.; Diggs, T.N.; Braunsdorf, N.R.; Birbiglia, G.V.; Kittridge, M.G.; Macaulay, C.I.; Espejo, I.S. Sandstone diagenesis and reservoir quality prediction: Models, myths, and reality. *Am. Assoc. Pet. Geol. Bull.* **2010**, *94*, 1093–1132. [CrossRef]
8. Leila, M.; Moscariello, A.; Šegvić, B. Depositional facies controls on the diagenesis and reservoir quality of the MessinianQawasim and Abu Madi formations, onshore Nile Delta, Egypt. *Geol. J.* **2019**, *54*, 1797–1813. [CrossRef]
9. Amel, H.; Jafarian, A.; Husinec, A.; Koeshidayatullah, A.; Swennen, R. Microfacies, depositional environment and diagenetic evolution controls on the reservoir quality of the Permian Upper Dalan Formation, Kish Gas Field, Zagros Basin. *Mar. Pet. Geol.* **2015**, *67*, 57–71. [CrossRef]
10. Lima, R.D.; De Ros, L.F. The role of depositional setting and diagenesis on the reservoir quality of Devonian sandstones from the Solimões Basin, Brazilian Amazonia. *Mar. Pet. Geol.* **2002**, *19*, 1047–1071. [CrossRef]
11. Henares, S.; Caracciolo, L.; Viseras, C.; Fernández, J.; Yeste, L.M. Diagenetic constraints on heterogeneous reservoir quality assessment: A Triassic outcrop analog of meandering fluvial reservoirs. *Am. Assoc. Pet. Geol. Bull.* **2016**, *100*, 1377–1398. [CrossRef]
12. Morad, S.; Al-Aasm, I.S.; Nader, F.H.; Ceriani, A.; Gasparrini, M.; Mansurbeg, H. Impact of diagenesis on the spatial and temporal distribution of reservoir quality in the Jurassic Arab D and C members, offshore Abu Dhabi oilfield, United Arab Emirates. *GeoArabia* **2012**, *17*, 17–56.
13. Oluwadebi, A.G.; Taylor, K.G.; Dowey, P.J. Diagenetic controls on the reservoir quality of the tight gas Collyhurst Sandstone Formation, Lower Permian, East Irish Sea Basin, United Kingdom. *Sediment. Geol.* **2018**, *371*, 55–74. [CrossRef]
14. Gutiérrez Paredes, H.C.; Catuneanu, O.; Hernández Romano, U. Controls on the quality of Miocene reservoirs, southern Gulf of Mexico. *J. South Am. Earth Sci.* **2018**, *81*, 45–65. [CrossRef]

15. Mansurbeg, H.; Morad, S.; Salem, A.; Marfil, R.; El-Ghali, M.A.; Nystuen, J.P.; Caja, M.A.; Amorosi, A.; Garcia, D.; La Iglesia, Λ. Diagenesis and reservoir quality evolution of palaeocene deep-water, marine sandstones, the Shetland-Faroes Basin, British continental shelf. *Mar. Pet. Geol.* **2008**, *25*, 514–543. [CrossRef]
16. Karmakar, R.; Manna, S.S.; Dutta, T. A geometrical model of diagenesis using percolation theory. *Phys. A Stat. Mech. Appl.* **2003**, *318*, 113–120. [CrossRef]
17. Hakimi, M.H.; Shalaby, M.R.; Abdullah, W.H. Diagenetic characteristics and reservoir quality of the Lower Cretaceous Biyadh sandstones at Kharir oilfield in the western central Masila Basin, Yemen. *J. Asian Earth Sci.* **2012**, *51*, 109–120. [CrossRef]
18. Bjørlykke, K. Open-system chemical behaviour of Wilcox Group mudstones. How is large scale mass transfer at great burial depth in sedimentary basins possible? A discussion. *Mar. Pet. Geol.* **2011**, *28*, 1381–1382. [CrossRef]
19. Thyne, G. A model of diagenetic mass transfer between adjacent sandstone and shale. *Mar. Pet. Geol.* **2001**, *18*, 743–755. [CrossRef]
20. Yuan, G.; Gluyas, J.; Cao, Y.; Oxtoby, N.H.; Jia, Z.; Wang, Y.; Xi, K.; Li, X. Diagenesis and reservoir quality evolution of the Eocene sandstones in the northern Dongying Sag, Bohai Bay Basin, East China. *Mar. Pet. Geol.* **2015**, *62*, 77–89. [CrossRef]
21. Zhang, Y.L.; Bao, Z.D.; Zhao, Y.; Jiang, L.; Zhou, Y.Q.; Gong, F.H. Origins of authigenic minerals and their impacts on reservoir quality of tight sandstones: Upper Triassic Chang-7 Member, Yanchang Formation, Ordos Basin, China. *Aust. J. Earth Sci.* **2017**, *64*, 519–536. [CrossRef]
22. Furong, W.; Sheng, H.; Yuguang, H.; Tian, D.; Zhiliang, H. The distribution and origin of carbonate cements in deep-buried sandstones in the central Junggar Basin, northwest China. *Geofluids* **2017**, *2017*, 18–23. [CrossRef]
23. Liu, M.; Shabaninejad, M.; Mostaghimi, P. Predictions of permeability, surface area and average dissolution rate during reactive transport in multi-mineral rocks. *J. Pet. Sci. Eng.* **2018**, *170*, 130–138. [CrossRef]
24. Poursoltani, M.R.; Gibling, M.R.; Pe-Piper, G. Diagenesis, burial history, and hydrocarbon potential of Cambrian sandstone in the northern continental margin of Gondwana: A case study of the Lalun Formation of central Iran. *J. Asian Earth Sci.* **2019**, *172*, 143–169. [CrossRef]
25. Christopher, B.A.; Kuiwu, L.; Oswald, G.W. Diagenesis and Reservoir Properties of the Permian Ecca Group Sandstones and Mudrocks in the Eastern Cape Province, South Africa. *Minerals* **2017**, *7*, 88. [CrossRef]
26. Feng, J.; Cao, J.; Hu, K.; Peng, X.; Chen, Y.; Wang, Y.; Wang, M. Dissolution and its impacts on reservoir formation in moderately to deeply buried strata of mixed siliciclastic-carbonate sediments, northwestern Qaidam Basin, northwest China. *Mar. Pet. Geol.* **2013**, *39*, 124–137. [CrossRef]
27. Sun, G.; Yin, J.; Zhang, S.; Lu, X.; Zhang, S.; Shi, J. Diagenesis and sedimentary environment of Miocene series in Eboliang III area. *Environ. Earth Sci.* **2015**, *74*, 5169–5179. [CrossRef]
28. Junwu, L.; Chengjin, Y.; Fengjie, L.; Yongliang, W.; Tingyong, D. Provenance analysis of the Neogene in Eboliang area Qaidam Basin. *J. Palaeogeogr.* **2015**, *17*, 186–197. (In Chinese)
29. Hong, L.; Bing, Z. Neogene sedimentary facies in the Eboliang area of the northern Qaidam Basin. *J. Stratigrap.* **2016**, *40*, 57–62. (In Chinese)
30. Jian, X.; Guan, P.; Zhang, D.W.; Zhang, W.; Feng, F.; Liu, R.J.; Lin, S.D. Provenance of Tertiary sandstone in the northern Qaidam basin, northeastern Tibetan Plateau: Integration of framework petrography, heavy mineral analysis and mineral chemistry. *Sediment. Geol.* **2013**, *290*, 109–125. [CrossRef]
31. Wang, T.; Yang, S.; Duan, S.; Chen, H.; Liu, H.; Cao, J. Multi-stage primary and secondary hydrocarbon migration and accumulation in lacustrine Jurassic petroleum systems in the northern Qaidam Basin, NW China. *Mar. Pet. Geol.* **2015**, *62*, 90–101. [CrossRef]
32. Miao, W.L.; Fan, Q.S.; Wei, H.C.; Zhang, X.Y.; Ma, H.Z. Clay mineralogical and geochemical constraints on late Pleistocene weathering processes of the Qaidam Basin, northern Tibetan Plateau. *J. Asian Earth Sci.* **2016**, *127*, 267–280. [CrossRef]
33. Wang, Y.; Zheng, J.; Zhang, W.; Li, S.; Liu, X.; Yang, X.; Liu, Y. Cenozoic uplift of the Tibetan Plateau: Evidence from the tectonic-sedimentary evolution of the western Qaidam Basin. *Geosci. Front.* **2012**, *3*, 175–187. [CrossRef]
34. Zongxing, L.; Chenglin, L.; Yinsheng, M. Geothermal Field and Tectono-Thermal Evolution since the Late Paleozoic of the Qaidam Basin, Western China. *Acta Geol. Sin.* **2015**, *89*, 678. (In Chinese) [CrossRef]

35. Zeqing, G.; Yinsheng, M.; Weihong, L.; Liqun, W.; Jixian, T.; Xu, Z.; Feng, M. Main factors controlling the formation of basement hydrocarbon reservoirs in the Qaidam Basin, western China. *J. Pet. Sci. Eng.* **2017**, *149*, 244–255. [CrossRef]

36. Folk, R.L. *Petrology of Sedimentary Rocks*; Hemphill Publishing Company: Austin, TX, USA, 1980; p. 182.

37. Rahman, M.J.J.; Worden, R.H. Diagenesis and its impact on the reservoir quality of Miocene sandstones (Surma Group) from the Bengal Basin, Bangladesh. *Mar. Pet. Geol.* **2016**, *77*, 898–915. [CrossRef]

38. Robinson, A.; Gluyas, J. Model calculations of loss of porosity in sandstones as a result of compaction and quartz cementation. *Mar. Pet. Geol.* **1992**, *9*, 319–323. [CrossRef]

39. Dutton, S.P. Calcite cement in Permian deep-water sandstones, Delaware Basin, west Texas: Origin, distribution, and effect on reservoir properties. *Am. Assoc. Pet. Geol. Bull.* **2008**, *92*, 765–787. [CrossRef]

40. Morad, S.; Al-Aasm, I.S.; Ramseyer, K.; Marfil, R.; Aldahan, A.A. Diagenesis of carbonate cements in Permo-Triassic sandstones from the Iberian Range, Spain: Evidence from chemical composition and stable isotopes. *Sediment. Geol.* **1990**, *67*, 281–295. [CrossRef]

41. Pszonka, J.; Wendorff, M. Carbonate cements and grains in submarine fan sandstones—the Cergowa Beds (Oligocene, Carpathians of Poland) recorded by cathodoluminescence. *Int. J. Earth Sci.* **2017**, *106*, 269–282. [CrossRef]

42. Guerra, N.C.; Kiang, C.H.; Sial, A.N. Carbonate cements in contemporaneous beachrocks, Jaguaribe beach, Itamaracáisland, northeastern Brazil: Petrographic, geochemical and isotopic aspects. *An. Acad. Bras. Cienc.* **2005**, *77*, 343–352. [CrossRef]

43. Sanyal, P.; Bhattacharya, S.K.; Prasad, M. Chemical diagenesis of Siwalik sandstone: Isotopic and mineralogical proxies from SuraiKhola section, Nepal. *Sediment. Geol.* **2005**, *180*, 57–74. [CrossRef]

44. Xiong, D.; Azmy, K.; Blamey, N.J.F. Diagenesis and origin of calcite cement in the Flemish Pass Basin sandstone reservoir (Upper Jurassic): Implications for porosity development. *Mar. Pet. Geol.* **2016**, *70*, 93–118. [CrossRef]

45. Yang, T.; Cao, Y.; Friis, H.; Liu, K.; Wang, Y.; Zhou, L.; Zhang, S.; Zhang, H. Genesis and distribution pattern of carbonate cements in lacustrine deep-water gravity-flow sandstone reservoirs in the third member of the Shahejie Formation in the Dongying Sag, Jiyang Depression, Eastern China. *Mar. Pet. Geol.* **2018**, *92*, 547–564. [CrossRef]

46. Odigi, M.I.; Amajor, L.C. Geochemistry of carbonate cements in Cretaceous sandstones, southeast Benue Trough, Nigeria: Implications for geochemical evolution of formation waters. *J. African Earth Sci.* **2010**, *57*, 213–226. [CrossRef]

47. Spotl, C.; Wrightt, V.P.; Institut, G. Groundwater dolocretes from the Upper Triassic of the Paris Basin, France: A case study of an arid, continental diageneticfacies. *Sedimentology* **1992**, *39*, 1119–1136. [CrossRef]

48. Drummond, C.N.; Wilkinson, B.H.; Lohmann, K.C.; Smith, G.R. Effect of regional topography and hydrology on the lacustrine isotopic record of Miocene paleoclimate in the Rocky Mountains. *Palaeogeogr. Palaeoclimatol. Palaeoecol.* **1993**, *101*, 67–79. [CrossRef]

49. Irwin, H.; Curtis, C.; Coleman, M. Isotopic evidence for source of diagenetic carbonates formed during burial of organic-rich sediments. *Nature* **1977**, *269*, 209–213. [CrossRef]

50. Travé, A.; Roca, E.; Playà, E.; Parcerisa, D.; Gómez-Gras, D.; Martín-Martín, J.D. Migration of Mn-rich fluids through normal faults and fine-grained terrigenous sediments during early development of the NeogeneVallès-Penedès half-graben (NE Spain). *Geofluids* **2009**, *9*, 303–320. [CrossRef]

51. Jin, Z.; Cao, J.; Hu, W.; Zhang, Y.; Yao, S.; Wang, X.; Zhang, Y.; Tang, Y.; Shi, X. Episodic petroleum fluid migration in fault zones of the northwestern Junggar Basin (northwest China): Evidence from hydrocarbon-bearing zoned calcite cement. *Am. Assoc. Pet. Geol. Bull.* **2008**, *92*, 1225–1243. [CrossRef]

52. Longstaffe, F.J. Stable isotope studies of diagnetic processes. *Mineral. Assoc. Can. Saskat.* **1987**, *13*, 187–257.

53. Craig, H. Isotopic variations in meteoric waters. *Science* **1961**, *133*, 1702–1703. [CrossRef] [PubMed]

54. Chen, Z.X.; Cheng, J.; Guo, P.W.; Lin, Z.Y.; Zhang, F.Y. Distribution characters and its control factors of stable isotope in precipitation over China. *Trans. Atmos. Sci.* **2010**, *33*, 667–679. (In Chinese)

55. Zhu, J.-J.; Chen, H.; Gong, G.-L. Hydrogen and oxygen isotopic compositions of precipitation and its water vapor sources in eastern Qaidam basin. *HuanjingKexue/Environ. Sci.* **2015**, *36*, 2748–2790. [CrossRef]

56. Worden, R.H.; Morad, S. Quartz Cementation in Oil Field Sandstones: A Review of the Key Controversies. *Quartz Cem. Sandstones* **2009**, 1–20. [CrossRef]

57. McBride, E.F. Quartz cement in sandstones: A review. *Earth Sci. Rev.* **1989**, *26*, 69–112. [CrossRef]

58. Bjørlykke, K.; Jahren, J. Open or closed geochemical systems during diagenesis in sedimentary basins: Constraints on mass transfer during diagenesis and the prediction of porosity in sandstone and carbonate reservoirs. *Am. Assoc. Pet. Geol. Bull.* **2012**, *96*, 2193–2214. [CrossRef]

59. Surdam, R.C.; Crossey, L.J.; Hagen, E.S.; Heasler, H.P. Organic-inorganic and sandstone diagenesis. *Am. Assoc. Pet. Geol. Bull.* **1989**, *73*, 1–23.

60. Liu, Y.; Hu, W.; Cao, J.; Wang, X.; Tang, Q.; Wu, H.; Kang, X. Diagenetic constraints on the heterogeneity of tight sandstone reservoirs: A case study on the Upper Triassic Xujiahe Formation in the Sichuan Basin, southwest China. *Mar. Pet. Geol.* **2018**, *92*, 650–669. [CrossRef]

61. Lynch, F.L.; Mack, L.E.; Land, L.S. Burial diagenesis of illite/smectite in shales and the origins of authigenic quartz and secondary porosity in sandstones. *Geochim. Cosmochim. Acta* **1997**, *61*, 1995–2006. [CrossRef]

62. Tang, L.; Gluyas, J.; Jones, S. Porosity preservation due to grain coating illite/smectite: Evidence from Buchan Formation (Upper Devonian) of the Ardmore Field, UK North Sea. *Proc. Geol. Assoc.* **2018**, *129*, 202–214. [CrossRef]

63. Friis, H.; Molenaar, N.; Varming, T. Chlorite meniscus cement—implications for diagenetic mineral growth after oil emplacement. *Terra Nov.* **2014**, *26*, 14–21. [CrossRef]

64. Thyberg, B.; Jahren, J.; Winje, T.; Bjørlykke, K.; Faleide, J.I.; Marcussen, Ø. Quartz cementation in Late Cretaceous mudstones, northern North Sea: Changes in rock properties due to dissolution of smectite and precipitation of micro-quartz crystals. *Mar. Pet. Geol.* **2010**, *27*, 1752–1764. [CrossRef]

65. Virolle, M.; Brigaud, B.; Bourillot, R.; Féniès, H.; Portier, E.; Duteil, T.; Nouet, J.; Patrier, P.; Beaufort, D. Detrital clay grain coats in estuarine clastic deposits: Origin and spatial distribution within a modern sedimentary system, the Gironde Estuary (south-west France). *Sedimentology* **2018**, *66*, 859–894. [CrossRef]

66. Morad, S.; Ketzer, J.M.; DeRos, F. Spatial and temporal distribution of diagenetic alterations in siliciclastic rocks: Implication for mass transfer in sedimentary basins. *Sedimentology* **2000**, *47*, 95–120. [CrossRef]

67. Grundtner, M.L.; Gross, D.; Linzer, H.G.; Neuhuber, S.; Sachsenhofer, R.F.; Scheucher, L. The diagenetic history of Oligocene-Miocene sandstones of the Austrian north Alpine foreland basin. *Mar. Pet. Geol.* **2016**, *77*, 418–434. [CrossRef]

68. Houseknecht, D.W. Influence of grain size and temperature on intergranular pressure solution, quartz cementation, and porosity in quartzose sandstone. *J. Sediment. Petrol.* **1984**, *54*, 348–361. [CrossRef]

69. Ehrenberg, S.N. Assessing the relative importance of compaction processes and cementation to reduction of porosity in sandstones: Discussion; compac-tion and porosity evolution of Pliocene sandstones, Ventura Basin, California: Discussion. *Am. Assoc. Pet. Geol. Bull.* **1989**, *73*, 1274–1276.

70. Cao, Z.; Liu, G.; Meng, W.; Wang, P.; Yang, C. Origin of different chlorite occurrences and their effects on tight clastic reservoir porosity. *J. Pet. Sci. Eng.* **2018**, *160*, 384–392. [CrossRef]

71. Haile, B.G.; Hellevang, H.; Aagaard, P.; Jahren, J. Experimental nucleation and growth of smectite and chlorite coatings on clean feldspar and quartz grain surfaces. *Mar. Pet. Geol.* **2015**, *68*, 664–674. [CrossRef]

72. Line, L.H.; Jahren, J.; Hellevang, H. Mechanical compaction in chlorite-coated sandstone reservoirs—Examples from Middle—Late Triassic channels in the southwestern Barents Sea. *Mar. Pet. Geol.* **2018**, *96*, 348–370. [CrossRef]

Article

Diagenetic Pore Fluid Evolution and Dolomitization of the Silurian and Devonian Carbonates, Huron Domain of Southwestern Ontario: Petrographic, Geochemical and Fluid Inclusion Evidence

Marco Tortola [1,*], Ihsan S. Al-Aasm [1] and Richard Crowe [2]

[1] School of the Environment, University of Windsor, Windsor, ON N9B 3P4, Canada; alaasm@uwindsor.ca
[2] Nuclear Waste Management Organization, Toronto, ON M4T 2S3, Canada; rcrowe@nwmo.ca
*   Correspondence: tortolam@uwindsor.ca

Received: 13 November 2019; Accepted: 5 February 2020; Published: 7 February 2020

Abstract: Core samples from two deep boreholes were analyzed for petrographic, stable and Sr isotopes, fluid inclusion microthermometry and major, minor, trace and rare-earth elements (REE) of different types of dolomite in the Silurian and Devonian carbonates of the eastern side of the Michigan Basin provided useful insights into the nature of dolomitization, and the evolution of diagenetic pore fluids in this part of the basin. Petrographic features show that both age groups are characterized by the presence of a pervasive replacive fine-crystalline (<50 μm) dolomite matrix (RD1) and pervasive and selective replacive medium crystalline (>50–100 μm) dolomite matrix (RD2 and RD3, respectively). In addition to these types, a coarse crystalline (>500 μm) saddle dolomite cement (SD) filling fractures and vugs is observed only in the Silurian rocks. Results from geochemical and fluid inclusion analyses indicate that the diagenesis of Silurian and Devonian formations show variations in terms of the evolution of the diagenetic fluid composition. These fluid systems are: (1) a diagenetic fluid system that affected Silurian carbonates and was altered by salt dissolution post-Silurian time. These carbonates show a negative shift in $\delta^{18}O$ values (dolomite $\delta^{18}O$ average: $-6.72‰$ VPDB), Sr isotopic composition slightly more radiogenic than coeval seawater (0.7078–0.7087), high temperatures (RD2 and SD dolomite T_h average: 110 °C) and hypersaline signature (RD2 and SD dolomite average salinity: 26.8 wt.% NaCl eq.); and (2) a diagenetic fluid system that affected Devonian carbonates, possibly occurred during the Alleghenian orogeny in Carboniferous time and characterized by a less pronounced negative shift in $\delta^{18}O$ values (dolomite $\delta^{18}O$ average: $-5.74‰$ VPDB), Sr isotopic composition in range with the postulated values for coeval seawater (0.7078–0.7080), lower temperatures (RD2 dolomite T_h average: 83 °C) and less saline signature (RD2 dolomite average salinity: 20.8 wt.% NaCl eq.).

Keywords: dolomitization; Huron Domain; Silurian; Devonian; fluid composition; Michigan Basin

---

1. Introduction

The Michigan Basin has been the subject of several diagenetic studies in the last three decades (e.g., [1–8]). However, very few comprehensive investigations regarding the nature and composition of diagenetic fluids, lateral extent of the diagenetic processes and paragenetic sequence of different types of dolomite within Silurian and Devonian successions in the Huron Domain have been carried out. Most previous studies focused on the Paleozoic carbonate succession in the area were limited in stratigraphic and/or spatial resolution (e.g., [1–4]). In addition, contrasting models have been proposed to explain the fluid flow evolution and migration in the Michigan Basin, such as migration of dolomitizing fluids from the center of the Michigan Basin toward the margins [5] or from the back reef downward through

Guelph limestones to the center of the Michigan basin (e.g., [6–8]). Numerous diagenetic processes such as dolomitization and Mississippi Valley Type (MVT) mineralization, salt dissolution, precipitation of late-stage cement and oil, and gas generation and migration have influenced the Paleozoic succession in the Michigan Basin [9]. Fractures and faults in carbonate bodies represent preferential pathways for fluids which in turn play a key role in the diagenesis and evolution of host rocks and hydrocarbon reservoirs [5]. The presence of fractures and faults in the Michigan intracratonic Basin represented an important control on the modification of host limestones and mobilization of dolomitizing fluids within the basin [4,10–12].

Coniglio et al. [8] performed a regional study of the Silurian Guelph Formation in southwestern Ontario. In this work, the authors underline how burial diagenesis produced the recrystallization of massive dolomites with homogeneous petrographic characteristics but with different geochemistry. According to their study, initial shallow burial dolomitization occurred due to the isolation of the Michigan Basin during the late Silurian. Isotopic values of $\delta^{13}C$ and $^{87}Sr/^{86}Sr$ suggest a Silurian seawater source. They also recorded a decrease of $\delta^{18}O$ values in dolomite from the platform to the lower slope which occurred with increasing burial depth. These authors suggested that seawater-derived dolomitizing fluids migrated from the margin toward the basin center with a progressive decrease in terms of the capability of the fluids to dolomitize the down-slope.

Luczaj et al. [13] conducted a study of fluid inclusion analysis in dolomite and calcite of the Middle Devonian Dundee Formation. Their work underlines that the sedimentary fabrics are modified by medium to coarse crystalline dolomite. They suggested that fluids characterized as dense Na–Ca–Mg–Cl brines interacted with host rock at high temperatures (around 120–150 °C). They invoked a model involving the transport of fluids and heat from deeper parts of the basin along with major fault and fracture zones systems.

Haeri-Ardakani et al. [5] evaluated diagenesis and fluid flow evolution within Ordovician, Silurian and Devonian carbonate successions in southwestern Ontario. Data from fluid inclusions analysis showed homogenization temperatures of replacive and saddle dolomite and late-stage calcite ranging between 75–145 °C. They suggested the involvement of hydrothermal fluids in replacement and recrystallization of dolomite and calcite cementation to explain homogenization temperatures higher than the peak expected during maximum burial for each interval (60–90 °C). Evidence of decreasing values in fluid inclusion homogenization temperatures from the center to the margin of the basin suggests that hydrothermal fluids originated from the center of the Michigan Basin. The authors suggest that the potential source of heat could be related to the buried mid-continental rift (MCR), while the preferential pathways for the migration of diagenetic fluids from the center of the Michigan Basin can be related to the Cambrian sandstones that unconformably overlay the Precambrian basement which could have behaved as regional aquifers [5].

The main purpose of this study was to conduct a thorough investigation of the characteristics and the occurrence of dolomitization within the Devonian and Silurian carbonate formations on the eastern side of the Michigan Basin (Figure 1). Dolomitization processes are examined in terms of dolomite types, distribution and geochemistry using an integrated approach utilizing petrography, multi-elements analysis, stable isotopes, Sr-isotopes and fluid inclusion analyses of rock core samples. Based on geologic, geochemical and petrographic investigations, a relative timing in terms of paragenesis of dolomitization is suggested, as well as relationships between existing regional structures and Devonian and Silurian strata-bound dolomite, nature and composition of mineralizing fluid during diagenesis, and extension of diagenetic events. This study represents an important contribution to the more extensive studies carried out by the Nuclear Waste Management Organization (NWMO) for the characterization of a suitable site to develop a Deep Geologic Repository (DGR) for low- and intermediate-level nuclear waste at the Bruce power station in Kincardine, Ontario.

2. Geologic and Tectonic Setting

The study area is located on the northeastern margin of the Michigan Basin (Figure 1a,b). It represents part of the northwestern flank of the Algonquin Arch that consists of Paleozoic sedimentary rocks that cover the subsurface basement [14]. The thickness of the Paleozoic succession ranges from a maximum of 4800 m at the center of the Michigan Basin to 850 m on the flank of the Algonquin Arch where the study area is located [15].

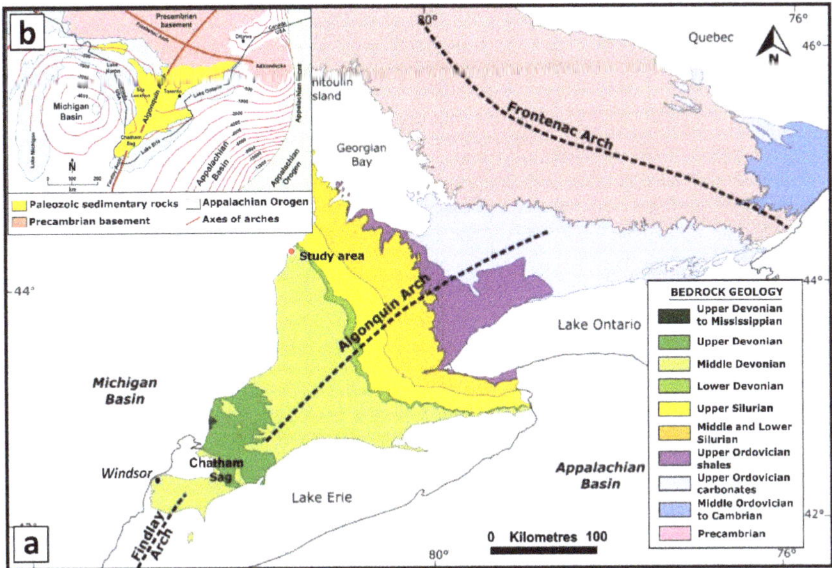

Figure 1. (**a**) Generalized Paleozoic Bedrock Geology Map of Southern Ontario (modified from OGS [16] and OPG [17]); (**b**) generalized basement structural contours (meters above the sea level datum) and location of structural arches and basins (modified from OPG [17] and Johnson et al. [18]).

On a regional scale, southern Ontario is located between two major Paleozoic sedimentary basins, the Appalachian Basin to the southeast and Michigan Basin to the west (Figure 1). The Algonquin Arch, which separates the two basins, extends from northeast of the Canadian shield to the southwest near the city of Chatham. Near the Windsor area, another structural high named the Findlay Arch starts extending to the southwest into Michigan and Ohio states [3].

The rifting that separated North America and Africa occurred during Precambrian to Cambrian represents the initial episode of subsidence (approximately 1100 Ma) and deposition within the Michigan Basin [10]. Thermal subsidence of the Precambrian basement followed this event approximately 580 Ma to 500 Ma because the cooled lithosphere becomes denser [19]. The subsequent and continuous deposition of sediments into the Michigan Basin caused the bending of the basement [20–22]. However, the development of the basin is not only the result of continuous subsidence but represents the result of a series of tectonic events that characterized the Paleozoic [23,24].

The Taconic and Acadian orogenies were a dominant control on the Paleozoic strata of the eastern flank of the Michigan Basin [25]. In fact, during the Taconic Orogeny (Ordovician) a large-scale eastward-tilting of the Laurentian margin governed the disappearance of the concentric depositional geometry of the Michigan Basin with the consequent deposition of Ordovician limestone and overlying shale successions [24,26,27]. The regional maximum principal stress at that time was northwesterly oriented the same as the main direction of the thrust motion along the Appalachian tectonic front [25,28].

Restricted marine conditions occurred during the late Silurian with the consequent deposition of evaporites of the Salina Group [15,18]. In addition, the presence of an unconformity reveals emergent conditions at the end of the Silurian [25]. The Middle Devonian Acadian Orogeny reactivated the Algonquin-Findlay Arch system with the return of the concentric geometry of the Michigan Basin [23]. The Lower and Middle Devonian units are dominated by limestones and dolostones, whereas, the Upper Devonian strata are mostly represented by shales [15].

The Caledonian (Devonian) and Alleghanian (Carboniferous) orogenies played an important role in terms of diagenetic fluid migration [29]. Late Paleozoic sediments have been removed from the entire study area. According to Wang et al. [30], the regional analysis of apatite fission track dating in the center of the Michigan Basin suggests that late to post Alleghenian erosion removed approximately 1000 m of carbonate successions from the margins and approximately 1500 m from the center of the basin.

The main characteristics of the Silurian and Devonian stratigraphy in the study area are shown in Figure 2. The Michigan Basin has been characterized by a typical tropical climate during mid to late Silurian [31] whereby shallow water reef complexes started to develop (e.g., Gasport, Lions Head, Fossil Hill and Guelph formations). Moreover, the Michigan Basin has been isolated and becoming an evaporative basin in which alternation of evaporites and carbonates (Salina Group) deposited during the rest of the Silurian [32]. Conversely, during the Middle Devonian, Michigan Basin has been characterized by an extremely arid climate with consequent deposition of shallow intertidal and supratidal lithofacies such as Amherstburg and Lucas formations [33].

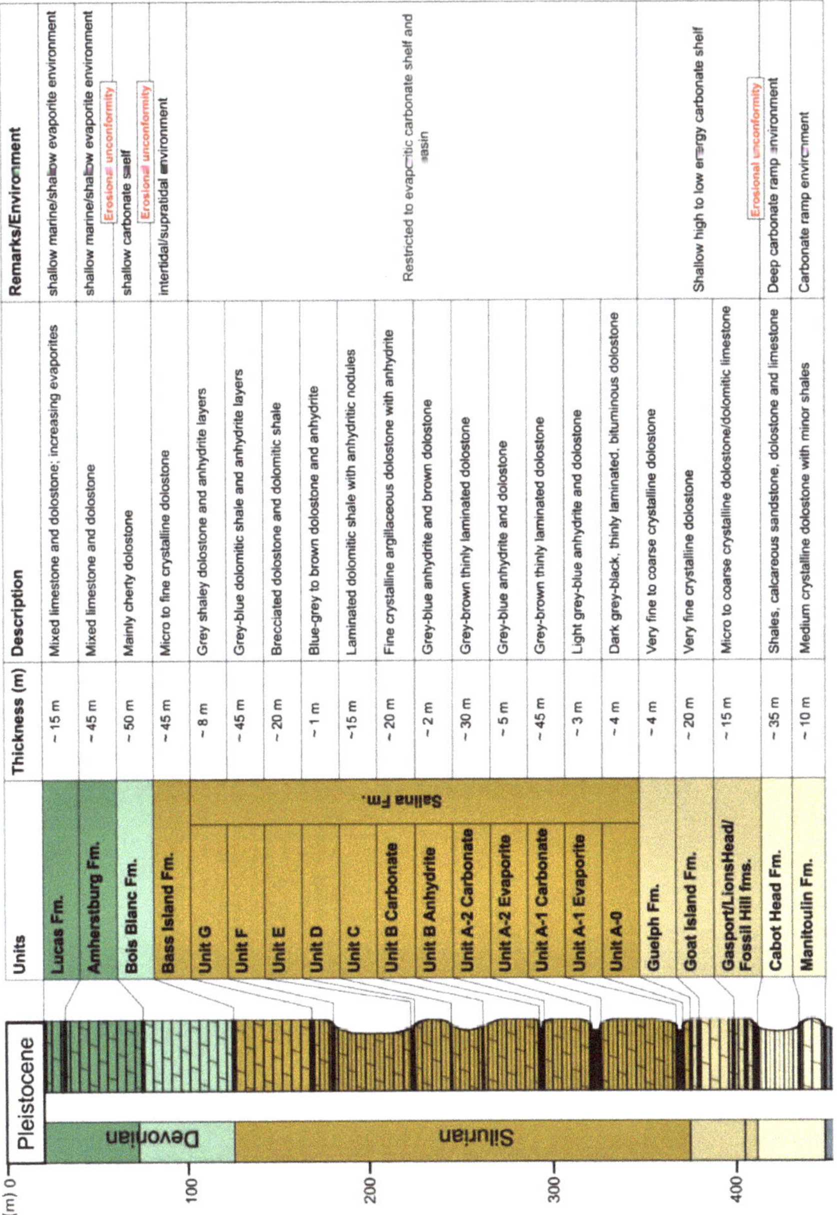

Figure 2. Summary of the Silurian and Devonian stratigraphy in the study area (modified from Johnson et al. [18]; AECOM [25]; Armstrong and Goodman [34]).

3. Materials and Methods

Core samples were collected from the Bruce Nuclear site, the proposed site for a deep geological repository for low- and intermediate-level nuclear waste, from two boreholes DGR1 and DGR8 (Figure 2) [17]. The samples were chosen based on dolomitized and fractured rocks in hand specimens. A summary of sample identification with relative age and depth is reported in Figure 3 and Table 1.

The selected core samples were photographed before making thin sections and wafers for fluid inclusions analysis. Petrographic examination of thin sections (n = 46) was performed using a standard petrographic microscope to provide lithofacies types, textural relationships and diagenetic fabric of the studied samples, to establish the paragenetic sequences. Subsequently, all thin sections were examined using cathodoluminescence microscopy (CL) to obtain data regarding the chemistry of the fluids, mineral zonation and diagenetic textures of host rock and fracture filling cement. The equipment used to carry out the cathodoluminescence analysis was a Technosyn Model 8200 MKII with a 12–15 Kv beam and a current intensity of 0.42–0.43 mA (Cambridge Image Technology Ltd., Hatfield, Hertfordshire, UK). The thin sections were also stained using a mixture of Alizarin Red S and potassium ferricyanide dissolved in a dilute solution of hydrochloric acid [35,36] to distinguish between ferroan or non-ferroan phases of calcite and dolomite.

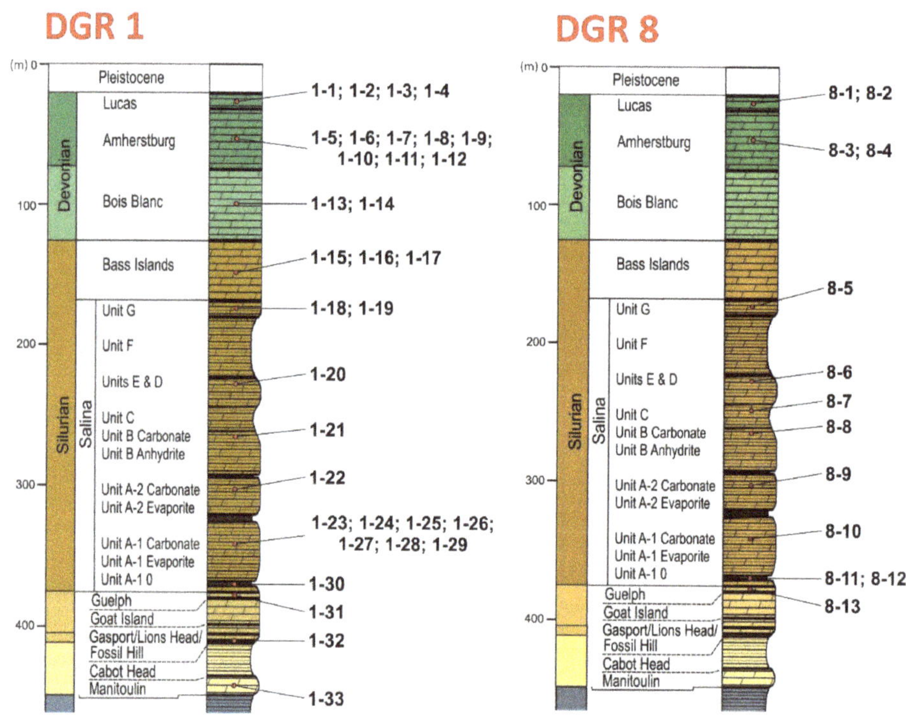

Figure 3. Stratigraphy at the Bruce Nuclear Site and sample locations (modified from AECOM [25]).

A total of forty-three (Silurian) and thirty-three (Devonian) samples were micro-sampled using a microscope-mounted drill assembly for oxygen and carbon isotopic composition from Silurian (n = 18 and n = 25) and Devonian (n = 12 and n = 21) specimens for calcitic and dolomitic components, respectively. Each sample was reacted in vacuo with 100% pure phosphoric acid four 4 h at 50 °C and 25 °C for dolomite and calcite respectively. The CO_2 produced during the reaction by the different diagenetic mineral phases was extracted using the method proposed by Al-Aasm et al. [37]

and measured employing a Delta-Plus mass spectrometer (Thermo Electron Corporation, Bremen, Germany). The delta values of oxygen and carbon isotopes have been reported in per mil (‰) relative to the PeeDee Belemnite (VPDB) standard and corrected for phosphoric acid fractionation. Precision was better than 0.05‰ for both ^{13}C and ^{18}O. Carbon dioxide extraction and isotopic composition measurements were performed at the Stable Isotopes Laboratory, University of Windsor (Windsor, ON, Canada).

Table 1. Lists of samples collected from the boreholes DGR1 and DGR8.

Cored Sample	Sample Code	Formation	Age	Depth (m)
DGR1-CR1	1-1	Lucas	Devonian	23.5
DGR1-CR2-(1)	1-2	Lucas	Devonian	26.25
DGR1-CR2-(2)	1-3	Lucas	Devonian	26.95
DGR1-CR2-(3)	1-4	Lucas	Devonian	27.75
DGR1-CR3-(1)	1-5	Amherstburg	Devonian	31.35
DGR1-CR3-(2)	1-6	Amherstburg	Devonian	31.85
DGR1-CR5	1-7	Amherstburg	Devonian	35.95
DGR1-CR8	1-8	Amherstburg	Devonian	42.58
DGR1-CR9	1-9	Amherstburg	Devonian	46.4
DGR1-CR10-(1)	1-10	Amherstburg	Devonian	48.65
DGR1-CR10-(2)	1-11	Amherstburg	Devonian	49.7
DGR1-CR11	1-12	Amherstburg	Devonian	51
DGR1-CR30-(1)	1-13	Bois Blanc	Devonian	104.05
DGR1-CR30-(2)	1-14	Bois Blanc	Devonian	105.08
DGR1-CR41	1-15	Bass Island	Silurian	128.93
DGR1-CR44	1-16	Bass Island	Silurian	133.03
DGR1-CR49	1-17	Bass Island	Silurian	142.28
DGR1-CR65	1-18	Salina—G-Unit	Silurian	176.13
DGR1-CR66	1-19	Salina—G-Unit	Silurian	178
DGR1-CR84	1-20	Salina—E-Unit	Silurian	229.32
DGR1-CR104	1-21	Salina—B-Unit (Carbonate)	Silurian	289.85
DGR1-CR105	1-22	Salina—A_2-Unit (Carbonate)	Silurian	294.14
DGR1-CR116-(1)	1-23	Salina—A_1-Unit (Carbonate)	Silurian	326.92
DGR1-CR116-(2)	1-24	Salina—A_1-Unit (Carbonate)	Silurian	328.17
DGR1-CR117	1-25	Salina—A_1-Unit (Carbonate)	Silurian	329.97
DGR1-CR118	1-26	Salina—A_1-Unit (Carbonate)	Silurian	333.48
DGR1-CR119	1-27	Salina—A_1-Unit (Carbonate)	Silurian	337.32
DGR1-CR121	1-28	Salina—A_1-Unit (Carbonate)	Silurian	342.27
DGR1-CR122	1-29	Salina—A_1-Unit (Carbonate)	Silurian	344.42
DGR1-CR130	1-30	Salina—A_0-Unit	Silurian	371.27
DGR1-CR133	1-31	Guelph	Silurian	378.1
DGR1-CR143	1-32	Fossil Hill	Silurian	410.1
DGR1-CR154	1-33	Manitoulin	Silurian	440.92
DGR8-CR10	8-1	Lucas	Devonian	22.71
DGR8-CR17	8-2	Lucas	Devonian	43.46
DGR8-CR18	8-3	Amherstburg	Devonian	48.61
DGR8-CR22	8-4	Amherstburg	Devonian	59.06
DGR8-CR64	8-5	Salina—G-Unit	Silurian	182.06
DGR8-CR90	8-6	Salina—E-Unit	Silurian	238.91
DGR8-CR100	8-7	Salina—C-Unit	Silurian	265.46
DGR8-CR110	8-8	Salina—B-Unit (Carbonate)	Silurian	295.91
DGR8-CR116	8-9	Salina—A_2-Unit (Carbonate)	Silurian	313.31
DGR8-CR122	8-10	Salina—A_1-Unit (Carbonate)	Silurian	332.26
DGR8-CR137	8-11	Salina—A_0-Unit	Silurian	377.21
DGR8-CR137	8-12	Salina—A_0-Unit	Silurian	378.75
DGR8-CR139	8-13	Guelph	Silurian	382.4

Around 10–50 mg of powdered samples (n = 22) were extracted to determine the Sr isotopic composition using a microscope-drill assembly, to be subsequently reacted with 2.5 N HCl for 24 h at room temperature. To avoid Strontium contamination from pore salts produced by drying the samples, they were washed twice using Milli-Q water. To measure the strontium isotope ratio in each sample, an automated Finnigan MAT 261TM mass spectrometer was employed, performing all the analysis in the static multicollector mode with Re filaments and normalizing $^{87}Sr/^{86}Sr$ ratios to $^{87}Sr/^{86}Sr$ = 8.375209. The mean standard error of the static multicollector was 0.00003 for NBS-987. These analyses were done at the University of Bochum, Bochum, Germany.

A thorough petrographic investigation of doubly polished thin sections was performed to identify and classify fluid inclusions trapped in diagenetic minerals. Measured ice-melting temperatures (Tmice) are reported as wt.% NaCl eq. using the relationship of Hall et al. [38] and Bodnar [39]. Detailed petrography was performed to determine the type of inclusions (primary or secondary/pseudosecondary) following the criteria proposed by Roedder [40]. For this study, attention was focused on primary liquid-vapor inclusion assemblages only. The microthermometric measurements were performed at the University of Windsor (Windsor, ON, Canada) employing a Linkam TH600 heating-freezing stage, coupled with the Olympus BX60. The device calibration was achieved using synthetic pure water and CO_2 inclusions as standard. Microthermometric analyses were completed on 126 primary fluid inclusions in selected carbonate samples from the Silurian (n = 85) and Devonian (n = 41) successions including pervasive replacive medium crystalline dolomite (RD2; n = 22 and n = 17 for Silurian and Devonian samples, respectively), saddle dolomite (SD; n = 31; observed in Silurian samples only) and late fracture-filling blocky calcite cement (BKC; n = 32 and n = 24 from Silurian and Devonian samples, respectively). Homogenization temperatures (T_h) and ice-melting temperatures (Tmice) were measured with a precision of ±1 °C and ±0.1 °C, respectively. Heating measurements were performed before freezing to avoid possible stretching of the inclusions during freezing [41].

Major, minor, trace and rare earth elements (REE) analyses were performed using inductively coupled plasma spectroscopy (ICP-MS) technique. Around 50–100 mg of powdered carbonate samples from different calcite and dolomite phases were digested utilizing 5% v/v acetic acid. Measurements were carried employing an Agilent 8800 inductively coupled mass spectrometer at the University of Waterloo (Waterloo, ON, Canada). All results are reported in part per million (ppm). The REEs' results obtained from ICP-MS analysis were normalized to PAAS (Post-Archean Australian Shales [42]). Rare-earth element anomalies such as Ce_{SN} [$(Ce/Ce^*)_{SN} = Ce_{SN}/(0.5La_{SN} + 0.5Pr_{SN})$], La_{SN} [$(Pr/Pr^*)_{SN} = Pr_{SN}/0.5Ce_{SN} + 0.5Nd_{SN})$], Eu_{SN} [$(Eu/Eu^*)_{SN} = Eu_{SN}/(0.5Sm_{SN} + 0.5Gd_{SN})$], and Gd_{SN} [$(Gd/Gd^*)_{SN} = Gd_{SN}/(0.33Sm_{SN} + 0.67Tb_{SN})$] were calculated employing the equations proposed by Bau and Dulski [43]. The proportions of LREE (La-Nd) over HREE (Ho-Lu) were calculated via $(La/Yb)_{SN}$ ratios as proposed by Kucera et al. [44]. The proportions of MREE (Sm-Dy) was calculated using $(Sm/La)_{SN}$ and $(Sm/Yb)_{SN}$ ratios [45] over LREE and HREE, respectively.

4. Results

4.1. Petrography of Silurian Formations

The diagenetic history of the Silurian formations includes processes such as dolomitization mechanical and chemical compaction, fracturing, dissolution, silicification, and calcite and evaporite cementation. The diagenetic processes above are described in Tortola [46], on which this paper is largely based.

4.1.1. Dolomitization

Four types of dolomite occur in the Silurian formations (Figure 4) and are classified following the criteria proposed by Sibley and Gregg [47]. These include including three types of replacive matrix dolomite and one of dolomite cement. These are a pervasive replacive micro to fine crystalline

(<50 μm) euhedral to subhedral dolomite matrix (RD1) (Figure 4A); a pervasive replacive medium (>50–100 μm) euhedral crystalline dolomite matrix (RD2) (Figure 4B), which represent the main constituents of the host rock; and a less abundant selective replacive medium (>50–100 μm) euhedral crystalline dolomite matrix (RD3) only observed along dissolution seams in partially dolomitized limestones, usually characterized by cloudy-dark cores and clear rims (Figure 4C). All dolomite types are non-ferroan and show dull red to non-luminescent (RD1 and RD2) and red to bright (RD3) under the CL (Figure 4D,E). In some cases, RD2 shows undulose extinction similar to saddle dolomite (e.g., [5,48]) and is characterized by cloudy non-ferroan cores followed by later clear and iron-rich rims (Figure 4B,F). In addition to these types, a coarse crystalline (>500 μm), ferroan saddle dolomite cement (SD; Figure 4G) filling fractures and vugs are observed, commonly predating late calcite cement (Figure 4E,H) and occasionally evaporite minerals (Figure 4I). Typically, SD is fluid inclusion-rich, zoned with cloudy cores and clear rims, and characterized by curved crystal boundaries.

Figure 4. Dolomitization in Silurian formations. (**A**) Photomicrograph (PPL) of planar-s, pervasive replacive micro to fine crystalline dolomite matrix (RD1). Sample 1-18; Well: DGR1-CR65; depth: 176.13 m. The blue arrow indicates the presence of framboidal pyrite in the matrix; (**B**) photomicrograph (XPL) of non-planar, pervasive replacive micro to fine crystalline dolomite matrix and planar-e pervasive replacive medium crystalline dolomite matrix (RD2). Sample 1-31; Well: DGR1-CR133; depth: 440.92 m; (**C**) photomicrograph (PPL—after staining) of planar-e, selective replacive medium crystalline dolomite matrix (RD3) associated with dissolution seams showing cloudy cores and clear rims (red arrow). Undolomitized limestone matrix (MC) shows a pink color after staining. Sample 8-12; well: DGR8-CR137 (2); depth: 378.75 m; (**D**) CL photomicrograph of isopachous calcite cement (ISC) and selective replacive medium crystalline dolomite matrix (RD3) both showing dully-red luminescence. Note the change from dull to red luminescent of ISC towards the center of the vug. Sample 8-10; well: DGR8-CR122; depth: 332.26 m; (**E**) CL photomicrograph of bright luminescent blocky calcite cement (BKC) postdating non-luminescent saddle dolomite cement (SD) surrounded by dull-red medium crystalline dolomite

matrix (RD2). Sample 8-13; Well: DGR8-CR139; Depth: 382.4 m; (**F**) photomicrograph (PPL—after staining) of planar-e to planar-s, pervasive replacive medium crystalline dolomite matrix (RD2) showing intercrystalline porosity (green arrows) and non-ferroan cores followed by later iron-rich rims (yellow arrow). Sample 1-24; well: DGR1-CR116 (2); depth: 328.17 m; (**G**) photomicrograph (XPL) of fracture filling saddle dolomite cement (SD) postdating pervasive replacive medium crystalline dolomite matrix (RD2) characterized by distinctive wavy extinction and curved crystal boundaries. Sample 1-31; Well: DGR1-CR133; Depth: 440.92 m. (**H**) CL photomicrograph of gypsum (GY) postdating non-luminescent with bright rims zoned saddle dolomite (SD). Sample 8-13; Well: DGR8-CR139; depth: 382.4 m; (**I**) photomicrograph (XPL) of gypsum cement (GY) postdating coarse saddle dolomite cement (SD). Sample 8-9; well: DGR8-CR116; depth: 313.31 m.

4.1.2. Fracturing

Three types of fractures are present in the studied formations. Thin randomly oriented hairline fractures (<0.5 cm) filled with drusy calcite cement (DC) represent the first generation of fractures (Figure 5A). Late wider subhorizontal and subvertical fractures (0.5–1 cm), both commonly occluded by saddle dolomite, late blocky calcite and evaporite cement representing the second and third generations, respectively (Figure 5B,C). Cross-cutting relationship of the last two generations of fractures shows that cementation took place after the formation of both fractures.

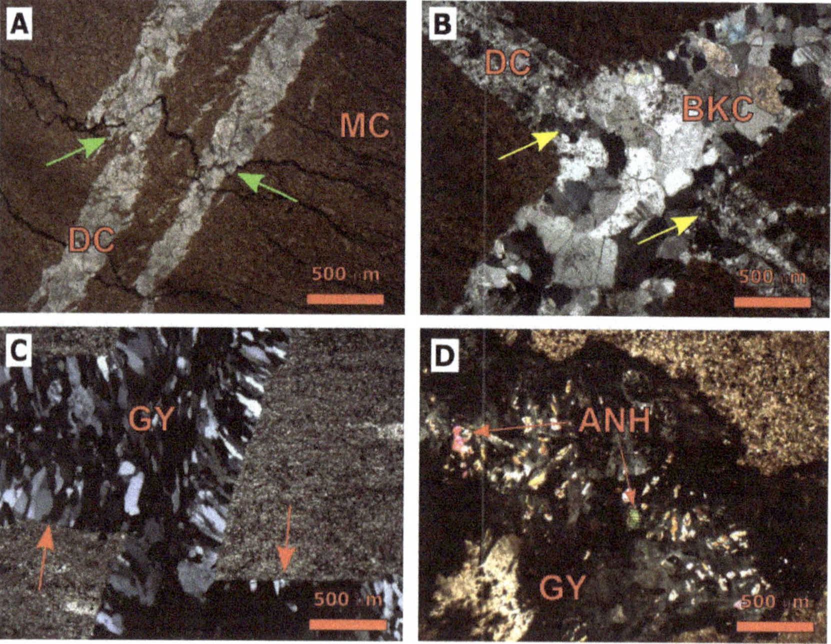

Figure 5. Fracturing in Silurian formations. (**A**) Photomicrograph (PPL) of multiple stylolites (green arrows) cross-cutting fracture-filling drusy calcite cement (DC). The surrounding matrix is micritic (MC). Sample 1-28; well: DGR1-CR121; depth: 342.27 m; (**B**) photomicrograph (XPL) of a sub-vertical fracture-filled by blocky calcite cement (BKC) cross-cutting and postdating (yellow arrows) a drusy mosaic calcite cement (DC) filling a randomly oriented fracture. Sample 1-28; well: DGR1-CR121; depth: 342.27 m; (**C**) photomicrograph (XPL) showing two generations of fractures; the first sub-horizontal (red arrows) displaced by the second sub-vertical, gypsum filled fracture. Sample 1-29; well: DGR1 - CR122; depth: 344.42 m; (**D**) photomicrograph (XPL) of a fracture filled by gypsum (GY) partially replaced by anhydrite (ANH) showing higher birefringence. Sample 8-7; well: DGR8-CR100; depth: 265.46 m.

4.1.3. Calcite Cementation

Early and late-stage calcite cement (Figure 6) have been distinguished including from early to late: isopachous, syntaxial overgrowth, drusy, and blocky calcite. Undolomitized calcite matrix (MC) from precursor limestone was also recognized (Figures 4C and 5A). Isopachous calcite cement (ISC) is characterized by cement rims growing around coated grains. It exhibits microcrystalline, bladed texture (Figure 6A) with non-ferroan, dully-red luminescence (Figure 4D) and crystal size ranging from 50 to 100 μm. Syntaxial calcite overgrowth cement (SXC) forms ferroan, bright-luminescent crystals ranging in size from 100 μm up to >500 μm, usually around echinoderm fragments (Figure 6B). Void-filling and pore-lining drusy calcite cement (DC) was mainly observed in intergranular and intraskeletal pores, molds and fractures. Common textural features show equant to elongate, anhedral to subhedral crystals ranging in size between 75–250 μm with dimension increasing towards the center of the pore space (Figure 6C). In most cases, DC is ferroan and exhibits dully-red luminescence under CL (Figure 6D).

A late fractures and void-filling blocky calcite cement (BKC) consisting of coarse non-ferroan to ferroan crystal (Figure 6E) ranging in size from 200 μm to >500 μm, shows zoned red to bright luminescence under CL (Figure 6F) and commonly postdate saddle dolomite cement.

In terms of paragenesis, early calcite cement (ISC, SXC and DC) predated RD1, RD2 and RD3 and predated fracture- and pore-filling saddle dolomite which is postdated by late blocky calcite (BKC) and evaporite (GY and ANH) cement.

4.1.4. Evaporite Cementation

Pore and fracture-filling fibrous gypsum (GY)/anhydrite (ANH) cements in the Silurian formations range in size from 50–150 μm (Figures 4I and 5C,D). They are the latest stage diagenetic mineral phases. Both gypsum and anhydrite are only present in samples from the Salina Group Formation, commonly postdating fracture-filling blocky calcite and saddle dolomite cement (Figure 4I).

4.2. Petrography of Devonian Formations

4.2.1. Dolomitization

Three types of dolomite occur in the Devonian formations (Figure 7), which are characterized by the presence of pervasive replacive micro to fine (<50 μm), euhedral to subhedral, crystalline dolomite matrix (RD1) (Figure 7A,B), a pervasive replacive medium (>50–100 μm) euhedral crystalline dolomite matrix (RD2) (Figure 7B–D), which represent the main constituents of the host dolostone, and a less abundant selective replacive medium (>50–100 μm) euhedral crystalline dolomite matrix (RD3 commonly associated with dissolution seams), and usually characterized by cloudy-dark core and clear rims (Figure 7E). All three types of dolomite are non-ferroan and show non-luminescent to dull red (RD1 and RD2) and red to bright red (RD3) under the CL (Figure 7B,F).

A schematic summary of the main features with relative nomenclatures of the main carbonate minerals observed and described for both age groups is presented in Table 2.

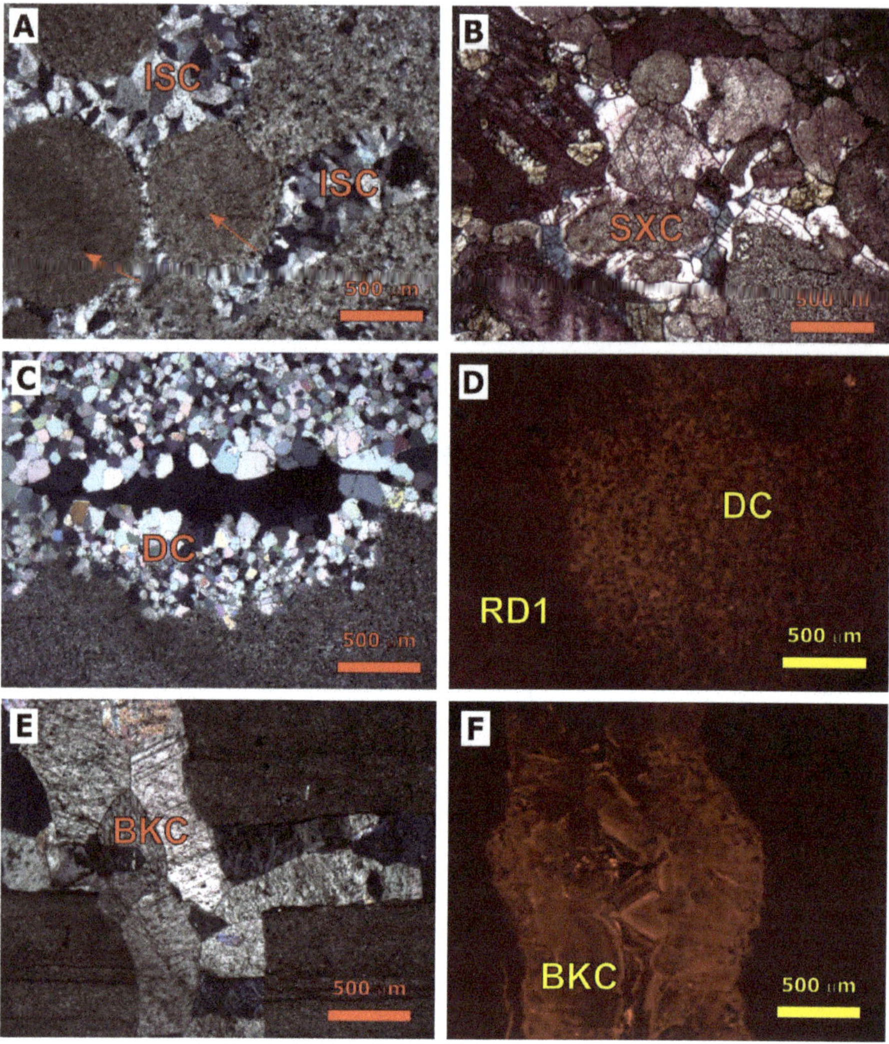

Figure 6. Calcite cementation in Silurian formations. (**A**) Photomicrograph (XPL) of isopachous calcite cement (ISC) filling interparticle pores in coated grains (red arrows). Sample 8-10; well: DGR8-CR122; depth: 332.26 m; (**B**) photomicrograph (PPL—after staining) of ferroan syntaxial overgrowth calcite cement (SXC) on crinoidal and echinoid skeletal fragments in a bioclastic grainstone. Sample 1-33; well: DGR1-CR154; depth: 440.92 m; (**C**) photomicrograph (XPL) of drusy calcite cement (DC) showing increasing crystal sizes towards the center of a vug in a fenestral dolostone. Sample 1-17; well: DGR1-CR49; depth: 142.28 m; (**D**) CL photomicrograph of pervasive replacive fine crystalline dolomite matrix (RD1) and vug-filling drusy calcite cement (DC) both showing dully-red luminescence. Sample 1-17; well: DGR1-CR49; depth: 142.28 m; (**E**) photomicrograph (XPL) of a single generation of blocky calcite cement (BKC) filling fractures in a fine crystalline, laminated dolostone. Sample 8-5; well: DGR8-CR64; depth: 182.06 m. (**F**) CL photomicrograph of fracture-filling, late blocky calcite cement (BKC) showing red to bright zoned luminescence. Sample 1-26; well: DGR1-CR119; depth: 333.48 m.

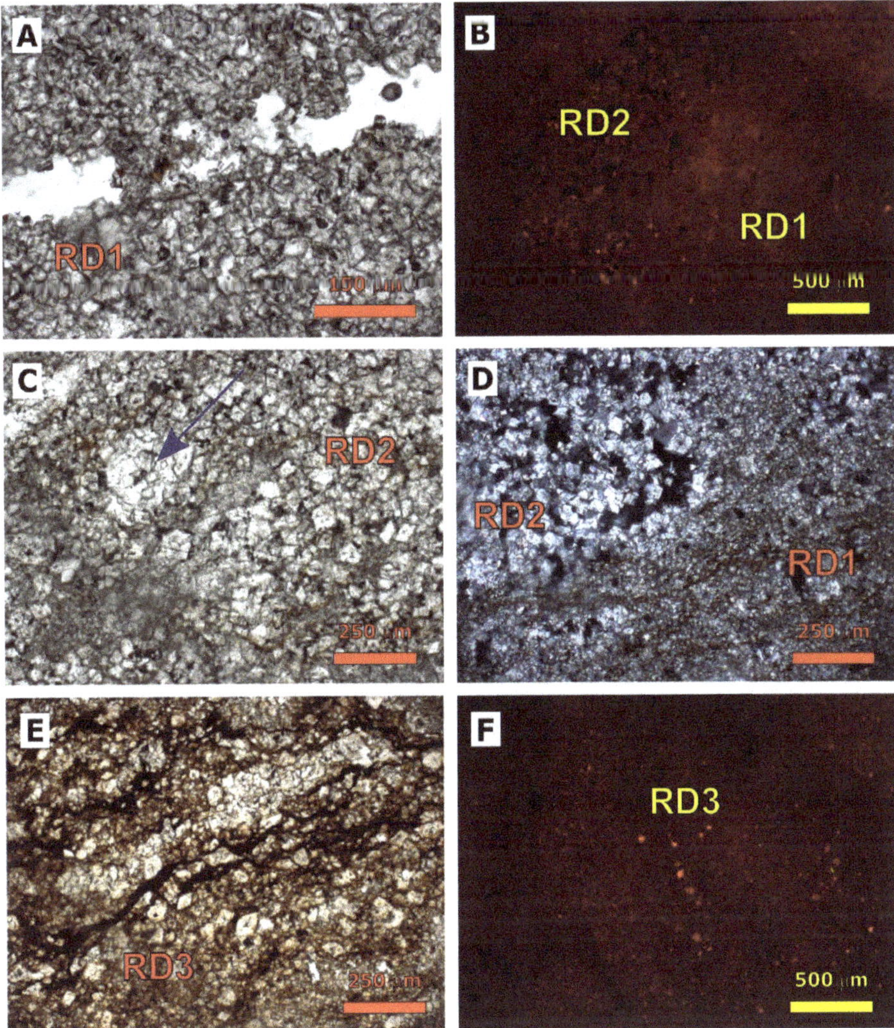

Figure 7. Dolomitization in Devonian formations. (**A**) Photomicrograph (PPL) of planar-e to planar-s, pervasive replacive micro to fine crystalline dolomite matrix (RD1). Sample 1-1; well: DGR1-CR1; depth: 23.5 m; (**B**) CL photomicrograph of pervasive replacive, fine-crystalline dolomite matrix (RD1) and pervasive replacive, medium crystalline dolomite matrix (RD2) showing red luminescence. Sample 8-4; well: DGR8-CR22; depth: 59.06 m; (**C**) photomicrograph (XPL) of planar-e to planar-s, medium crystalline dolomite matrix (RD2) pervasively replacing precursor limestone and partially replacing allochems fragments (blue arrow). Sample 1-4; well: DGR1-CR2 (3); depth: 27.75 m; (**D**) photomicrograph (XPL) of planar-e to planar-s, pervasive replacive medium crystalline dolomite matrix (RD2) with higher intercrystalline porosity (black areas) compared to the pervasive replacive micro to fine-crystalline dolomite matrix (RD1). Sample 1-3-2; well: DGR1-CR2 (2); depth: 26.95 m; (**E**) photomicrograph (PPL) of planar-e to planar-s, selective replacive medium crystalline dolomite matrix (RD3) associated with dissolution seams showing dark cores and clear rims. Sample 8-4; well: DGR8-CR22; depth: 59.06 m. (**F**) CL photomicrograph of selective replacive medium crystalline dolomite matrix (RD3) showing dull to red luminescence. Sample 1-3-2; well: DGR1-CR2 (2); depth: 313.31 m.

Table 2. Summary of the main petrographic characteristics of the carbonate minerals observed in Silurian and Devonian samples.

Age	Mineral	Description	Texture	Size	Luminescence	Ferroan/ Non-Ferroan
Silurian	RD1	Pervasive replacive micro to fine crystalline dolomite matrix	Non-planar to planar-s	<50 µm	Dully red/bright	Non-ferroan
Silurian	RD2	Pervasive replacive medium crystalline dolomite matrix	Planar-e/planar-s	>50 µm up to 150 µm	Dully red/ non-luminescent	Non-ferroan
Silurian	RD3	Selective replacive medium crystalline dolomite matrix (commonly associated with dissolution seams)	Planar-e	>50 µm up to 150 µm	Dull/red	Non-ferroan
Silurian	SD	Coarse crystalline saddle dolomite cement filling fractures and vugs	Curved crystal faces and cleavage planes and undulose extinction	>500 µm	Dull/ non-luminescent	Ferroan
Silurian	ISC	Isopachous calcite cement characterized by cement rims growing around coated grains	bladed	50–100 µm	Dully red	Non-ferroan
Silurian	SXC	Syntaxial calcite overgrowth cement usually around echinoderm fragments	-	100 µm to >500 µm	Bright red	Ferroan
Silurian	DC	Void-filling and pore-lining cement in intergranular and intraskeletal pores, molds and fractures	Equant to elongate, anhedral to subhedral	75–250 µm Size increase towards the center	Dull red	Ferroan
Silurian	BKC	Blocky calcite cement consisting of coarse-grained crystals without a preferred orientation mainly filling fractures and voids	-	200 µm to >500 µm	Red/bright orange zoned	Non-ferroan/ ferroan
Devonian	RD1	Pervasive replacive micro to fine crystalline dolomite matrix	Non-planar to planar-s	<50 µm	Dull/red	Non-ferroan
Devonian	RD2	Pervasive replacive medium crystalline dolomite matrix	Planar-e/ planar-s	>50 µm up to 100 µm	Dull/ non-luminescent	Non-ferroan
Devonian	RD3	Selective replacive medium crystalline dolomite matrix (commonly associated with dissolution seams)	Planar-e	>50 µm up to 100 µm	Red/bright	Non-ferroan
Devonian	SXC	Syntaxial calcite overgrowth cement usually around echinoderm fragments	-	100 µm to >500 µm	Non-luminescent	Non-ferroan
Devonian	DTC	Dogtooth calcite cement characterized by sharply pointed acute crystals growing normal to the substrate (mainly skeletal grains)	Elongate scalenohedral or rhombohedral	50 µm	Red/bright mostly zoned	Non-ferroan
Devonian	DC	Void-filling and pore-lining cement in intergranular and intraskeletal pores, molds and fractures	Equant to elongate, anhedral to subhedral	75–250 µm Size increase towards the center	Dull red to bright, zoned in most cases	Non-ferroan
Devonian	BKC	Blocky calcite cement consisting of coarse-grained crystals without a preferred orientation mainly filling fractures and voids	-	200 µm to >500 µm	Red/bright zoned	Non-ferroan

4.2.2. Fracturing

The fractures observed in the Devonian formations can be divided in two types. The first is represented by sub-horizontal fractures (0.5–1 cm) filled by blocky calcite cement (Figure 8A).

The second type is represented by sub-vertical fractures (0.1–0.5 cm) commonly filled by blocky calcite (Figure 8A,B) crosscutting chemical compactional features (Figure 8C,D).

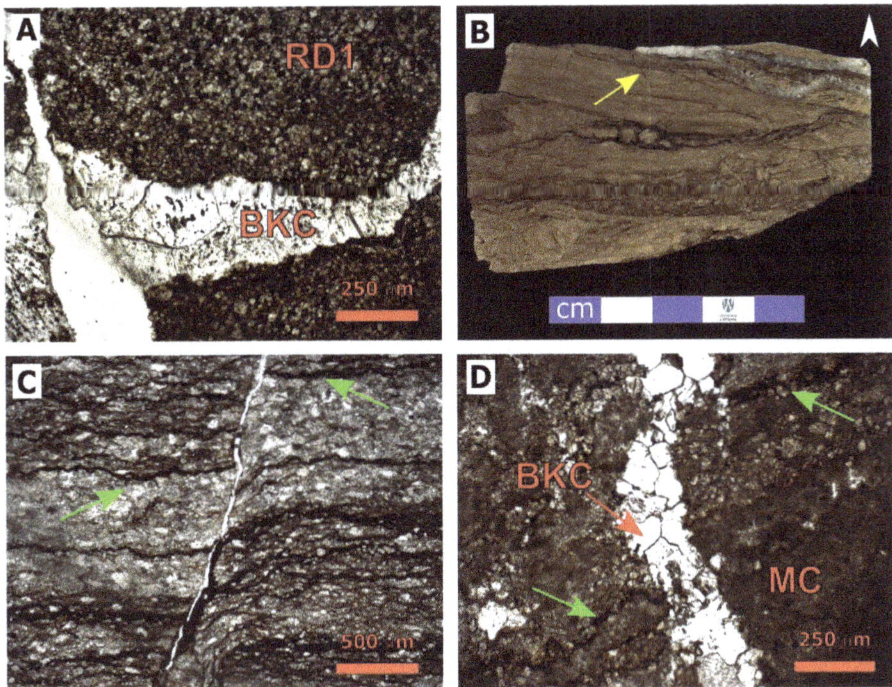

Figure 8. Fracturing in Devonian formation. (**A**) Photomicrograph (PPL) of a sub-vertical, unfilled hairline fracture cross-cutting sub-horizontal blocky calcite (BKC) vein in a fine crystalline dolostone. Sample 8-1; well: DGR8-CR10; depth: 22.71 m; (**B**) Core photograph showing two systems of fractures in a dolostone from the Lucas Formation (Devonian). A sub-vertical unfilled hairline fracture (yellow arrow) crosscut a sub-horizontal fracture-filled by late calcite cement. The white arrow indicates the top of the sedimentary sequence. Sample 8-1; well: DGR8-CR10; depth: 22.71 m. (**C**) Photomicrograph (PPL) of a sub-vertical, unfilled hairline fracture crosscutting and displacing dissolution seams (green arrows) and sedimentary structures in a partially dolomitized grainstone. Sample 1-9; well: DGR1-CR9; depth: 46.4 m; (**D**) photomicrograph (PPL) of sub-vertical, blocky calcite (BKC) vein cross-cutting dissolution seams and stylolites (green arrows) in a partially dolomitized limestone. Note the presence of undolomitized micritic matrix (MC) from the precursor limestone. Sample 1-2; well: DGR1-CR2 (1); depth: 26.25 m.

4.2.3. Calcite Cementation

Four types of calcite cement, ranging from early to late-stage, (Figure 9) have been distinguished, including syntaxial overgrowth, dogtooth, drusy and blocky calcite. Undolomitized calcite matrix (MC) from the precursor limestone was also observed in some specimens (Figure 8D).

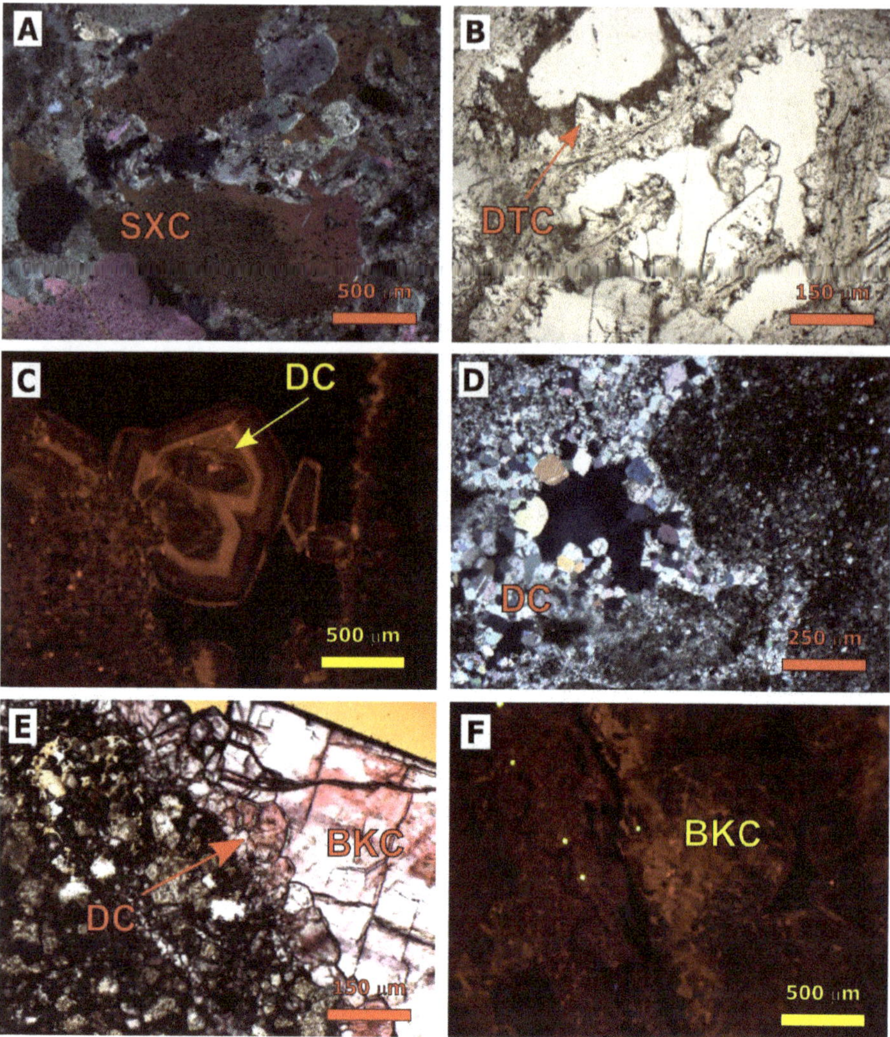

Figure 9. Calcite cementation in Devonian formations. (**A**) Photomicrograph (XPL) of syntaxial overgrowth calcite cement (SXC) on echinoderm skeletal fragments in a bioclastic grainstone. Sample 1-10; well: DGR1-CR10 (1); depth: 48.65 m; (**B**) photomicrograph (PPL) of dogtooth calcite cement (DTC) growing normally with respect to the surface of the intraskeletal chambers in a coral from a bioclastic grainstone. Sample 1-11; well: DGR1-CR10 (2); depth: 49.7 m; (**C**) CL photomicrograph of fracture-filling drusy calcite cement (DC) showing dull to bright zoned luminescence. Sample 1-13; well: DGR1-CR30 (1); depth: 26.95 m; (**D**) photomicrograph (XPL) of drusy calcite cement (DC) showing increasing crystal sizes towards the center of a fracture in a brecciated cherty dolostone. Sample 1-13; well: DGR1-CR30 (1); depth: 104.05 m; (**E**) photomicrograph (PPL—after staining) of non-ferroan, drusy calcite cement (DC) lining a fracture subsequently filled by non-ferroan blocky calcite (BKC) in fine crystalline dolostone. Sample 1-13; well: DGR1-CR30 (1); depth: 104.05 m; (**F**) CL photomicrograph of bright luminescent fracture-filling blocky calcite cement (BKC). Sample 8-1; well: DGR8-CR10; depth: 22.71 m.

Syntaxial calcite overgrowth cement (SXC) forms non-ferroan, non-luminescent crystals, usually around echinoderm fragments, ranging in size from 100 μm up to >500 μm (Figure 9A).

Dogtooth calcite cement (DTC), is characterized by non-ferroan, non-luminescent, sharply pointed acute crystals (50 μm) growing normal to the substrate. It is mainly observed lining intraskeletal chambers, and commonly presenting textures from elongate scalenohedral to rhombohedral (Figure 9B).

Pore-lining and void-filling drusy calcite cement (DC) is mainly observed in intergranular and intraskeletal pores, molds and fractures (Figure 9C–E). Common textural features show equant to elongate, in some cases anhedral to subhedral crystals ranging in size between 75–250 μm with dimension increasing towards the center of the pore space (Figure 9D). In most cases, DC is non-ferroan (Figure 9E) and shows zoned, dull red to bright orange luminescence (Figure 9C)

A late fracture and void-filling blocky calcite cement (BKC) consisting of the coarse-grained, non-ferroan crystal (Figure 9E) without preferred orientation, ranges in size from 200 μm to >500 μm, shows zoned red to bright orange luminescence under CL (Figure 9F).

Paragenetically, early calcite cement (SXC, DTC and DC) predate replacive dolomite matrix RD1, RD2 and RD3, and predate the late fracture- and pore-filling blocky calcite (BKC) cement.

4.3. Geochemistry of Silurian and Devonian Formations

4.3.1. Oxygen and Carbon Stable Isotopes

The isotopic composition of RD1 from the two age groups shows more negative $\delta^{18}O$ and $\delta^{13}C$ average values in the Silurian samples than in the Devonian samples. $\delta^{18}O$ isotopic composition of Silurian and Devonian pervasive replacive micro to fine crystalline dolomite matrix (RD1) range from −7.83 to −4.46‰ VPDB (average: −6.43‰ ± 1.1‰) and from −7.01 to −4.46‰ VPDB (average: −5.95‰ ± 0.71‰), respectively. $\delta^{13}C$ isotopic values in RD1 samples range from −2.73 to 3.78‰ VPDB (average: 0.18‰ ± 1.9‰) for Silurian formations and from 2.01 to 4.28‰ VPDB (average: 3.05‰ ± 0.88‰) for Devonian formations (n = 20; Figure 10; Table 3).

The isotopic compositions of RD2 from Silurian formations show more negative $\delta^{18}O$ and $\delta^{13}C$ average values than values of RD2 from Devonian samples. $\delta^{18}O$ isotopic composition of Silurian and Devonian pervasive replacive medium crystalline dolomite matrix (RD2) range from −7.96 to −3.85‰ VPDB (average: −6.62‰ ± 1.37‰) and from −7.66 to −4.75‰ VPDB (average: −6.11‰ ± 1.07‰), respectively. $\delta^{13}C$ isotopic concentrations in RD2 samples range from −0.32 to 3.84‰ VPDB (average: 2.15‰ ± 1.37‰) for Silurian formations and from 0.99 to 3.93 ‰ VPDB (average: 2.75‰ ± 1.21‰) for Devonian successions (n = 12; Figure 10; Table 3).

RD3 from Silurian samples is characterized by more negative $\delta^{18}O$ average values but comparable $\delta^{13}C$ average values Devonian samples. $\delta^{18}O$ isotopic composition of Silurian and Devonian selective replacive medium crystalline dolomite matrix (RD3) range from −7.53 to -5.91‰ VPDB (average: −6.95‰ ± 0.9‰) and from −6.20 to −3.91‰ VPDB (average: −5.31‰ ± 0.71‰), respectively. $\delta^{13}C$ isotopic concentrations in RD3 samples range from 0.65 to 4.02‰ VPDB (average: 2.66‰ ± 1.77‰) for Silurian formations and from 1.09 to 4.63‰ VPDB (average: 3.08‰ ± 1.07‰) for Devonian successions (n = 11; Figure 10; Table 3).

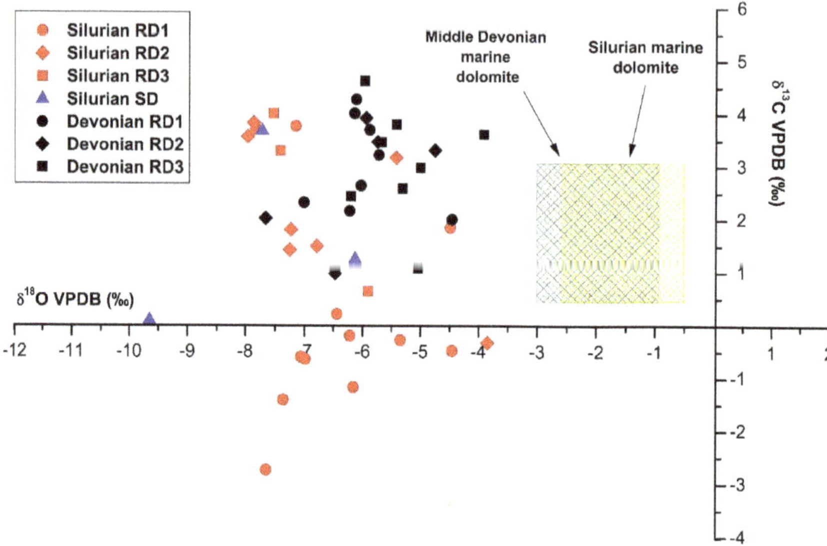

Figure 10. Cross plot of $\delta^{13}C$ vs $\delta^{18}O$ of different types of dolomite in Silurian and Devonian formations. The data shows an overlap between $\delta^{13}C$ and $\delta^{18}O$ values in Silurian and Devonian dolomites. In both age groups, $\delta^{13}C$ values fall in the range of values estimated for the marine calcite of equivalent age except for the more negative values in some Silurian RD1 samples possibly related to bacterial sulfate reduction (BSR). The negative shift in $\delta^{18}O$ values, more pronounced in Silurian dolomites, can be related to dolomite recrystallization during burial and increasing temperature. The green box represents isotopic values of the Middle Devonian marine calcite [49] and the yellow box represents the estimated isotopic values for the Silurian marine dolomite [50].

Table 3. Oxygen and carbon stable isotopes, Sr isotopes results from dolomite and calcite samples.

Sample ID	Age	Phase	$\delta^{13}C_{VPDB}$ (‰)	$\delta^{18}O_{VPDB}$ (‰)	$^{87}Sr/^{86}Sr$
1-15a	Silurian	RD1	−0.47	−4.46	
1-16a	Silurian	RD1	−0.27	−5.35	
1-17a	Silurian	RD1	1.86	−4.49	0.708705
1-18a	Silurian	RD1	−2.73	−7.65	0.708877
1-20a	Silurian	RD1	−0.59	−7.06	
1-21a	Silurian	RD1	0.22	−6.44	
1-31a	Silurian	RD1	3.78	−7.15	
8-6a	Silurian	RD1	−1.4	−7.36	0.708919
8-5a	Silurian	RD1	−0.63	−6.99	
8-8a	Silurian	RD1	−1.16	−6.16	
8-9a	Silurian	RD1	−0.19	−6.22	
8-13a	Silurian	RD1	3.72	−7.83	
1-16b	Silurian	RD2	−0.32	−3.85	
1-24a	Silurian	RD2	1.51	−6.79	0.708469
1-23b	Silurian	RD2	1.44	−7.25	
1-31b	Silurian	RD2	3.58	−7.96	0.708587
8-11a	Silurian	RD2	3.18	−5.42	
8-10b	Silurian	RD2	1.82	−7.23	
8-13b	Silurian	RD2	3.84	−7.86	
1-30a	Silurian	RD3	4.02	−7.53	0.708905
1-33b	Silurian	RD3	0.65	−5.91	
8-12	Silurian	RD3	3.31	−7.41	0.708743

Table 3. *Cont.*

Sample ID	Age	Phase	$\delta^{13}C_{VPDB}$ (‰)	$\delta^{18}O_{VPDB}$ (‰)	$^{87}Sr/^{86}Sr$
1-31c	Silurian	SD	3.69	−7.72	0.708648
8-13c	Silurian	SD	0.10	−9.65	
1-28a	Silurian	SD	1.27	−6.13	0.708516
1-24b	Silurian	MC	1.32	−5.55	
1-27b	Silurian	MC	0.14	−8.16	
1-30b	Silurian	MC	3.53	−8.09	
1-33c	Silurian	MC	0.21	−4.73	
8-11b	Silurian	MC	2.89	−5.35	
1-23a	Silurian	ISC	1.37	−6.76	
8-10a	Silurian	ISC	1.27	−7.77	
1-32	Silurian	SXC	0.12	−4.81	
1-33a	Silurian	SXC	−0.21	−5.54	
1-17b	Silurian	DC	−0.88	−8.85	0.708154
1-25	Silurian	BKC	0.02	−4.54	
1-26a	Silurian	BKC	0.00	−6.78	
1-27a	Silurian	BKC	0.03	−7.52	0.708383
1-28b	Silurian	BKC	0.06	−7.51	
1-31d	Silurian	BKC	3.17	−6.41	
8-5b	Silurian	BKC	−3.75	−9.19	0.708054
8-9c	Silurian	BKC	−1.06	−10.19	
8-13d	Silurian	BKC	3.10	−8.09	
8-9d	Silurian	GY	-	-	0.708486
1-29a	Silurian	GY	-	-	0.708434
1-1a	Devonian	RD1	2.65	−6.03	0.708103
1-1b	Devonian	RD1	2.17	−6.22	
1-3-1a	Devonian	RD1	3.23	−5.72	
1-5a	Devonian	RD1	4.02	−6.14	
1-5b	Devonian	RD1	4.28	−6.11	
1-7a	Devonian	RD1	3.71	−5.88	
8-1a	Devonian	RD1	2.01	−4.46	0.707884
8-4a	Devonian	RD1	2.33	−7.01	
1-3-2b	Devonian	RD2	3.32	−4.75	0.708089
1-4a	Devonian	RD2	3.93	−5.94	
1-7b	Devonian	RD2	3.48	−5.74	
1-14a	Devonian	RD2	0.99	−6.47	
8-4b	Devonian	RD2	2.03	−7.66	0.708920
1-2a	Devonian	RD3	3.47	−5.67	
1-3-2c	Devonian	RD3	3.62	−3.91	
1-6a	Devonian	RD3	4.63	−5.97	0.708146
1-10a	Devonian	RD3	2.60	−5.31	
1-13b	Devonian	RD3	1.09	−5.04	
8-2a	Devonian	RD3	3.81	−5.41	
8-3a	Devonian	RD3	2.99	−5.00	
8-4c	Devonian	RD3	2.45	−6.20	0.708509
1-2b	Devonian	MC	2.52	−5.87	
1-4b	Devonian	MC	2.81	−5.77	
1-6b	Devonian	MC	4.12	−4.53	
1-8b	Devonian	MC	1.92	−6.47	
8-2b	Devonian	MC	2.83	−6.59	
8-3b	Devonian	MC	1.99	−6.51	
1-11a	Devonian	SXC	0.97	−4.44	
1-12a	Devonian	SXC	1.55	−4.42	
1-13a	Devonian	DC	−0.59	−6.22	0.708051
1-14b	Devonian	DC	−1.15	−6.49	
1-2c	Devonian	BKC	−1.55	−8.01	
8-1b	Devonian	BKC	−2.40	−8.61	0.707982

The $\delta^{18}O$ and $\delta^{13}C$ isotopic values of Silurian Saddle dolomite cement (SD) range from -9.65 to $-6.13‰$ VPDB (average: $-7.84‰ \pm 1.76‰$) and from 0.10 to 3.69‰ VPDB (average: $1.69‰ \pm 1.83‰$), respectively (n = 3; Figure 10 and Table 3).

Early calcite shows more negative $\delta^{18}O$ and more positive $\delta^{13}C$ average values in Devonian formations compared with Silurian samples. Silurian and Devonian early calcite matrix and cement (MC, ISC, SXC, DTC and DC) show $\delta^{18}O$ isotopic composition ranging from -8.85 to $-4.73‰$ VPDB (average: $-6.56‰ \pm 1.47‰$) and from -6.59 to $-4.42‰$ VPDB (average: $-5.73‰ \pm 0.87‰$), respectively. $\delta^{13}C$ isotopic values in early calcite samples (MC, ISC, SXC, DTC and DC) range from -0.88 to 3.53‰ VPDB (average: $0.97‰ \pm 1.32‰$) for Silurian formations and from -1.15 to 4.12‰ VPDB (average: $1.70‰ \pm 1.52‰$) for Devonian successions (n = 20; Figure 11; Table 3).

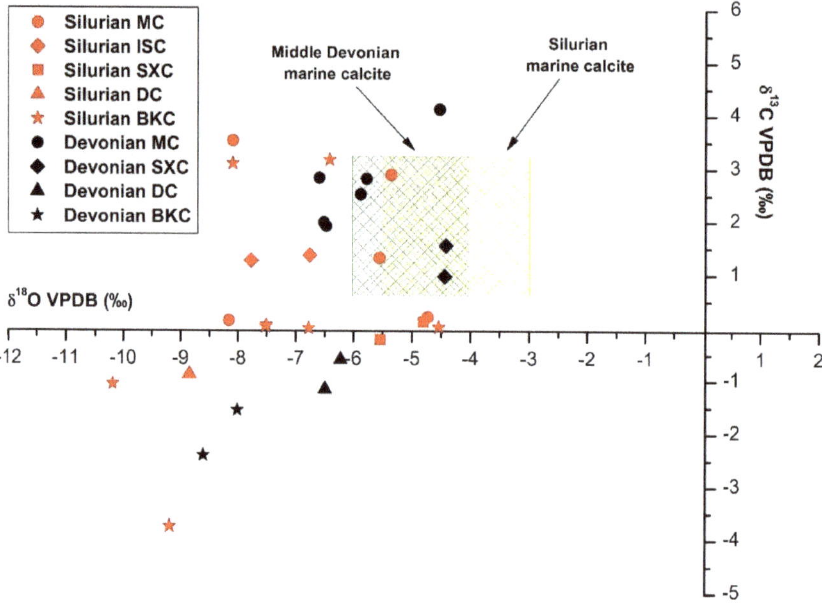

Figure 11. Crossplot of $\delta^{13}C$ vs $\delta^{18}O$ of different generations of calcite cement in Silurian and Devonian formations. Most of the calcitic samples show $\delta^{13}C$ values, in range with postulated values for marine calcite of the respective age with the exception for late BKC which clearly shows a negative shift from the equilibrium values in both age groups. This may suggest a separate input of oxidized organic carbon possibly related to interaction with hydrocarbons. The green box represents isotopic values of the Middle Devonian marine calcite [49] and the yellow box represents the estimated isotopic values for the Silurian marine dolomite [50].

Late calcite cement (BKC) shows more negative $\delta^{18}O$ and $\delta^{13}C$ average values in Devonian formations compared with the Silurian samples. The $\delta^{18}O$ isotopic composition of Silurian and Devonian late fracture-filling calcite cement (BKC) range from -10.19 to $-4.54‰$ VPDB (average: $-7.53‰ \pm 1.62‰$) and from -8.61 to $-8.01‰$ VPDB (average: $-8.31‰ \pm 0.3‰$), respectively. $\delta^{13}C$ isotopic concentrations in BKC samples range from -3.75 to 3.17‰ VPDB (average: $0.20‰ \pm 2.08‰$) for Silurian formations and from -2.40 to $-1.55‰$ VPDB (average: $-1.98‰ \pm 0.43‰$) for Devonian successions (n = 10; Figure 11; Table 3).

4.3.2. Strontium Isotopes

Due to the mixed nature of the samples, only one pure sample of drusy calcite cement (DC), for each age group, were microsampled avoiding then contamination with other carbonate minerals. The Sr isotopic values of DC were 0.708154 and 0.708051 for the Silurian and Devonian samples, respectively (Figure 12). The Silurian late calcite cement (BKC) show $^{87}Sr/^{86}Sr$ isotopic ratio values ranging from 0.708054 to 0.708383 (n = 2) and 0.707982 for Devonian BKC (n = 1). $^{87}Sr/^{86}Sr$ ratios of matrix dolomite samples (RD1, RD2 and RD3) from the Silurian (n = 7) and Devonian (n = 6) carbonate successions range from 0.708469 to 0.708919 and 0.707884 to 0.708920, respectively. Saddle dolomite sampled from Silurian formations has Sr isotopic values ranging from 0.708516 to 0.708648 (n = 2; Figure 12). Gypsum samples from Silurian formations show Sr isotopic values ranging from 0.708404 to 0.708486.

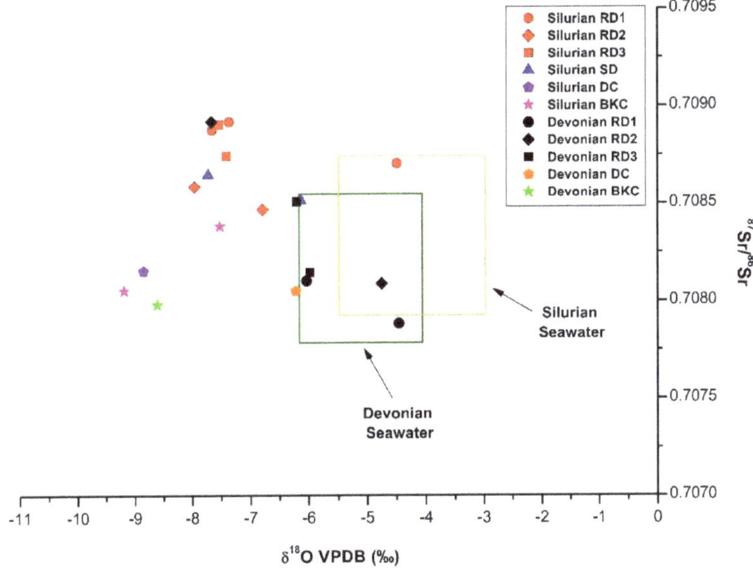

Figure 12. Cross plot of $\delta^{18}O$ vs $^{87}Sr/^{86}Sr$ ratios of different dolomitic and calcitic components from the two age groups. In both age groups, Sr isotopic ratios show seawater composition of their respective age as the primary source of diagenetic fluids with minor rock-water interactions as indicated by some exceptions of more radiogenic signatures in the Silurian formations. The green box represents Devonian seawater [51] and the yellow box represents Silurian seawater [51,52].

$^{87}Sr/^{86}Sr$ isotopic ratios for dolomite and calcite by age are presented in Table 3.

4.3.3. Fluid Inclusions Microthermometry

Fluid inclusions measurements were focused only on dolomite and blocky calcite cement. Due to the small size of crystal and/or fluid inclusions, microthermometric measurements in mineral phases such as RD1and RD3 were not achievable in both age groups. The melting (Tm) and homogenization temperatures (T_h) were measured from two-phase, liquid-rich primary fluid inclusions which ranged in size from 1 μm up to 20 μm in diameter with variable shapes from irregular to nearly circular (Figure 13). In some cases, measurements of the melting temperatures for fluid inclusions hosted in RD2 were not possible due to their small size in both Silurian and Devonian successions (measured n = 9 and n = 7 for Silurian and Devonian samples, respectively). Hydrocarbon fluid inclusions were not identified in the selected samples from the studied formations.

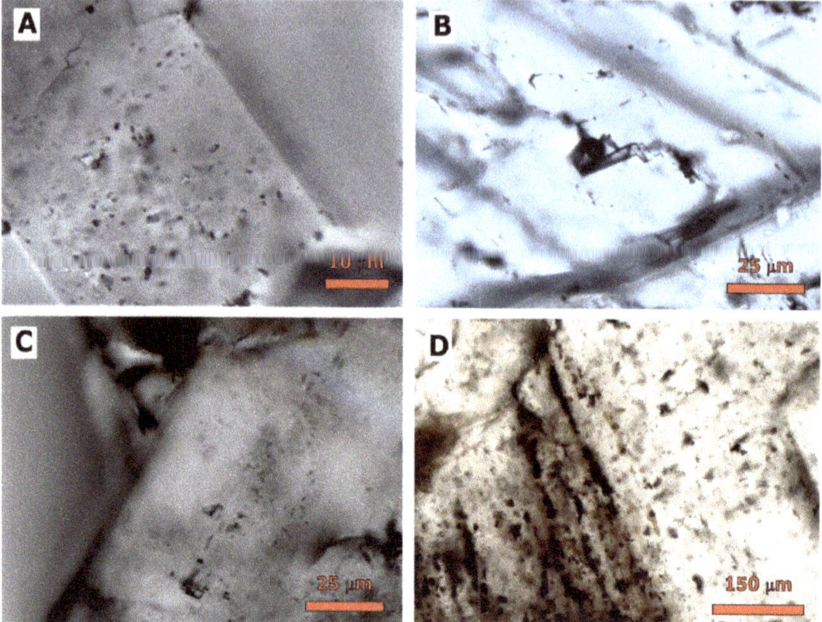

Figure 13. Fluid Inclusion assemblages. (**A**) Photomicrograph (PPL) of fluid inclusion assemblage in an RD2 crystal from Lucas Formation. Sample 1-3-2; Well: DGR1-CR2 (2); depth: 26.95 m; (**B**) photomicrograph (PPL) of isolated two-phase, primary fluid inclusions (liquid-rich with vapor bubble) hosted within blocky calcite cement occluding fractures in the Lucas Formation. Sample 8-1; Well: DGR8-CR10; depth: 22.71 m; (**C**) photomicrograph (PPL) of fluid inclusions assemblage in RD2 along the crystal growth showing two-phase primary fluid inclusions. Sample 1-31; well: DGR1-CR133; depth: 378.1 m; (**D**) photomicrograph (PPL) of fluid inclusion assemblage in SD along the crystal growth, showing a trail of two-phase primary fluid inclusions with irregular and elongated shapes. Sample 1-28; well: DGR1-CR121; depth: 342.27 m.

Fluid inclusions hosted in RD2 from Silurian samples show higher T_h and salinity compared to fluid inclusions hosted in RD2 from Silurian formations. Microthermometric results of pervasive, replacive medium crystalline dolomite matrix (RD2) vary from T_h: 49.7 to 134.1 °C (average: 89.1 ± 20.9), 22.03 to 25.2 wt.% NaCl eq. (average: 23.3 ± 1.1) and T_h: 69.9 to 102.3 °C (average: 82.7 ± 9.8), 18.9 to 21.8 wt.% NaCl eq. (average: 20.8 ± 0.9) for the Silurian and Devonian samples, respectively.

Saddle dolomite (SD) hosted in Silurian successions shows values ranging from T_h: 101.2 to 193.4 °C (average: 124.8 ± 20.5), 25.97 to 32.6 wt.% NaCl eq. (average: 28.5 ± 1.8).

Fluid inclusions hosted in BKC from Silurian samples show lower T_h and higher salinity compared to fluid inclusions hosted in BKC from Silurian formations. In late calcite cement (BKC), measured values range from T_h: 70.2 to 194.3 °C (average: 120.9 ± 35.4), 21.4 to 32.3 wt.% NaCl eq. (average: 27.6 ± 3.2) and T_h: 146.20 to 237.9 °C (average: 194.87 ± 26.9), 15.17 to 23.95 wt.% NaCl eq. (average: 20.21 ± 2.31) for the mentioned age groups, respectively.

The summaries of the fluid inclusion results are presented in Figures 14–16 and Table 4. Examples of the main characteristics of fluid inclusion assemblages hosted within the different calcite and dolomite phases are illustrated in Figure 13.

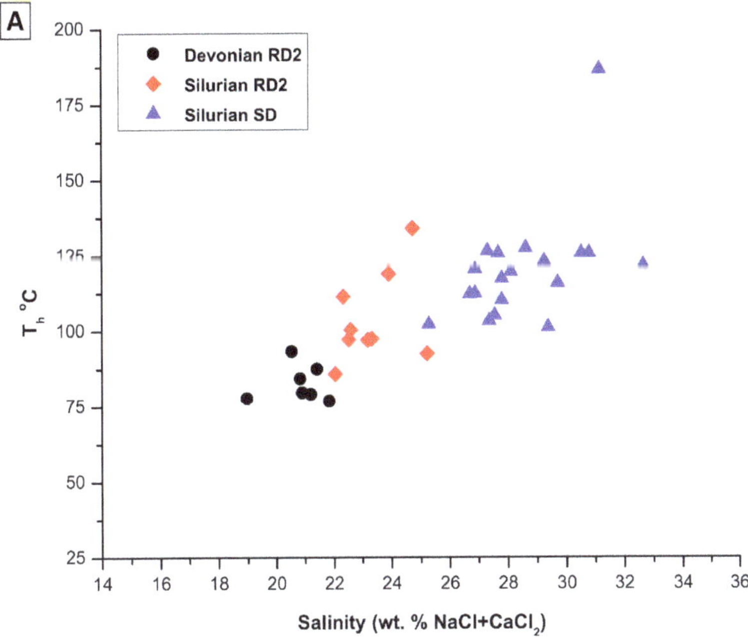

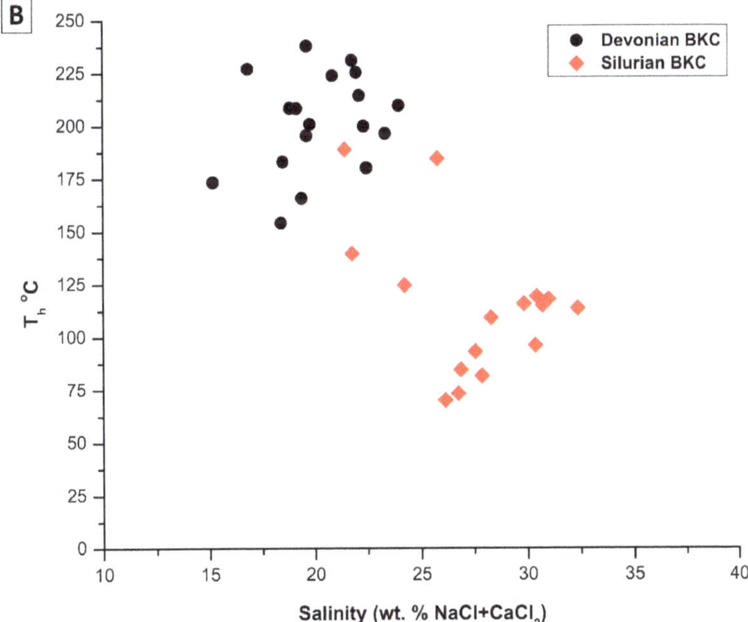

Figure 14. Homogenization temperature vs. salinity cross plots for analyzed (**A**) dolomite and (**B**) calcite phases.

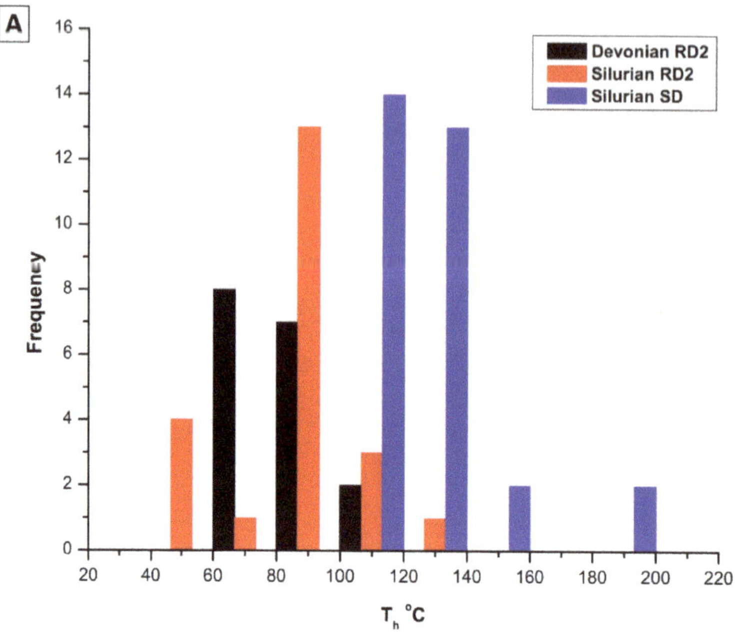

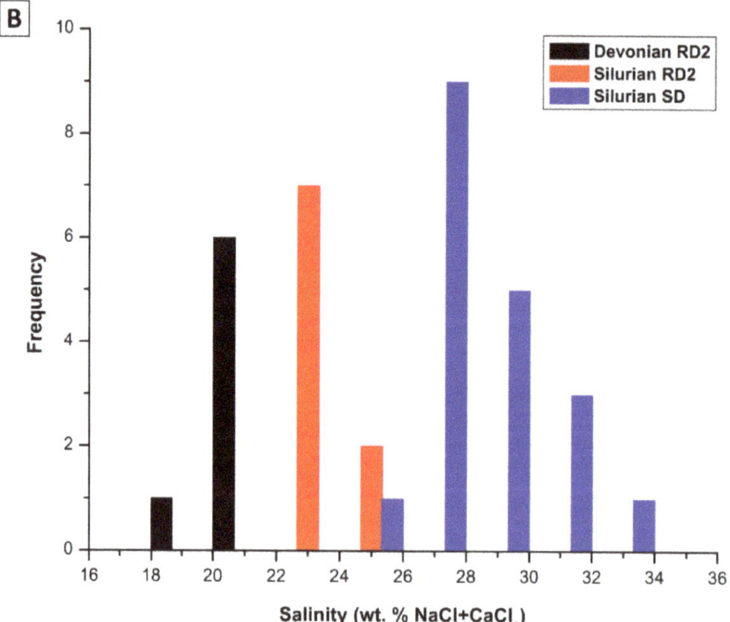

Figure 15. Histogram plots of (**A**) homogenization temperature and (**B**) salinity for analyzed dolomite phases in Silurian and Devonian formations.

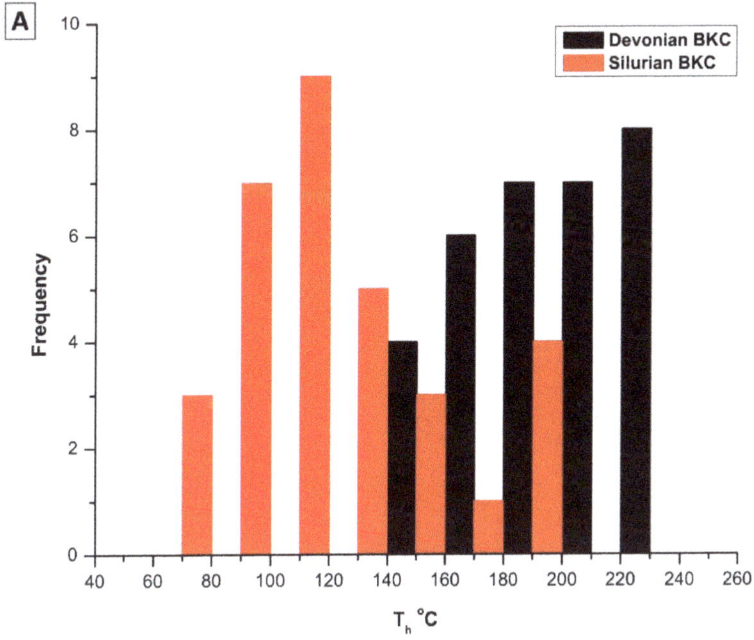

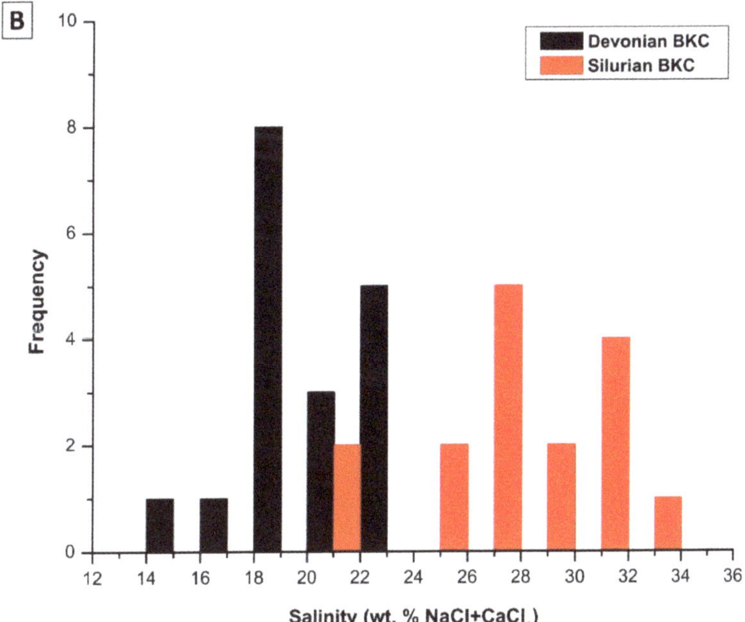

Figure 16. Histogram plots of (**A**) homogenization temperature and (**B**) salinity for calcite cement (BKC) in Silurian and Devonian formations.

Table 4. Fluid inclusions results.

Sample	Age	Host Mineral	Occurrence	Size (µm)	T_h (°C)	Tm_{ice} (°C)	Salinity
1-31-1	Silurian	RD2	Cluster	2	54.2		
1-31-2	Silurian	RD2	Cluster	3	55.1		
1-31-3	Silurian	RD2	Cluster	1	57.5		
1-31-4	Silurian	RD2	Cluster	7	100.4	−20.3	22.50
1-31-5	Silurian	RD2	Cluster	1	49.7		
1-31-6	Silurian	RD2	Cluster	6	118.9	−22.3	23.89
1-31-7	Silurian	RD2	Cluster	5	92.6	−24.4	25.21
1-31-8	Silurian	RD2	Cluster	5	97.5	−21.4	23.31
1-31-9	Silurian	RD2	Cluster	4	88.9		
1-31-10	Silurian	RD2	Isolated	5	85.9	−19.5	22.03
1-31-11	Silurian	RD2	Isolated	2	71.4		
1-31-12	Silurian	RD2	Isolated	10	134.1	−23.6	24.71
1-31-13	Silurian	RD2	Crystal growth	2	91.5		
1-31-14	Silurian	RD2	Crystal growth	1	92.8		
1-31-15	Silurian	RD2	Crystal growth	1	93.6		
1-31-16	Silurian	RD2	Crystal growth	1	84.7		
1-31-17	Silurian	RD2	Crystal growth	1	89.6		
1-31-18	Silurian	RD2	Crystal growth	4	97.1	−21.2	23.18
1-31-19	Silurian	RD2	Crystal growth	2	96.4		
1-31-20	Silurian	RD2	Crystal growth	1	99.8		
1-31-21	Silurian	RD2	Crystal growth	6	97.3	−20.2	22.51
8-13-31	Silurian	RD2	Crystal growth	3	111.4	−19.9	22.31
1-31-22	Silurian	BKC	Isolated	6	159.9		
1-31-23	Silurian	BKC	Isolated	5	184.7	−25.3	25.76
8-13-1	Silurian	BKC	Short trail	7	119.3	−32.7	30.42
8-13-2	Silurian	BKC	Short trail	8	118.1	−33.5	30.97
8-13-3	Silurian	BKC	Short trail	6	109.3	−29.4	28.28
8-13-4	Silurian	BKC	Cluster	4	106.4		
8-13-5	Silurian	BKC	Cluster	4	145.4		
8-13-6	Silurian	BKC	Cluster	2	79.4		
8-13-7	Silurian	BKC	Cluster	2	95.7		
8-13-8	Silurian	BKC	Cluster	3	130.1		
8-13-9	Silurian	BKC	Cluster	2	149.7		
8-13-10	Silurian	BKC	Cluster	2	172.3		
8-13-11	Silurian	BKC	Isolated	10	115.9	−31.8	29.82
8-13-12	Silurian	BKC	Crystal growth	13	81.8	−28.7	27.85
8-13-13	Silurian	BKC	Crystal growth	18	96.2	−32.6	30.35
8-13-14	Silurian	BKC	Crystal growth	15	93.3	−28.2	27.54
8-13-15	Silurian	BKC	Crystal growth	6	89.4		
8-13-16	Silurian	BKC	Crystal growth	5	81.6		
8-13-17	Silurian	BKC	Isolated	8	73.4	−26.9	26.74
8-13-18	Silurian	BKC	Isolated	6	84.7	−27.1	26.86
8-13-19	Silurian	BKC	Isolated	2	104.7		
8-13-20	Silurian	BKC	Isolated	3	103.5		
8-13-21	Silurian	BKC	Isolated	2	122.6		
8-13-22	Silurian	BKC	Crystal growth	10	113.8	−35.4	32.33
1-28-6	Silurian	BKC	Isolated	12	115.2	−33.1	30.69
1-28-7	Silurian	BKC	Cluster	3	123.3		
1-28-8	Silurian	BKC	Cluster	4	189	−18.6	21.40
1-28-9	Silurian	BKC	Cluster	4	139.7	−19.1	21.75
1-28-10	Silurian	BKC	Isolated	6	182.8		
1-28-11	Silurian	BKC	Isolated	6	194.3		
1-28-12	Silurian	BKC	Isolated	4	124.7	−22.8	24.21
1-28-13	Silurian	BKC	Isolated	10	70.2	−25.9	26.13
8-13-19	Silurian	SD	Crystal growth	5	122.6		
8-13-20	Silurian	SD	Crystal growth	5	123.4		
8-13-21	Silurian	SD	Crystal growth	3	142.7		

Table 4. *Cont.*

Sample	Age	Host Mineral	Occurrence	Size (μm)	T_h (°C)	Tm_{ice} (°C)	Salinity
8-13-22	Silurian	SD	Crystal growth	4	193.4		
8-13-23	Silurian	SD	Scattered	5	121.8	−35.8	32.63
8-13-24	Silurian	SD	Scattered	6	125.9	−33.2	30.76
8-13-25	Silurian	SD	Scattered	2	117.16		
8-13-26	Silurian	SD	Scattered	3	145.1		
8-13-27	Silurian	SD	Scattered	2	115.5		
8-13-28	Silurian	SD	Scattered	2	118.1		
8-13-29	Silurian	SD	Scattered	4	119.5	−29.1	28.09
8-13-30	Silurian	SD	Isolated	10	102.3	−24.5	25.27
8-13-31	Silurian	SD	Isolated	15	110.2	−28.6	27.78
8-13-32	Silurian	SD	Crystal growth	6	112.5	−27.1	26.86
8-13-33	Silurian	SD	Crystal growth	7	123.2	−30.9	29.23
8-13-34	Silurian	SD	Crystal growth	11	105.2	−28.2	27.54
8-13-35	Silurian	SD	Crystal growth	5	115.9	−31.6	29.69
8-13-36	Silurian	SD	Crystal growth	4	108.9		
8-13-37	Silurian	SD	Crystal growth	10	117.3	−28.6	27.78
8-13-38	Silurian	SD	Crystal growth	9	103.4	−27.9	27.35
8-13-39	Silurian	SD	Crystal growth	6	101.2	−31.1	29.36
8-13-40	Silurian	SD	Crystal growth	13	127.5	−29.9	28.60
8-13-41	Silurian	SD	Crystal growth	3	125.9	−32.8	30.49
8-13-42	Silurian	SD	Crystal growth	4	132.7		
1-28-1	Silurian	SD	Crystal growth	6	112.3	−26.8	26.68
1-28-2	Silurian	SD	Crystal growth	8	120.5	−27.1	26.86
1-28-3	Silurian	SD	Crystal growth	5	126.6	−27.8	27.29
1-28-4	Silurian	SD	Crystal growth	3	130.3		
1-28-5	Silurian	SD	Crystal growth	3	136.4		
1-28-6	Silurian	SD	Crystal growth	5	125.8	−28.4	27.66
1-28-7	Silurian	SD	Crystal growth	7	186.6	−33.7	31.11
8-4-1.	Devonian	RD2	Crystal growth	3	101.1		
8-4-2.	Devonian	RD2	Isolated	5	87.5	−18.6	21.40
8-4-2.	Devonian	RD2	Cluster	2	85.3		
8-4-2.	Devonian	RD2	Cluster	4	79.6	−17.9	20.89
8-4-2.	Devonian	RD2	Cluster	1	88.1		
8-4-2.	Devonian	RD2	Cluster	6	77.8	−15.4	18.96
8-4-2.	Devonian	RD2	Cluster	4	84.3	−17.8	20.82
8-4-2.	Devonian	RD2	Cluster	2	75.4		
1-3-2-1	Devonian	RD2	Isolated	5	76.9	−19.2	21.82
1-3-2-2	Devonian	RD2	Isolated	3	85.2		
1-3-2-3	Devonian	RD2	Cluster	4	93.3	−17.4	20.52
1-3-2-4	Devonian	RD2	Isolated	6	79.1	−18.3	21.19
1-3-2-5	Devonian	RD2	Isolated	1	71.4		
1-3-2-6	Devonian	RD2	Isolated	2	102.3		
1-3-2-7	Devonian	RD2	Isolated	1	76.3		
1-3-2-8	Devonian	RD2	Isolated	1	69.9		
1-3-2-9	Devonian	RD2	Isolated	2	81.1		
8-1-1.	Devonian	BKC	Isolated	8	173.5	−11.2	15.17
8-1-2.	Devonian	BKC	Cluster	10	205.1		
8-1-3.	Devonian	BKC	Cluster	2	166.9		
8-1-4.	Devonian	BKC	Crystal growth	8	175		
8-1-5.	Devonian	BKC	Crystal growth	8	208.4	−15.6	19.13
8-1-6.	Devonian	BKC	Trail	2	154.1		
8-1-7.	Devonian	BKC	Trail	2	157.9		
8-1-8.	Devonian	BKC	Trail	2	163.2		
8-1-9.	Devonian	BKC	Scattered	8	183.1	−14.8	18.47
8-1-10.	Devonian	BKC	Scattered	8	208.6	−15.2	18.80
8-1-11.	Devonian	BKC	Scattered	4	221.2		
8-1-12.	Devonian	BKC	Scattered	3	182.4		
8-1-13.	Devonian	BKC	Cluster	6	199.8	−19.9	22.31
8-1-14.	Devonian	BKC	Scattered	3	168.3		
8-1-15.	Devonian	BKC	Cluster	5	237.8		

Table 4. *Cont.*

Sample	Age	Host Mineral	Occurrence	Size (µm)	T_h (°C)	Tm_{ice} (°C)	Salinity
8-1-16.	Devonian	BKC	Cluster	7	180.2	−20.1	22.44
8-1-17.	Devonian	BKC	Cluster	8	225.4	−19.4	21.96
8-1-18.	Devonian	BKC	Cluster	5	225.8		
8-1-19.	Devonian	BKC	Isolated	7	196.4	−21.4	23.31
8-1-20.	Devonian	BKC	Trail	5	207.1		
8-1-21.	Devonian	BKC	Isolated	8	165.9	−15.9	19.37
8-1-22.	Devonian	BKC	Trail	8	237.9	−16.2	19.60
8-1-23.	Devonian	BKC	Isolated	0	231.1	−19.1	21.75
8-1-24.	Devonian	BKC	Isolated	7	209.8	−22.4	23.95
8-1-25.	Devonian	BKC	Scattered	8	227.1	−12.9	16.80
8-1-26.	Devonian	BKC	Scattered	4	146.2		
8-1-27.	Devonian	BKC	Scattered	4	188.6		
8-1-28.	Devonian	BKC	Isolated	6	200.8	−16.4	19.76
8-1-29.	Devonian	BKC	Isolated	6	195.5	−16.2	19.60
8-1-30.	Devonian	BKC	Isolated	7	223.9	−17.8	20.82
8-1-31.	Devonian	BKC	Isolated	4	154.3	−14.7	18.38
8-1-32.	Devonian	BKC	Isolated	9	214.5	−19.6	22.10

4.3.4. Major, Minor and Trace Elements Analysis

Major Elements and Dolomite Stoichiometry

The concentrations of the major elements, calcium and magnesium, were determined by ICP-MS analysis for dolomite and calcite phases to establish their stoichiometry. Ideal or stoichiometric dolomite would be characterized by the equal molar percentage of $MgCO_3$ and $CaCO_3$ and would be less soluble and more stable compared to non-stoichiometric dolomite [53]. In contrast, non-stoichiometric dolomite will have more Ca than Mg and because of its heterogeneous structure and composition is unstable and dissolve more rapidly.

All the dolomite types recognized petrographically were sampled and analyzed without any contamination from calcite components for both age groups with the only exception for the selective replacive dolomite (RD3) which could not be sampled without possible contamination by calcite or other dolomite phases in the Silurian specimens. With rare exceptions, all types of dolomite analyzed in this study were non-stoichiometric in both age groups (Table 5) with variable proportions of $MgCO_3$ and $CaCO_3$.

In both age group RD1 in non-stoichiometric but results less stoichiometric in Silurian formation compared to Devonian samples. Pervasive replacive micro to fine crystalline dolomite (RD1) ranges from 48.34 to 59.36 mole % $CaCO_3$ (average: 55.19 ± 3.73) and from 51.78 to 60.77 mole % $CaCO_3$ (average: 57.71 ± 3.04) for Silurian and Devonian formations, respectively (n_S = 11 and n_D = 7).

RD2 shows comparable values of mole % $CaCO_3$, but it is slightly more calcian in Devonian formations. Pervasive replacive medium crystalline dolomite (RD2) ranges from 56.11 to 62.12 mole % $CaCO_3$ (average: 58.04 ± 2.4) and from 58.39 to 61.57 mole % $CaCO_3$ (average: 59.71 ± 1.16) for Silurian and Devonian formations, respectively (n_S = 4 and n_D = 4).

Selective replacive medium crystalline dolomite (RD3) ranges from 47.93 to 59.96 mole % $CaCO_3$ (average: 53.55 ± 4.94) in Devonian formations (n_D = 3).

Saddle dolomite cement (SD), only observed and sampled from Silurian formations (n_S = 2), ranges from 63.47 to 67.00 mole % $CaCO_3$ (average: 65.23 ± 1.77).

Table 5. Major, minor and trace elements results

Sample	Age	Type	MgCO$_3$ (mole %)	CaCO$_3$ (mole %)	Mn (ppm)	Fe (ppm)	Sr (ppm)
1-15a	Silurian	RD1	43.0	57.0	42.5	678.9	89.3
1-16a	Silurian	RD1	41.6	58.4	62.1	599.1	90.1
1-17a	Silurian	RD1	42.0	58.0	34.2	606.3	70.2
1-18	Silurian	RD1	46.8	53.2	197.6	3458.5	71.9
1-20	Silurian	RD1	40.6	59.4	130.2	2564.8	72.4
1-21	Silurian	RD1	48.7	51.3	67.9	339.3	71.9
1-31a	Silurian	RD1	49.8	50.2	446.7	2984.0	97.7
8-5a	Silurian	RD1	51.7	48.3	208.9	3515.8	94.8
8-6	Silurian	RD1	41.2	58.8	157.3	3276.2	81.2
8-8	Silurian	RD1	45.6	54.4	100.1	617.7	107.9
8-13a	Silurian	RD1	41.8	58.2	552.2	4020.5	80.2
1-16b	Silurian	RD2	43.9	56.1	66.3	809.7	81.2
1-24	Silurian	RD2	37.9	62.1	103.7	6261.1	166.3
1-31b	Silurian	RD2	42.6	57.4	702.5	5183.0	70.7
8-13b	Silurian	RD2	43.5	56.5	748.4	5929.5	71.9
1-28a	Silurian	SD	33.0	67.0	214.6	20,374.0	81.8
8-13c	Silurian	SD	36.5	63.5	1436.5	18,005.3	111.2
1-1a	Devonian	RD1	48.2	51.8	30.1	81.3	46.8
1-3-1a	Devonian	RD1	39.5	60.5	29.1	80.6	58.4
1-5a	Devonian	RD1	39.3	60.7	46.1	171.2	81.9
1-5b	Devonian	RD1	43.4	56.6	40.7	148.4	66.3
1-7a	Devonian	RD1	43.7	56.3	35.7	112.2	72.6
8-1a	Devonian	RD1	39.2	60.8	25.8	90.0	62.0
8-4a	Devonian	RD1	42.7	57.3	54.2	690.5	67.3
1-3-2a	Devonian	RD2	40.4	59.6	32.9	83.0	60.3
1-4a	Devonian	RD2	40.7	59.3	39.8	89.6	93.9
1-7b	Devonian	RD2	41.6	58.4	37.4	117.4	71.7
8-4b	Devonian	RD2	38.4	61.6	59.9	756.1	79.4
1-3-2b	Devonian	RD3	47.2	52.8	40.8	360.9	65.7
1-6a	Devonian	RD3	40.0	60.0	42.5	173.0	89.1
8-4c	Devonian	RD3	52.1	47.9	59.0	724.8	78.9

Minor and Trace Elements: Iron, Manganese and Strontium

The results of trace elements in dolomitic phases are presented in Table 5.

RD1 sampled from Silurian formations show higher contents of Mn, Fe and Sr to Devonian samples. Pervasive replacive fine crystalline dolomite matrix RD1 in Silurian rocks (n = 11) is characterized by Mn contents between 34.2 to 552.2 ppm (average: 181.5 ± 161.5 ppm), Fe between 339.3 to 4020.5 ppm (average: 2060.1 ± 1404.6) and Sr between 70.2 to 107.9 ppm (average: 85.6 ± 11.9), whereas in Devonian rocks (n = 7) Mn ranges between 25.8 to 54.2 ppm (average: 37.4 ± 9.5), Fe between 80.6 to 690.5 ppm (average: 196.3 ± 204.3) and Sr between 46.8 to 81.9 ppm (average: 65 ± 10.2).

Silurian RD2 samples show higher contents of Mn, Fe and Sr compared to Devonian samples. Pervasive replacive medium crystalline dolomite matrix RD2 in Silurian rocks (n = 4) is characterized by Mn contents between 66.3 to 748.4 ppm (average: 405.2 ± 320.9), Fe between 809.7 to 6261.1 ppm (average: 4545.8 ± 2192.1) and Sr between 70.7 to 166.3 ppm (average: 97.5 ± 39.9), whereas in Devonian rocks (n = 4) Mn ranges between 32.9 to 59.9 ppm (average: 42.5 ± 10.3), Fe between 83 to 756.1 ppm (average: 261.5 ± 285.8) and Sr between 60.2 to 93.8 ppm (average: 76.3 ± 12.2). Selective replacive medium crystalline dolomite matrix (RD3) sampled from Devonian formations only (n = 3) and is characterized by Mn contents between 59 to 40.8 ppm (average: 47.5 ± 8.2), Fe between 724.8 to 173 ppm (average: 419.6 ± 229), and Sr between 89.1 to 65.7 ppm (average: 77.9 ± 9.6)

Values of trace elements in saddle dolomite cement (SD) encountered exclusively in Silurian formations (n = 2) is characterized by Mn contents between 214.5 to 1436.5 ppm (average: 825.5 ± 611),

Fe between 18005.3 to 20373.9 ppm (average: 19189.7 ± 1184.3), and Sr between 81.8 to 111.2 ppm (average: 96.5 ± 14.7).

Silurian dolomites have higher Mn, Fe and Sr content than Devonian dolomites and also the progressive increase in Fe and Mn in the successive generations.

In addition to the previously mentioned results, samples (n = 10 and n = 5) from early and late calcite cement were analyzed to establish their trace elements concentrations.

Isopachous calcite cement (ISC) observed only in Silurian formations (n = 2) is characterized by Mn contents between 41.9 to 48.2 ppm (average: 45.1 ± 3.2), Fe between 164.8 to 230.5 ppm (average: 197.7 ± 32.8) and Sr between 177.1 to 244.3 ppm (average: 210.7 ± 33.6).

Syntaxial overgrowth calcite cement (SXC) in Silurian rocks (n = 2) show higher contents of Mn, Fe compared with respective values obtained from Devonian samples, but lower Sr concentrations. It is characterized by Mn contents between 272.1 to 589.23 ppm (average: 430.7 ± 158.6), Fe between 617.7 to 1124.5 ppm (average: 871.1 ± 253.4) and Sr between 230.8 to 244.3 ppm (average: 237.6 ± 6.7), whereas in Devonian rocks (n = 3) Mn ranges between 15.7 to 20.3 ppm (average: 17.9 ± 1.9), Fe between 41 to 51.4 ppm (average: 47.4 ± 4.6) and Sr between 239.1 to 498.7 ppm (average: 333 ± 117.5).

A single measurement in each age group of drusy calcite cement (DC) show values of Mn: 36.5 ppm and 116.2 ppm, Fe: 401 ppm and 217.67 ppm, and Sr: 116.8 ppm and 1425 ppm for the Silurian (n = 1) and Devonian (n = 1) samples, respectively.

Late-stage blocky calcite cement (BKC) sampled from Silurian formations show significantly higher contents of Mn, Fe, and Sr compared to Devonian samples. Trace element values from Silurian samples (n_S = 5) is characterized by contents of Mn between 82.1 and 754.2 ppm (average: 225.9 ± 264.6), Fe between 79.2 to 2097.6 ppm (average: 1036.5 ± 811.4) and Sr between 164.6 to 538.6 ppm (average: 317.3 ± 144), whereas results from the single pure sample available of blocky calcite cement from Devonian formations show values of Mn: 35.9 ppm, Fe: 8.7 ppm and Sr: 643.9 ppm. BKC

4.3.5. Rare Earth Elements (REE)

Selected samples of pervasive replacive fine crystalline dolomite matrix (RD1), pervasive replacive medium crystalline dolomite matrix (RD2), saddle dolomite cement (SD) and blocky calcite cement (BKC) were analyzed for REEs concentrations from Silurian formations. In addition to the selected samples of pervasive replacive fine-crystalline dolomite matrix (RD1), pervasive replacive, medium crystalline dolomite matrix (RD2), selective replacive medium crystalline dolomite matrix (RD3) and blocky calcite cement (BKC) were analyzed for REEs concentrations from Devonian successions (Table 6).

Replacive dolomite RD1 and RD2 sampled from Silurian formations show sub-parallel trends compared to the pattern of Silurian brachiopods shown by Azmy et al. [54] but with substantial differences from Devonian patterns, Silurian dolomite trends are characterized by a minor negative La anomaly and both cases of positive and negative Ce anomalies (Figure 17A,B). Both types of dolomite RD1 and RD2 exhibit higher average ΣREE (11.88 ± 6.25 ppm, and 10.62 ± 8.92 ppm, respectively) compared with the average of Silurian brachiopods (2.21 ppm) shown by Azmy et al. [55]. Saddle dolomite (SD) and blocky calcite cement (BKC) show similar REE shale-normalized trends (Figure 17A,B) with average ΣREE (47.06 ± 41.95 ppm and 28.83 ± 24.47 ppm, respectively) and both significantly differ from the Silurian Brachiopods REE patterns. As shown in Figure 17, SD and BKC present both negative La and Ce anomalies (Figure 17B) as well as a slight negative Eu anomaly (Figure 17A).

All the dolomite and late calcite samples from Devonian formations exhibit shale-normalized patterns comparable to those of modern warm water brachiopods (Figure 18A).

Table 6. Rare earth elements (REE) results.

Sample	Age	Type	La (ppm)	Ce (ppm)	Pr (ppm)	Nd (ppm)	Sm (ppm)	Eu (ppm)	Gd (ppm)	Tb (ppm)	Dy (ppm)	Ho (ppm)	Er (ppm)	Tm (ppm)	Yb (ppm)	Lu (ppm)	ΣREE (ppm)
1-15a	Silurian	RD1	1.51	3.42	0.44	1.65	0.36	0.09	0.31	0.05	0.26	0.05	0.16	0.02	0.12	0.02	8.46
1-16a	Silurian	RD1	3.80	9.17	1.16	4.22	0.84	0.16	0.76	0.11	0.62	0.11	0.33	0.04	0.26	0.04	21.62
1-17a	Silurian	RD1	0.78	1.42	0.15	0.52	0.10	0.02	0.10	0.01	0.06	0.01	0.04	0.00	0.03	0.00	3.25
1-18	Silurian	RD1	1.76	4.02	0.51	2.07	0.43	0.08	0.41	0.05	0.28	0.05	0.12	0.01	0.07	0.01	9.88
1-20	Silurian	RD1	3.97	9.62	1.16	4.58	1.05	0.25	1.06	0.15	0.80	0.15	0.41	0.05	0.27	0.04	23.55
1-21	Silurian	RD1	0.87	1.99	0.23	0.90	0.16	0.05	0.17	0.02	0.13	0.02	0.07	0.01	0.06	0.00	4.68
1-31a	Silurian	RD1	2.06	4.23	0.54	2.04	0.41	0.10	0.42	0.06	0.32	0.06	0.19	0.02	0.14	0.02	10.62
8-5a	Silurian	RD1	2.19	5.21	0.64	2.40	0.47	0.11	0.50	0.06	0.35	0.07	0.18	0.02	0.12	0.02	12.33
8-6	Silurian	RD1	2.75	6.99	0.87	3.46	0.75	0.19	0.82	0.12	0.64	0.12	0.30	0.04	0.20	0.03	17.27
8-8	Silurian	RD1	1.17	2.82	0.35	1.34	0.27	0.06	0.30	0.04	0.22	0.04	0.11	0.01	0.12	0.01	6.86
8-13a	Silurian	RD1	2.42	4.90	0.60	2.19	0.48	0.10	0.46	0.07	0.39	0.08	0.25	0.02	0.16	0.02	12.13
1-16b	Silurian	RD2	3.64	9.45	1.23	4.86	0.88	0.18	0.86	0.12	0.65	0.12	0.36	0.05	0.23	0.03	22.67
1-23b	Silurian	RD2	0.28	0.50	0.05	0.26	0.05	0.01	0.05	0.01	0.04	0.01	0.02	0.00	0.02	0.00	1.30
1-24	Silurian	RD2	0.32	0.69	0.09	0.35	0.06	0.01	0.07	0.01	0.07	0.02	0.04	0.01	0.02	0.01	1.77
1-31b	Silurian	RD2	2.87	6.52	0.80	2.80	0.49	0.12	0.48	0.06	0.37	0.06	0.21	0.03	0.15	0.02	14.98
8-10b	Silurian	RD2	0.29	0.53	0.06	0.23	0.06	0.01	0.06	0.01	0.05	0.01	0.03	0.00	0.02	0.00	1.37
8-13b	Silurian	RD2	3.32	7.82	0.97	3.79	0.73	0.13	0.67	0.08	0.48	0.09	0.24	0.04	0.23	0.03	18.61
1-28a	Silurian	SD	0.36	0.77	0.09	0.39	0.07	0.02	0.04	0.01	0.05	0.01	0.02	0.00	0.02	0.00	1.85
1-31c	Silurian	SD	5.75	16.24	2.14	8.07	1.39	0.23	1.06	0.13	0.68	0.11	0.33	0.04	0.21	0.03	36.41
8-13c	Silurian	SD	19.39	47.98	5.92	20.35	3.42	0.44	2.30	0.27	1.34	0.23	0.69	0.08	0.45	0.06	102.93
1-23a	Silurian	ISC	0.40	0.71	0.08	0.33	0.05	0.01	0.06	0.01	0.05	0.01	0.05	0.00	0.02	0.00	1.80
8-10a	Silurian	ISC	0.20	0.39	0.05	0.20	0.04	0.01	0.04	0.01	0.03	0.01	0.02	0.00	0.02	0.00	1.01
1-32	Silurian	SXC	6.07	10.30	1.44	5.74	1.09	0.24	1.17	0.14	0.76	0.14	0.35	0.04	0.24	0.03	27.75
1-33a	Silurian	SXC	3.86	7.09	0.94	3.92	1.08	0.26	1.42	0.20	1.39	0.29	0.93	0.12	0.80	0.10	22.42
1-17b	Silurian	DC	1.22	2.08	0.23	0.82	0.15	0.04	0.12	0.02	0.10	0.01	0.05	0.01	0.03	0.01	4.88
1-26	Silurian	BKC	2.99	7.48	0.95	4.01	1.08	0.24	1.19	0.19	1.15	0.22	0.60	0.08	0.48	0.06	20.73
1-27	Silurian	BKC	2.07	5.09	0.67	2.76	0.53	0.13	0.56	0.08	0.42	0.07	0.22	0.03	0.15	0.02	12.78
1-28b	Silurian	BKC	0.70	1.24	0.13	0.49	0.10	0.02	0.13	0.02	0.13	0.03	0.09	0.01	0.07	0.01	3.17
1-31d	Silurian	BKC	9.35	26.81	3.53	13.19	2.43	0.34	1.75	0.21	1.03	0.19	0.53	0.06	0.40	0.05	59.87
8-5b	Silurian	BKC	1.72	4.60	0.62	2.28	0.40	0.10	0.42	0.06	0.32	0.08	0.24	0.03	0.20	0.03	11.09
8-13d	Silurian	BKC	14.53	32.22	3.47	10.91	1.60	0.21	1.18	0.12	0.53	0.09	0.24	0.03	0.20	0.03	65.36
1-1a	Devonian	RD1	0.13	0.25	0.03	0.14	0.03	0.01	0.03	0.00	0.02	0.00	0.01	0.00	0.01	0.00	0.69
1-3-1a	Devonian	RD1	0.23	0.42	0.06	0.18	0.05	0.01	0.04	0.01	0.04	0.01	0.02	0.00	0.03	0.00	1.10
1-5a	Devonian	RD1	0.13	0.18	0.02	0.09	0.02	0.00	BDL	0.00	0.03	0.01	0.03	0.00	0.02	0.00	0.54

Table 6. *Cont.*

Sample	Age	Type	La (ppm)	Ce (ppm)	Pr (ppm)	Nd (ppm)	Sm (ppm)	Eu (ppm)	Gd (ppm)	Tb (ppm)	Dy (ppm)	Ho (ppm)	Er (ppm)	Tm (ppm)	Yb (ppm)	Lu (ppm)	ΣREE (ppm)
1-5b	Devonian	RD1	0.11	0.17	0.02	0.07	0.01	0.00	BDL	BDL	0.02	0.00	0.02	0.00	0.02	0.00	0.44
1-7a	Devonian	RD1	0.14	0.20	0.02	0.11	0.03	0.01	BDL	0.00	0.02	0.00	0.01	0.00	0.02	0.00	0.56
8-1a	Devonian	RD1	0.31	0.53	0.07	0.24	0.04	0.01	0.04	0.01	0.05	0.01	0.02	0.00	0.02	0.00	1.35
8-4a	Devonian	RD1	0.76	0.94	0.15	0.61	0.17	0.03	0.18	0.03	0.16	0.04	0.12	0.01	0.07	0.01	3.29
1-3-2a	Devonian	RD2	0.25	0.45	0.06	0.23	0.04	0.01	0.04	0.01	0.05	0.01	0.03	0.00	0.02	0.01	1.22
1-4a	Devonian	RD2	0.33	0.40	0.05	0.15	0.03	0.01	BDL	0.00	0.03	0.00	0.01	0.00	0.02	0.00	1.05
1-7b	Devonian	RD2	0.14	0.20	0.03	0.10	0.03	0.00	BDL	0.01	0.02	0.01	0.01	0.00	0.01	0.00	0.56
8-4b	Devonian	RD2	0.89	1.09	0.18	0.72	0.16	0.03	0.22	0.03	0.19	0.04	0.14	0.02	0.09	0.01	3.80
1-2a	Devonian	RD3	0.43	0.66	0.08	0.29	0.06	0.01	0.07	0.01	0.04	0.01	0.02	0.00	0.02	0.00	1.72
1-3-2b	Devonian	RD3	0.42	0.76	0.11	0.42	0.12	0.02	0.08	0.01	0.08	0.02	0.05	0.01	0.04	0.01	2.15
1-6a	Devonian	RD3	0.20	0.23	0.03	0.13	0.03	0.00	0.03	0.00	0.03	0.01	0.02	0.00	0.01	0.00	0.73
1-10a	Devonian	RD3	1.67	1.53	0.20	0.70	0.13	0.03	0.14	0.02	0.14	0.03	0.10	0.01	0.06	0.01	4.78
1-13b	Devonian	RD3	6.74	6.77	1.32	5.40	1.07	0.25	1.25	0.16	0.96	0.20	0.56	0.06	0.38	0.05	25.17
8-4c	Devonian	RD3	0.91	1.12	0.18	0.75	0.16	0.04	0.23	0.03	0.21	0.05	0.12	0.02	0.11	0.01	3.94
1-10b	Devonian	SXC	0.82	0.74	0.11	0.38	0.04	0.01	0.06	0.01	0.07	0.02	0.06	0.01	0.04	0.00	2.38
1-11a	Devonian	SXC	1.13	1.16	0.16	0.64	0.09	0.02	0.14	0.02	0.12	0.02	0.07	0.01	0.05	0.01	3.63
1-12a	Devonian	SXC	1.20	0.90	0.15	0.61	0.12	0.03	0.21	0.03	0.22	0.05	0.16	0.02	0.11	0.02	3.83
1-13a	Devonian	DC	5.61	5.71	1.07	4.07	0.71	0.14	0.70	0.11	0.60	0.13	0.39	0.05	0.36	0.06	19.70
8-1b	Devonian	BKC	0.88	1.22	0.14	0.48	0.07	0.01	0.07	0.01	0.06	0.02	0.04	0.01	0.05	0.01	3.08

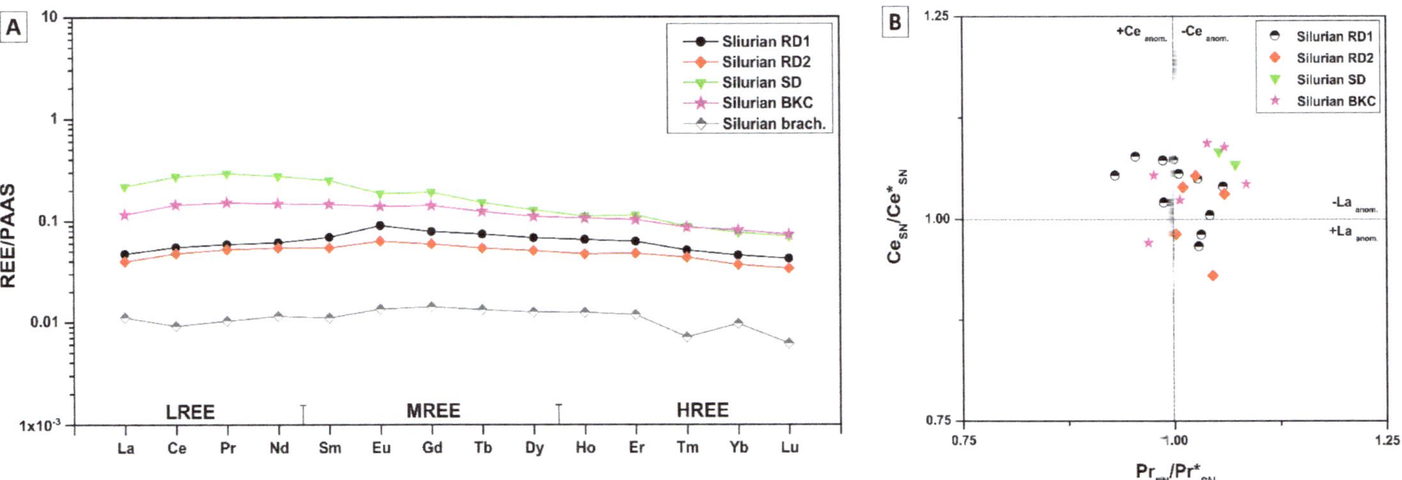

Figure 17. (**A**) Post-Archean Australian Shale (PAAS) normalized rare-earth element (REE) pattern for average values of fine (RD1) and medium (RD2) crystalline dolomite, saddle dolomite (SD) and fracture-filling calcite (BKC) samples, compared with average PAAS normalized REE pattern of well-preserved Silurian brachiopods [54]. (**B**) Ce $(Ce/Ce^*)_{SN}$–La $(Pr/Pr^*)_{SN}$ anomaly crossplot of RD1, RD2, SD and BKC samples.

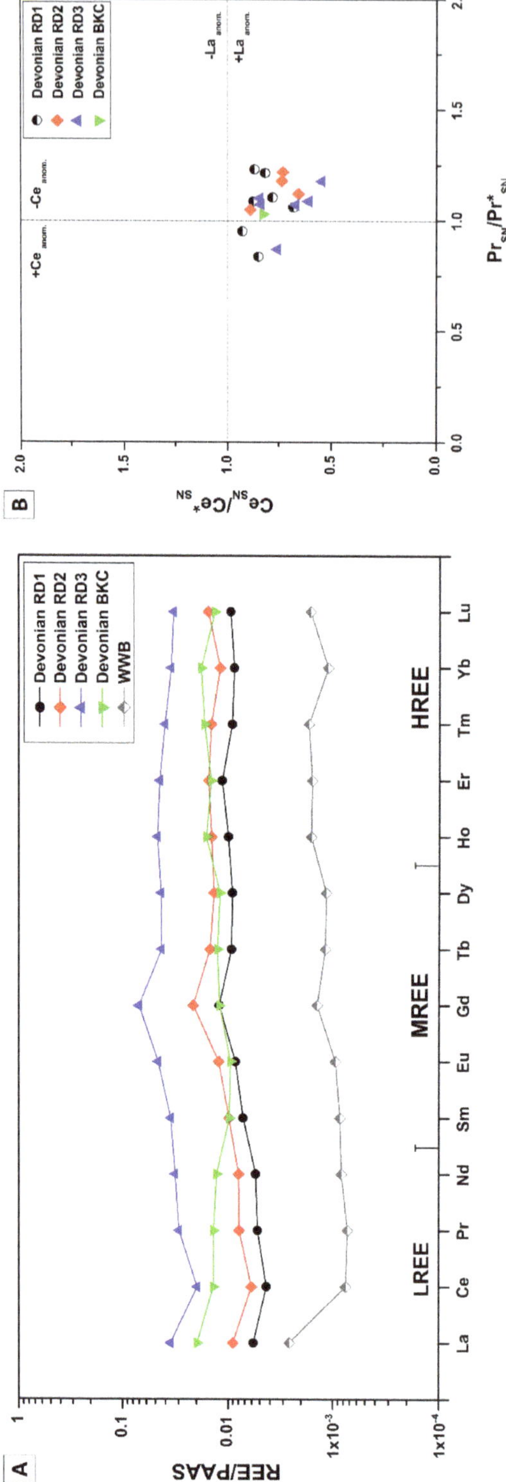

Figure 18. (**A**) PAAS normalized REE pattern for average values of fine (RD1) and medium (RD2) crystalline dolomite, dissolution seams dolomite (RD3) and fracture-filling calcite (BKC) samples, compared with average PAAS normalized REE pattern of modern warm water brachiopods [?]. (**B**) Ce (Ce/Ce*)$_{SN}$–La (Pr/Pr*)$_{SN}$ anomaly cross plot of RD1, RD2, RD3 and BKC samples.

The average shale-normalized REEs concentrations in samples from Devonian formations present all the typical characteristic of seawater patterns including depletion of LREE over HREE (averages: RD1: $(La/Yb)_{SN}$ = 0.66; RD2: $(La/Yb)_{SN}$ = 0.70; RD3: $(La/Yb)_{SN}$ = 1.03), enrichment of MREE (averages: RD1: $(Sm/La)_{SN}$ = 1.25; RD2: $(Sm/La)_{SN}$ = 1.13; RD3: $(Sm/La)_{SN}$ = 1.0), negative Ce anomaly (Figure 18A,B), and slightly positive La and Gd anomalies (Figure 18A). Replacive dolomite RD1, RD2 and RD3 exhibit higher and progressively increasing average ΣREE (1.14 ± 0.93 ppm, 1.66 ± 1.26 ppm and 6.41 ± 8.50 ppm, respectively) compared with the average of modern warm water brachiopods (0.25 ± 0.2 ppm) proposed by Azmy et al. [55]. Slightly different patterns, which are still comparable to the trend of modern warm water brachiopods, characterize REEs of BKC with ΣREE of 3.08 ppm which falls between those of RD1 and RD3 (Figure 18A).

5. Discussion and Interpretations

The petrographic and geochemical results obtained in this study indicate that diagenesis of the Silurian and Devonian dolomitized successions have been subjected to fluids of variable origins and compositions, as described below.

5.1. Constraints from Petrography

5.1.1. Silurian

The diagenetic history of the Silurian formations includes dolomitization, mechanical and chemical compaction, fracturing, dissolution, silicification and calcite and evaporite cementation (Figure 19). The diagenetic processes above listed are described in depth in Tortola [46], on which this paper is largely based.

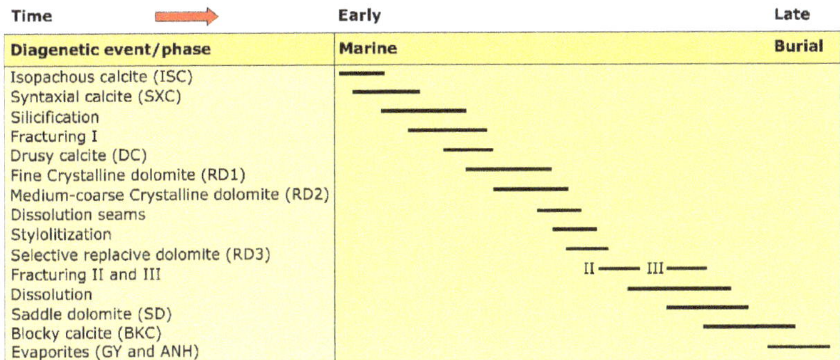

Figure 19. Paragenesis of Silurian formations (modified from Tortola [46]).

Dolomitization

Replacive dolomite crystal size is often influenced by the texture of the precursor limestone [55]. Hence, fine-crystalline dolomite may represent the replacement of fine-grained limestone, whereas coarser replacive dolomite may either represent replacement of coarse-grained limestone or recrystallization of precursor dolomite matrix [55].

RD1 is commonly characterized by closely packed unimodal mosaics of micro to fine-crystalline dolomite which preserved the finely laminated texture of the precursor mudstone. The characteristics showed by RD1, are comparable to fine crystalline dolomite observed in Silurian formations from different areas of southwestern Ontario [5,7,8,56]. This dolomite has been interpreted to be formed from evaporative reflux of hypersaline brines in a sabkha-like setting in a near-surface or shallow burial setting, in the condition of relatively high salinity and low temperature [47,57,58]. This interpretation is also supported by a cross-cutting relationship with chemical compactional features, such as dissolution seams and stylolites, which evidently postdated RD1. However, there is no evidence of supratidal lithofacies and abundant evaporites.

As mentioned by Zheng [7], the relatively coarser size of replacive fine crystalline dolomite (20–50 μm) compared to modern dolomites (<4 μm) can be explained by an increase of crystal size due to early recrystallization. This early recrystallization is possibly supported by the negative shift in oxygen isotopic composition discussed in the next section. Cathodoluminescence of mainly dull to dark red luminescence for RD1 suggests replacement in a reducing environment at shallow burial conditions [59].

Medium replacive dolomite (RD2) shows slightly coarser crystal size but similar texture compared to RD1. Dolomite crystal coarsening can occur in different scenarios: (1) slow precipitation rate [60]; (2) increase of temperature due to progressive increasing burial conditions [47]; (3) recrystallization of precursor dolomite or overgrowth of later stage dolomite on existent cores of earlier dolomite [47,61–64].

Many studies (e.g., [62,64,65]) suggest that non-planar crystal boundaries, coarsening of crystal size and absence of zonation under CL can represent evidence of recrystallization. According to Machel [66], the most accredited definition of recrystallization is the one proposed by the Glossary of Geology [67], which states that recrystallization of a mineral determines the formation of a new mineral commonly characterized by coarser crystals and with either same or different composition. The coarsening of crystal size coupled with the lack of zonation under CL observed in RD2 may indicate recrystallization [64,68,69]. The occurrence of strictly non-planar crystal boundaries was not observed in RD2 from both Silurian and Devonian samples, possibly because early dolomite, formed from seawater can preserve unchanged geochemistry and texture after a certain depth of burial and several million years [70].

According to Machel [66], dolomite can be defined "significantly recrystallized" if it is characterized by the variation of at least one of the following parameters: (1) increase of crystal size or in non-planar crystal boundaries; (2) progressive ordering; (3) change in composition including stable and radiogenic isotopes, stoichiometry, trace and rare earth elements, fluid inclusions and zonation; (4) change in paleomagnetic properties.

The increase of crystal size, the negative shift of $\delta^{18}O$ from postulated values of marine dolomite formed during Silurian and changes in trace and rare earth elements (discussed in the next sections), make RD2 "significantly recrystallized", but also "insignificantly recrystallized" with respect to the increase of non-planar crystal boundaries and progressive ordering [1,5,66]. Crystal coarsening of fabric-destructive pervasive replacive medium (50–100 μm) crystalline dolomite (RD2) may suggest recrystallization from the precursor dolomite RD1, but the presence of cloudy, non-ferroan cores and clear, ferroan rims as a common characteristic of RD2 crystals (Figure 4F) may also suggest overgrowth on earlier dolomite [47]. Under CL, medium dolomite RD2 shows slightly more red luminescence compared to RD1, which indicates that its formation took place in slightly deeper burial conditions in a more reduced environment [59]. As the hypothesis of overgrowth of RD2 on early dolomite is not quite confirmed by CL observation, no brighter luminescent zones have been observed in correspondence of the clear rims, the interpretation of recrystallization seems more likely. Selective replacive medium crystalline dolomite (RD3) occurred in association with dissolution seams during early chemical compaction. Similar textures of patchy dolomite formed in association with chemical compactional features have been previously described (e.g., [71,72]) as dolomite formed from marine or connate waters flowed through dissolution seams which are believed to have behaved as preferential transport path for the dolomitizing fluids. Alternatively, the magnesium required for the formation of RD3 may be supplied by the remobilization of the Mg released in solution from the dissolution of high magnesium calcite [72,73]. The association with dissolution seams suggests that RD3 formed in shallow to intermediate burial realm postdating RD1 and RD2. The common presence of dark cores and clear rims as distinctive characteristics of RD3 crystals (Figure 4C) may also suggest overgrowth on the previous dolomite phases [47]. This evidence, as mentioned for RD2, is not confirmed by cathodoluminescence which shows luminescence comparable with those of RD2 without any zonation or difference in terms of luminescence between cores and rims in RD3 crystals (Figure 4D).

The less abundant saddle dolomite (SD) is observed as fracture and vug-filling cement occluding second and third generation of fracture, postdating RD1, RD2 and RD3 and predating blocky calcite (BKC) and evaporite cement (GY and ANH) (Figure 4E,G–I). This petrographic evidence suggests that saddle dolomite formed during a late diagenetic event possibly in intermediate burial conditions.

Calcite Cementation

Calcite cementation represents an important diagenetic process that affected Silurian carbonate rocks in the Michigan Basin. It started in the early stage, during marine environmental conditions, to continue during progressively deeper burial.

Isopachous calcite (ISC) was only observed in Silurian formations as cement characterized by bladed crystals surrounding coated grains (Figure 6A) formed prior to mechanical compaction and consequently prior to SXC, DC and BKC. It is interpreted to have formed from seawater immediately after deposition in a near-surface marine environment, inhibiting and hence predating mechanical compaction. Under CL this cement is dull-red (Figure 4D), occasionally non-luminescent, indicating precipitation in a slightly reducing environment where Fe^{2+} and Mn^{2+} were available to be incorporated in the crystal lattice [59].

Syntaxial overgrowth calcite (SXC) is mainly cloudy and inclusion-rich, evidence of precipitation in near-surface marine, meteoric or mixing-zone environments [74]. The presence of dissolution seams deflecting around and cross-cutting SXC indicates that its precipitation took place before chemical compaction. Under CL this cement is red to bright red indicating precipitation in a redox environment where Fe^{2+} and Mn^{2+} were available to be incorporated in the crystal lattice [59]. Isotopic values for this cement from both Silurian and Devonian rocks (Figure 11) show overlapping and/or slightly depleted values relative to the postulated ratios for their respective ages; hence evidence for their formation form seawater parentage fluids.

Drusy mosaic calcite (DC), was mainly observed as vug, mould and fracture-filling cement (Figure 6C), which can form in near-surface meteoric as well as in burial environment [75,76] but its formation is not commonly related to the marine environment [62]. Under CL it mainly shows dull-red luminescence (Figure 6D), evidence of precipitation in a reducing condition [59]. Paragenetically, DC postdate silica and predated dissolution seams and stylolites (Figure 5A) as well as late blocky calcite cement (Figure 5B). Hence, its formation represents a relatively early diagenetic event predating chemical compaction.

Late diagenetic blocky calcite cement (BKC) is characterized by coarse crystals (>500 μm) and it is mainly observed as vug and fracture filling cement which possibly formed in a burial environment (e.g., [77]). Petrographically, BKC shows dull to bright orange zoned luminescence under CL indicating formation in a progressively reducing environment. Based on petrographic evidence, BKC is interpreted to be the latest phase of calcite cementation, postdating saddle dolomite cement (SD) (Figure 4I) and predating evaporite cementation.

5.1.2. Devonian

The diagenetic history of Devonian successions includes dolomitization mechanical and chemical compaction, fracturing, dissolution, silicification, and calcite cementation (Figure 20). These sequences are based on petrographic cross-cutting relationships supplemented by geochemical evidence.

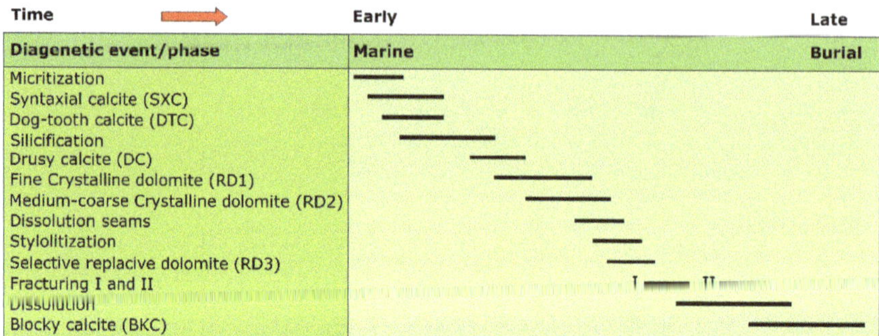

Figure 20. Paragenesis of Devonian Formations (modified from Tortola [46]).

Dolomitization

Devonian replacive dolomites RD1, RD2 and RD3 (Figure 7) show similar characteristics in terms of timing and environments compared to the respective dolomite phases observed in the Silurian formations.

RD1 is commonly characterized by closely packed unimodal mosaics of micro to fine-crystalline dolomite (Figure 7A) which preserved the original texture of the precursor mudstone. Fine-crystalline dolomites distinguished in Devonian formations in previous studies (e.g., [2,5,33,56]) show similar texture if compared to RD1 and are interpreted to be formed from evaporative reflux of hypersaline brines in a sabkha-like setting. Hence, the mimetic pervasive replacive micro to fine crystalline dolomite (RD1) is possibly formed in a low temperature, near-surface or shallow burial environment. However, there is no evidence of abundant evaporites commonly associated with this model. Slightly coarser size of RD1 crystals, in comparison with recent dolomites, suggest early recrystallization [7], which is supported by oxygen isotopic composition discussed in Section 5.2. Cathodoluminescence characteristics of RD1 showing mainly dull to dark red luminescence (Figure 7B) suggest replacement in a shallow reducing environment [53]. Cross-cutting relationship with dissolution seams and stylolites suggests that RD1 formed prior to chemical compaction.

Crystal coarsening of fabric-destructive pervasive replacive medium crystalline dolomite (RD2) (Figure 7C,D) suggests possible recrystallization of the precursor dolomite RD1 [47,61–64]. The lack of zonation under CL coupled with the coarsening of crystal size observed in RD2 may represent additional evidence of recrystallization [64,68]. Devonian medium replacive dolomite shows slightly brighter red luminescence compared to RD1 (Figure 7B), possible evidence of a slightly deeper burial formation in a more reducing environment.

Selective replacive medium crystalline dolomite (RD3) has similar characteristics to RD3 observed in Silurian formations. It is commonly associated with dissolution seams and stylolites, which act as preferential paths for dolomitizing fluids [72]. Hence, it is interpreted to have formed in a shallow to intermediate burial regime postdating RD1 and RD2 (Figure 7E,F) and alongside with chemical compaction. The common presence of dark cores and clear rims as distinctive characteristics of RD3 crystals (Figure 7E) may also indicate overgrowth on the previous dolomite phases [47]. This hypothesis is not confirmed by cathodoluminescence results which shows luminescence slightly brighter with those of RD2 but without any zonation or difference in terms of luminescence between cores and rims within the crystals (Figure 7B,F).

Calcite Cementation

Calcite cementation represents an important process within the paragenesis of Devonian carbonate rocks in the Michigan Basin (Figure 9). It started in the early-stage, during marine environmental conditions, to proceed during progressively deeper burial stages.

Syntaxial overgrowth calcite cement (SXC) is commonly observed in bioclastic grainstone, usually as nucleation over echinoderm plates or crinoid ossicles that act as single crystals (Figure 9A). Syntaxial overgrowth is mainly cloudy and inclusion-rich, evidence of precipitation in near-surface marine, meteoric or mixing-zone environments [74]. The presence of dissolution seams deflecting around and crosscutting SXC indicates that its precipitation took place before chemical compaction. Under CL this cement is dull-red indicating precipitation in a reducing environment where Fe^{2+} and Mn^{2+} were available to be incorporated in the crystal lattice [59].

Dogtooth calcite cement (DTC), which is only present in bioclastic grainstones from Devonian formations (Figure 9B), can form in different environments. According to Reinhold [78] this cement can form: (1) in a marine-phreatic environment when dogtooth calcite occurs in intraskeletal voids; (2) in shallow burial environment when observed in stromatactis cavities succeeding radiaxial fibrous calcite cement. DTC is observed growing normally to the substrate (mainly in intraskeletal pores) with sharply pointed acute crystals. It formed early within the diagenetic history of the Devonian formations, in a near-surface or shallow burial environment, predating the formation of silica and consequently predating RD1. The presence of zoned dull to red luminescence under CL indicates possible formation in a progressively reducing environment [53]. As sampling of a pure specimen for geochemical analyses was not possible, the conclusion concerning time and environment of DTC's occurrence are based exclusively on petrographic observations.

Drusy mosaic calcite (DC), mainly observed as vug, mould and fracture-filling cement (Figure 9D,E), can form in near-surface meteoric as well as in burial environment [75,76]. Its formation is not commonly related to the marine environment [71]. CL observations reveal luminescent zones ranging from non- to orange bright luminescent (Figure 9C) indicating fluctuation from well oxidizing environment, which inhibited the incorporation of Fe^{2+} and Mn^{2+}, to a slightly more reducing condition, respectively [59]. DC postdated silica and predated dissolution seams, stylolites and late blocky calcite cement (Figure 9E). Hence, its formation is thought to represent a relatively early diagenetic event which possibly took place in a shallow burial environment, predating chemical compaction.

Late diagenetic blocky calcite cement (BKC) possibly formed in the burial environment. BKC shows dull to bright zoned luminescence under CL (Figure 9F) indicating formation in a progressively reducing environment. Based on petrographic evidence, BKC is interpreted to be the latest phase of calcite cementation (Figure 20) postdating fractures.

5.2. Constraints from Stable O, C and Radiogenic Sr Isotopes

The conditions of precipitation and/or recrystallization of diagenetic minerals can be predicted using their oxygen isotopic composition, which is a function of the temperature as well as the $\delta^{18}O$ composition of the diagenetic fluids involved [50]. The recrystallization of dolomite, and associated chemical and textural changes, is considered a frequent process in carbonates associated with increasing burial depth [50,64,66,69,79]. Isotopic data show a general overlap between $\delta^{18}O$ and $\delta^{13}C$ values of replacive dolomites (RD2 and RD3) from Silurian and Devonian successions (Figure 10). In both age groups, for these dolomite samples display similar $\delta^{13}C$ compared to those of carbonates formed in equilibrium with seawater of the respective age (for Silurian: 0.5 to 2.5‰ VPDB [80]; and for Middle Devonian: 0.5 to 2.5‰ VPDB [49]) (Figure 10). The $\delta^{13}C$ values possibly reflect the carbon isotopic composition of the precursor limestones (e.g., [81]). However, many $\delta^{13}C$ values in Silurian RD1 departed from the postulated values for Silurian carbonates. In contrast, in both age groups, the $\delta^{18}O$ isotopic composition shows a negative shift from the postulated values for their respective age.

5.2.1. Silurian

Dolomite formed in equilibrium with seawater should display $\delta^{18}O$ isotopic composition enriched of about 3 ± 1‰ (e.g., [82]). Calcitic matrix is characterized by $\delta^{18}O$ values varying from −8.16 to −4.73‰ VPDB and hence with some exceptions, mostly falling into the range of the postulated values for Silurian marine calcite (−3 to −5.5‰ VPDB [80]). Assuming that the Silurian seawater with an isotopic composition of −3.5 ‰ SMOW [83] represents the source of dolomitizing fluids for the formation of RD1, its range of $\delta^{18}O$ values should vary from −0.25 to −2.5‰ VPDB. The calculated average value (−1.4‰ VPDB) would correspond to an estimated value of 24 °C employing the fractionation equation proposed by Land [82]. However, values obtained from RD1 show a negative shift of about 2–6‰ from the $\delta^{18}O$ average value (−1.4‰ VPDB) expected for the Silurian dolomite (Figure 10). RD2, RD3 and SD exhibit a negative shift of about 2–8‰ compared with the $\delta^{18}O$ average value of the Silurian marine dolomite (Figure 10). An explanation to justify this negative shift of $\delta^{18}O$ values can be the interaction with meteoric water or formation in the burial regime at relatively high temperatures (e.g., [69]). The interaction with meteoric water should be demonstrated by evidence of subaerial exposure such as karst phenomena, presence of vadose cement or other features associated with meteoric diagenesis. Although previous studies underline the presence of an erosional unconformities (between Cabot Head and the overlying Fossil Hill formations and between Bass Island and Bois Blanc formations [34]), the lack of evidence of karstification in the studied samples (e.g., [69]), make the meteoric diagenesis an improbable scenario for matrix dolomites. Hence, the hypothesis of recrystallization with burial seems to be more likely.

RD1 shows negative $\delta^{13}C$ values, possibly related to the formation of this dolomite in a shallow burial environment associated with bacterial sulphate reduction (BSR) [84]. This hypothesis is supported by the presence of minute framboidal pyrite crystals (Figure 4A) embedded within RD1 [85].

Even though, there is no evidence of the corresponding variation in terms of crystal size coarsening and depletion of $\delta^{18}O$ values (e.g., [8]) comparing data from RD1 and coarser dolomites (RD2 and RD3), the negative shift of $\delta^{18}O$ values could be associated to recrystallization and/or to interaction with hydrothermal fluids at relatively high temperature (e.g., [5,69,77,86,87]).

Most of the dolomite samples (both, matrix and cement) from Silurian formations are characterized by $^{87}Sr/^{86}Sr$ ratios falling in the range of postulated values for Silurian seawater (0.7078–0.7087 [52,53]) (Figure 12). However, some samples show slightly more radiogenic values than those of coeval seawater. This variation from the postulated values can be related to the recrystallization of early dolomite and to water/rock interaction with feldspar-bearing rocks [5]. In an overall view, the $^{87}Sr/^{86}Sr$ ratios for dolomite matrix (RD1, RD2 and RD3), dolomite cement (SD) and late-stage calcite cement (BKC) suggest for both a seawater-dominated parental fluid.

Most of the calcitic samples show $\delta^{13}C$ values, in range with postulated values for Silurian marine calcite with the exception for late blocky calcite which clearly shows a negative shift from the equilibrium values (Figure 11). This may suggest a separate input of oxidized organic carbon possibly related to interaction with hydrocarbons (e.g., [77]).

5.2.2. Devonian

Calcitic matrix is characterized by $\delta^{18}O$ values varying from −8.16 to −4.73‰ VPDB and hence with some exceptions (Figure 11), mostly falling into the range of the postulated values for Devonian marine calcite (−4 to −6 ‰ VPDB [49]). Most of the calcitic samples show $\delta^{13}C$ values, in range with postulated values for Devonian marine calcite with the exception for late BKC which clearly shows a negative shift from the equilibrium values (Figure 11). This, as already mentioned for Silurian BKC, may suggest a separate input of oxidized organic carbon possibly related to interaction with hydrocarbons (e.g., [77]).

Considering that dolomite formed in equilibrium with seawater, it should display $\delta^{18}O$ isotopic composition enriched of about 3 ± 1‰ (e.g., [82]), and assuming that Devonian seawater, with an isotopic composition of −2‰ SMOW [51,88], represents the source of dolomitizing fluids for the formation of RD1, its range of $\delta^{18}O$ values should vary from −1 to −3‰ VPDB. The calculated average value (−2‰ VPDB) would correspond to an estimated value of 33 °C employing the fractionation equation proposed by Land [82]. However, RD1 shows a negative shift of about 1–5‰ from the average value (−2‰ VPDB) of $\delta^{18}O$ expected for Devonian dolomite (Figure 10). RD2 and RD3 also exhibit a similar negative shift compared with the $\delta^{18}O$ average value of Devonian marine dolomite (Figure 10).

This negative shift of $\delta^{18}O$ values can be also related to the interaction with meteoric water or formation in burial regime at relatively high temperatures (e.g., [69]). As already mentioned in the previous section, the interaction with meteoric water should be demonstrated by evidence of subaerial exposure such as karst features, such as the presence of vadose cement or other evidence associated with meteoric diagenesis. Although there is no evidence of karstification in the samples collected and analyzed in this study, a previous study [89] demonstrated the presence of dissolution processes mostly active in the shallow subsurface, usually <200 m depth in south-western Ontario. At the local scale, approximately 180 m of Devonian bedrock beneath the Bruce nuclear site are karstic with karstification and higher-permeability confined intervals also found within the Salina A1 dolostone and the Guelph Formation [89]. Hence, the possibility of meteoric water influence cannot be discounted completely.

However, the absence of a negative shift of $\delta^{13}C$ values, commonly associated with the meteoric diagenesis (e.g., [69]), coupled with the marked $\delta^{18}O$ isotopic negative values for matrix dolomite may favour the hypothesis of recrystallization.

The majority of the dolomite matrix, as well as the late calcite (BKC) from Devonian formations, are characterized by $^{87}Sr/^{86}Sr$ ratios falling in the range of postulated values for Devonian seawater (0.7078–0.7080 [51]) (Figure 12). However, one sample shows more radiogenic isotopic composition than those of coeval seawater. This variation from the postulated values, even if isolated, can be related to the recrystallization of early dolomite and to water/rock interaction with the overlying Upper Devonian sandstones and shales [56]. Hence, the results $^{87}Sr/^{86}Sr$ ratios for dolomite matrix (RD1, RD2 and RD3), and late-stage calcite cement (BKC) suggest for both seawater-dominated parental fluids.

5.3. Constraints from Fluid Inclusions

As shown in Figure 15 Silurian replacive dolomite (RD2) is characterized by slightly higher homogenization temperature (89.1 °C) compared to those hosted in Devonian RD2 (83.2 °C). These values are in both cases lower than T_h values given by Barnes et al. [90] (Th: 155 °C) for the central part of the Michigan Basin, but comparable T_h values and salinity with those reported by Haeri-Ardakani et al. [5] (T_h: 104 °C; 20.3 wt.% NaCl eq.) for different areas of southwestern Ontario. Fluid inclusions hosted in saddle dolomite in Silurian formation also show slightly higher values of T_h and salinity (T_h: 124 °C; 28.5 wt.% NaCl eq.) compared to matrix dolomite and to saddle dolomite fluid inclusions analyzed in the study carried out by Haeri-Ardakani et al. [5] (T_h: 108 °C; 25.5 wt.% NaCl eq.).

Late calcite cement (BKC) fluid inclusions results show significantly higher T_h values but lower salinity for the Devonian samples compared to those hosted in the Silurian calcite. Fluid inclusions hosted in BKC of the Devonian formation show T_h values significantly higher and lower salinity if compared with values obtained in the central part of the Michigan Basin by Luczaj et al. [13] (T_h: 140 °C; 29.2 wt.% NaCl eq.), whereas it shows similar salinity and higher T_h if compared with the values proposed by Haeri-Ardakani et al. [5] (T_h: 126 °C; 19.7 wt.% NaCl eq.).

These results coupled with the negative shift of $\delta^{18}O$ values may suggest, in both age groups, the formation of RD2 was at relatively high temperatures from saline brines (e.g., [77]). Similar ranges of homogenization temperatures characterize blocky calcite cement and saddle dolomite, in the Silurian formations, and may represent evidence of similarity in terms of mineralizing fluid responsible for their formation. This assumption is also supported by evidence from REE analysis (discussed in the following section). The overall higher salinity (Figures 15B and 16B) which characterizes RD2 and BKC from Silurian formations compared to those of the same mineral phases from Devonian formations, suggests that the diagenetic fluids responsible for their formation were saline basinal brines resulted from the dissolution of evaporites hosted in Silurian formations (e.g., [5,77]). The high homogenization temperatures of late calcite cement and saddle dolomite and the relatively high T_h of fluid inclusions hosted in replacive dolomite (Figures 14A and 15A) are significantly higher than the maximum burial temperatures suggested for the study area (ca 60–90 °C [91]) for both age groups and hence, it may suggest the involvement of hydrothermal fluids [5,13,90].

The distinct patterns of salinity and homogenization temperature (Figures 14–16) that characterize Devonian and Silurian dolomite may suggest that dolomitization/dolomite alteration and other related diagenetic processes are the result of two different diagenetic fluid systems: (1) a diagenetic fluid system that affected Silurian carbonates characterized by high temperatures and high salinity and (2) a diagenetic fluid system that affected Devonian carbonates characterized by slightly lower temperatures and salinity (e.g., [77]).

5.4. Constraints from Geochemistry

5.4.1. Stoichiometry

Most dolomite samples, with some exceptions for RD1, show Ca-rich, non-stoichiometric composition in both Silurian and Devonian formation. Several studies on dolomite [50,68,79,92,93] suggested that the recrystallization of fine crystalline dolomite forms coarser, more stable and more stoichiometric dolomite. However, coarser pervasive dolomites are overall less stoichiometric than fine crystalline dolomite. A possible explanation for such a trend is that these results reflect the average composition of the cores and rims of the crystals. Separate measurements for cores and rims, which were not performed for these dolomites, may have shown very substantial stoichiometric differences. The model proposed by Sibley [92] suggests that progressive recrystallization of non-stoichiometric dolomite should result in an increasingly stoichiometric dolomite characterized by Ca-rich cores and stoichiometric rims. Even though, this phenomenon has been observed and documented in many studies (e.g., [72,94]), stoichiometry has not been achieved in the Silurian and Devonian formations investigated. An experimental study by Malone et al. [95], demonstrated that dolomite subjected to temperatures of 50 °C–200 °C lead to its recrystallization with a rapid increase of stoichiometry in the first part of the experiment, but with a slowing down as reactions progressed. Despite complete recrystallization at 200 °C, dolomite stoichiometry was not achieved.

The interpretation that better explains this trend is that earlier-formed dolomite has been recrystallized by later, Ca-rich, and warmer basinal fluids, under conditions that did not allow these dolomites to achieve stoichiometry (e.g., [69]).

5.4.2. Trace Elements: Iron, Manganese and Strontium

The distribution of iron and manganese in solution is strictly controlled by redox conditions (Eh) [96] as well as their concentrations in precursor limestone and precipitation rate [97]. In the condition of positive Eh (oxidizing conditions), Fe^{2+} and Mn^{2+} oxidize passing from a divalent to a trivalent state inhibiting their incorporation in carbonates crystal lattice [59].

RD1, RD2 and SD in Silurian formations and RD1, RD2 and RD3 in Devonian formations show a progressive increase in Fe and Mn concentration (Table 5). Previous studies on dolomite [65,98] suggested that the relative enrichment of Fe and Mn in later dolomites can be attributed to dolomite recrystallization. Higher content of Fe and Mn in coarser dolomites could be related to either formation under relatively more reducing conditions [99] or increasing Fe and Mn distribution coefficients resulting from higher temperatures or slower precipitation rate [100]. Despite this increase of Fe and Mn there is no evidence of a corresponding variation in terms of Sr concentration between RD1 and younger dolomite phases in both Silurian and Devonian formations. This evidence may suggest that the different phases of dolomite observed within both age groups are perhaps linked to their parental mineralizing fluids (in this case seawater-dominated fluids). The low Sr concentrations (mostly in the range between 50 and 100 ppm for all dolomite samples in both age groups) are comparable to those of ancient dolomite precipitated from marine waters (e.g., 40–150 ppm [101]), but significantly lower than modern protodolomites (500–700 ppm) formed in evaporitic marine settings [79,102]. Decrease of Sr concentration with increasing crystal size is interpreted to be associated with dolomite recrystallization in presence of Sr-depleted meteoric waters (e.g., [50,79,98,103]. However, matrix dolomites from both age groups do not show evidence of a decrease of Sr concentrations with increasing crystal size.

Despite, the lack of evidence of co-variation of Sr with either Fe and Mn, their higher concentrations in coarser dolomite (RD2 and SD in Silurian samples and RD2 and RD3 in Devonian samples) compared to RD1, can be related to their formation from Fe- and Mn-enriched fluids under more reducing conditions [7].

5.4.3. Rare Earth Elements (REE)

Previous studies focused on REE compositions of carbonate rocks suggest controversial scenarios In terms of preservation [104,105] or modification [54], after diagenesis, of rare earth element patterns and concentrations compared with those of their precursor seawater. Concentrations of REE in ancient carbonate rocks can be useful tools to establish the nature of the mineralizing source [106–108]. Chaudhuri and Cullers [109], Webb et al. [104] and Allwood et al. [105] suggested that during the post-depositional processes such as diagenesis and metamorphism REEs patterns can remain essentially unaffected, considering that a large amount of diagenetic fluids is required to record substantial changes [106]. In other words, the REE concentrations in carbonate rocks are principally influenced by the depositional environment [110,111]. However, according to recent studies (e.g., [5,54]) diagenetic processes may sensibly alter REEs patterns and concentrations, making them an ideal tool for the understanding of REEs relocation in carbonate rocks.

The distinctive Post-Archean Australian Shales (PAAS) normalized seawater REE concentrations have patterns characterized by the following features: depletion of light REE, negative Ce anomaly and slight positive La and Gd anomalies [43,112–114].

Silurian

Average shale-normalized REE patterns of RD1 and RD2 show similar trends compared to those of Silurian brachiopods [54] but with significantly higher ΣREE (Figure 17A). All dolomite and calcite samples are characterized by a slight enrichment of LREE over HREE and a more significant enrichment of MREE. The comparison between the trend of Silurian brachiopods [54] and those of RD1 and RD2 (Figure 17A), possibly suggests an initial common seawater-dominated source of diagenetic fluids partially masked afterward by a different evolution of the diagenetic fluids [5]. Differently from Devonian patterns, Silurian dolomite trends are, in fact, characterized by a minor negative La anomaly and both cases of positive and negative Ce anomalies (Figure 17A,B) possibly implying their formation from evolved seawater. Both types of dolomite RD1and RD2 exhibit higher average ΣREE (11.07 ± 0.93 ppm, and 9.79 3 ± 7.8 ppm, respectively) compared with the average of Silurian brachiopods (2.21 ppm) proposed by Azmy et al. [54].

Saddle dolomite (SD) and Blocky calcite cement (BKC) show similar REE shale-normalized trends (Figure 17A) with average ΣREE (47.06 ± 41.95 ppm and 23.74 ± 23.00 ppm, respectively) and both significantly differ from the Silurian Brachiopods REE patterns. SD and BKC both show major negative La and Ce (Figure 17B) anomalies as well as a slight negative Eu anomaly (Figure 17A). The differences between replacive dolomite matrix RD1 and RD2 compared with Saddle dolomite and blocky calcite cement include progressively higher ΣREE and a more pronounced enrichment in LREE, and hence may suggest a different source of REEs (Figure 17).

Devonian

Average shale-normalized REE patterns and their comparison with those of modern warm water brachiopods [54] show that all dolomite types and late calcite cement sampled from Devonian formations have maintained the original seawater REE characteristics with a negative Ce anomaly (Figure 18A,B), and slightly positive La and Gd anomalies [43,112–114]. Proceeding from older (RD1) to younger (RD3) dolomite, shale-normalized REE show similar patterns coupled with a progressive enrichment in average ΣREE (Figure 18). These characteristics may indicate a common seawater-dominated diagenetic fluid slightly modified during progressive deeper burial because of the increasing water/rock interaction [5].

5.5. Evolution of Diagenetic Fluids

The evolution of diagenetic fluids involved during dolomitization in the eastern side of the Michigan Basin can be deduced by integrating petrographic, isotopic and fluid inclusions data.

The relationships between the oxygen isotopic composition of the mineral phase, of the fluid ($\delta^{18}O_{fluid}$) and temperature of formation (T_h from fluid inclusions studies) are shown in Figure 21.

Figure 21A shows the relationship between the measured $\delta^{18}O$ values of dolomite from Silurian and Devonian formations vs. temperature. It also shows the $\delta^{18}O$ values of formation fluids as related to the dolomite-water oxygen fractionation equation [82]. Isotopic compositions of fluid involved in the formation of RD2 in Silurian and Devonian formations show an overlap but Devonian RD2 shows slightly more enriched $\delta^{18}O_{SMOW}$ likely related to the presence of more evolved basinal fluids (e.g., [77]) and a narrow range of temperature compared to the Silurian RD2. On the other hand, Saddle dolomite shows more enriched $\delta^{18}O_{SMOW}$ and higher temperatures than the RD2 matrix dolomite. Considering the most negative $\delta^{18}O$ value from Silurian RD2 (−7.96‰ VPDB) and assuming a Silurian seawater value of −3.5‰ SMOW [83], results in a precipitation temperature of 56 °C, which is significantly lower than the average temperature obtained from fluid inclusion microthermometry (89 °C). Considering the most negative $\delta^{18}O$ value from Devonian RD2 (−7.66‰ VPDB) and assuming a Devonian seawater value of −2‰ SMOW [51], results in a precipitation temperature of 33 °C, which also results significantly lower than the average temperature obtained from fluid inclusion microthermometry (83 °C). Thermal history reconstructions of the Michigan Basin have been carried out by many authors using different methods [4,6,7,13,91,115,116]. Considering the burial curve reconstructed by Zheng [7] the maximum burial temperature expected for Silurian formations would be ca. 63 °C, assuming the atmospheric temperature of 20 °C, a thickness of 1700 m for the overlaying sedimentary cover and a geothermal gradient of 25 °C /km. Recrystallization with the involvement of hydrothermal fluids (e.g., [5]) can be assumed considering the significantly higher (SD: 125 °C; RD2: 89 °C) temperature obtained from fluid inclusions microthermometric analysis. Considering the burial reconstruction proposed by Legall et al. [91] for Devonian formation using conodont and acritarch alteration index, the maximum burial temperature is estimated to be 60 °C, and hence, considerably lower than the temperatures obtained from fluid inclusions microthermometry. It follows that in both age groups, higher values of homogenization temperatures compared to those expected for maximum burial cannot be justified by the burial dolomitization model only [77] but must involve the presence of hydrothermal fluids [5,56].

Figure 21B shows the relationship between the measured $\delta^{18}O$ values of late calcite cement (BKC) from Silurian and Devonian formations vs. temperature. It also shows the $\delta^{18}O$ values of formation fluids as related to the calcite-water oxygen fractionation equation [117]. Isotopic compositions of fluid involved in the formation of BKC in Silurian and Devonian formations show an overlap. Devonian BKC, however, shows slightly more enriched $\delta^{18}O_{SMOW}$ values compared to the Silurian BKC. This trend reflects a slight change in fluid chemistry in the two age groups that can be attributed to more evolved basinal fluids (e.g., [77]).

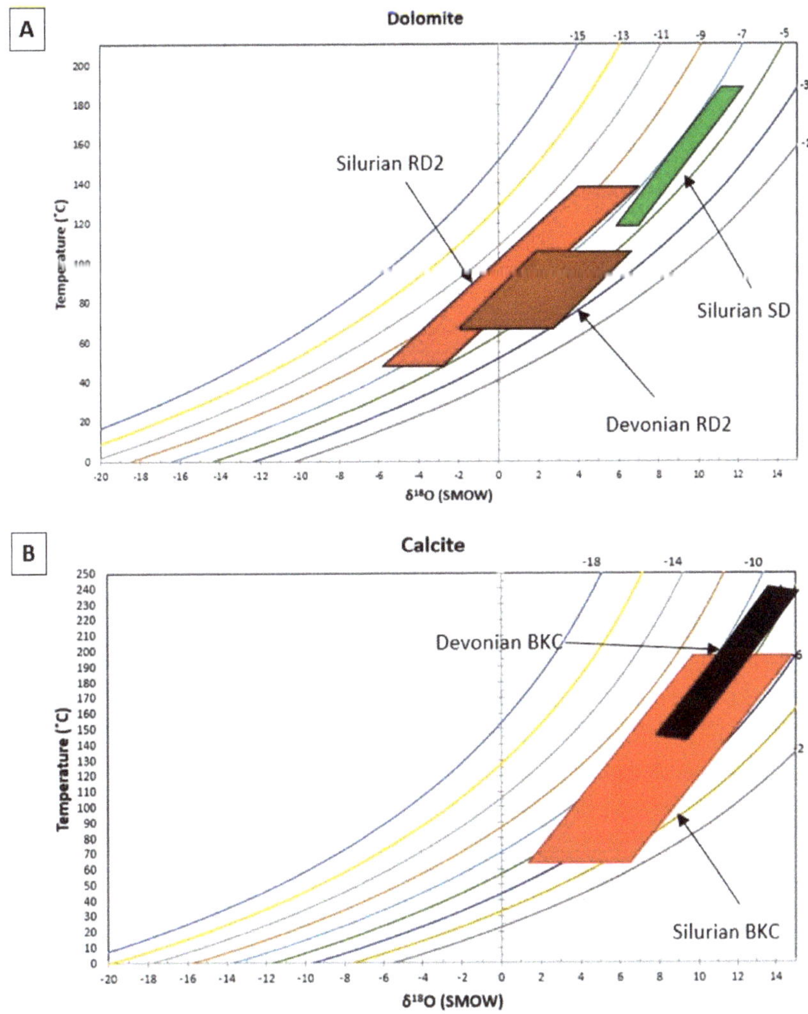

Figure 21. (**A**) Oxygen isotope values of dolomite from Silurian and Devonian formations plotted on the temperature-dependent, dolomite-water oxygen fractionation curve [82]. (**B**) Oxygen isotope values of calcite cement from Silurian and Devonian formations plotted on the temperature-dependent, calcite-water oxygen fractionation curve [117].

The fluid flow in intracratonic basins, such as the Michigan Basin, can be influenced by topography, compaction and convection related to density gradients (e.g., [118]).

The most likely fluid flow model, considering the low topographic relief and the carbonate rocks-dominated characteristic of the Michigan Basin, is the buoyancy-driven model related to temperature and density gradient [5]. Fractures and faults in sedimentary basins characterized by alternating carbonate rocks with variable permeability can promote the migration of cross-formational fluids [119]. In the Michigan Basin, faults and fracture systems in Paleozoic successions are controlled by the reactivation of basement structures and represent preferential pathways that promote fluid migration through low permeability sedimentary rocks such as shales and evaporites [5]. Homogenization temperatures of fluid inclusions hosted in dolomite samples from both Silurian and Devonian formations are significantly higher than temperatures estimated for maximum burial but lower than temperatures of fluid inclusions hosted in dolomites from previous studies carried out in the central part of the Michigan Basin (e.g., [13,91]). This evidence suggests that dolomite formation can be related to the presence of hydrothermal fluids that originated in the central part of the Michigan Basin migrated towards the margins [5]. The potential source of heat could be related to the reactivation of the buried mid-continental rift (MCR) during late Devonian to Mississippian time [120,121], while the preferential path for the migration of diagenetic fluids from the center of the Michigan Basin can be attributed to the Cambrian sandstones unconformably overlying the Precambrian basement [5,77]. Late calcite cement in sub-horizontal fractures near the base of the Silurian sequence dated in a recent study [120] show U–Pb ages of 318 ± 10 Ma determined by LA ICP-MS and 313 ± 1 Ma by ID-TIMS. Hence, the authors suggest that the fluid flow of hydrothermal brines was influenced by the Alleghanian orogenies [120]. According to Sutcliff et al. [122] absolute dating of Cambrian and Silurian secondary minerals shows the same age and are contemporaneous with Alleghanian magmatism and uplift which caused the migration of hydrothermal fluids westwards from high-level Alleghanian plutons through the basal Cambrian.

6. Conclusions

Examination of cores, petrographic studies, microthermometric and geochemical analyses of Silurian and Devonian carbonates of the eastern side of the Michigan Basin (Huron Domain) led to the following conclusions regarding dolomitization and other related diagenetic processes observed in these successions:

- Silurian and Devonian formations are characterized by the presence of three types of replacive dolomite matrix RD1 (formed in shallow burial conditions), RD2 and RD3 (formed in intermediate burial conditions). In addition to these types, a coarse crystalline ferroan saddle dolomite cement (formed in intermediate burial conditions) filling fractures and vugs is observed only in the Silurian rocks. Early- and late-stage calcite cement have been distinguished in both groups of formations including isopachous, syntaxial overgrowth, dogtooth, drusy and blocky calcite.
- Isotopic data show an overlap between $\delta^{13}C$ and $\delta^{18}O$ values in Silurian and Devonian dolomites. In both Devonian and Silurian, many samples $\delta^{13}C$ values fall in the range of values estimated for the marine dolomite of equivalent age. The negative shift in $\delta^{18}O$ values, however, show evidence of dolomite recrystallization during burial and increasing temperature. Several other lines of evidence for recrystallization have been observed such as increasing crystal size, zonation, etched surfaces between cores and rims of crystals, and an increase in radiogenic strontium ratios.
- Silurian and Devonian Sr isotopic ratios show the seawater composition of their respective age as the primary source of diagenetic fluids with minor rock-water interactions. However, Silurian samples show a more radiogenic signature that documents more intense interactions of water with feldspar-bearing rocks
- REE shale-normalized patterns suggest that in both age groups the diagenetic fluids were originally of coeval seawater composition subsequently modified via water-rock interaction. The different evolution of the diagenetic fluids is more prominent in REE_{SN} patterns from Silurian samples and it is possibly related to brines, which were modified by the dissolution of Silurian evaporites from the Salina series.
- Homogenization temperatures characterizing fluid inclusions hosted in RD2 from the Devonian samples, and RD2 and saddle dolomite in the Silurian successions exceed the temperature estimated for maximum burial. Hence, it may suggest the involvement of hydrothermal fluids during their formation.
- The potential source of heat could be related to the reactivation of the buried mid-continental rift (MCR), whereas the preferential path for the migration of diagenetic fluids from the central part of the Michigan Basin towards the margins via faults and fractures can be attributed to the Cambrian

sandstones unconformably overlying the Precambrian basement that could have behaved as regional aquifers.

- Geochemical and fluid inclusion investigations denote two distinct fluid systems related to the diagenetic and tectonic history of Michigan Basin.

Author Contributions: Conceptualization, M.T.; writing—original draft preparation, M.T.; supervision, I.S.A.-A.; writing—review and editing, I.S.A-A.; review and editing, R.C. All authors have read and agreed to the published version of the manuscript.

Funding: Funding for this research was provided by NWMO to I.S.A.-A.

Acknowledgments: M.T. and I.S.A.-A. acknowledge the help from M. Price during isotope analysis. The critical review by M. Hobbs is greatly appreciated. The anonymous reviewers are also acknowledged for their improvements to the paper.

Conflicts of Interest: The authors declare no conflict of interest.

References

1. Cercone, K.R.; Lohmann, K.C. Late burial diagenesis of Niagaran (Middle Silurian) pinnacle reefs in Michigan Basin. *Am. Assoc. Geol. Bull.* **1987**, *71*, 156–166.

2. Hamilton, G.D. Styles of reservoir development in middle Devonian carbonates of southwestern Ontario. In *Diagenetic History of Ordovician and Devonian Oil and Gas Reservoirs in Southwestern Ontario*; Open File Report 5822; Coniglio, M., Frape, S.K., Eds.; Ontario Geological Survey: Sudbury, ON, Canada, 1991.

3. Hamilton, D.G.; Coniglio, M. Reservoir development in the middle Devonian of southwestern Ontario. In Proceedings of the Ontario Petroleum Institute Thirty-First Annual Conference, Niagara Falls, ON, Canada, 28–30 October 1992; Technical Paper No. 9. Volume 32, p. 20.

4. Coniglio, M.; Sherlock, R.; Williams-Jones, A.E.; Middleton, K.; Frape, S.K. Burial and hydrothermal diagenesis of Ordovician carbonates from the Michigan Basin, Ontario, Canada. *Int. Assoc. Sedimentol.* **1994**, *21*, 231–254.

5. Haeri-Ardakani, O.; Al-Aasm, I.S.; Coniglio, M. Fracture mineralization and fluid flow evolution: An example from Ordovician-Devonian carbonates, southwestern Ontario, Canada. *Geofluids* **2013**, *13*, 1–20. [CrossRef]

6. Cercone, K.R. Evaporative sea-level drawdown in the Silurian Michigan Basin. *Geology* **1988**, *16*, 109–130. [CrossRef]

7. Zheng, Q. Carbonate Diagenesis and Porosity Evolution in the Guelph Formation, Southwestern Ontario. Ph.D. Thesis, University of Waterloo, Waterloo, ON, Canada, 1999.

8. Coniglio, M.; Zheng, Q.; Carter, T.R. Dolomitization and recrystallization of middle Silurian reefs and platformal carbonates of the Guelph formation, Michigan basin, Southwestern Ontario. *Bull. Can. Pet. Geol.* **2003**, *51*, 177–199. [CrossRef]

9. Coniglio, M.; William-Jones, A.E. Diagenesis of Ordovician carbonates from the northeast Michigan Basin, Manitoulin island area, Ontario: Evidence from petrography, stable isotopes and fluid inclusions. *Sedimentology* **1992**, *39*, 813–836. [CrossRef]

10. Sanford, B.V.; Thompson, F.J.; McFall, G.H. Plate tectonics—A possible controlling mechanism in the development of hydrocarbon traps in southwestern Ontario. *Bull. Can. Pet. Geol.* **1985**, *33*, 52–71.

11. Carter, T.R.; Trevail, R.A.; Easton, R.M. Basement controls on some hydrocarbon traps in southern Ontario, Canada. In *Basement and Basins of Eastern North America*; Van der Pluijm, B.A., Catacosinos, P.A., Eds.; Geological Society of America: Boulder, CO, USA, 1996; Volume 308, pp. 95–107.

12. Boyce, J.J.; Morris, W.A. Basement-controlled faulting of Paleozoic strata in southern Ontario, Canada: New evidence from geophysical lineament mapping. *Tectonophysics* **2002**, *353*, 151–171. [CrossRef]

13. Luczaj, J.A. Evidence against the Dorag (mixing-zone) model for dolomitization along the Wisconsin arch-A case for hydrothermal diagenesis. *Am. Assoc. Pet. Geol. Bull.* **2006**, *90*, 1719–1738. [CrossRef]

14. Carter, T.R.; Easton, R.M. Extension of Grenville basement beneath southwestern Ontario: lithology and tectonic subdivisions. In *Subsurface Geology of Southwestern Ontario: A Core Workshop*; Ontario Petroleum Institute: London, ON, Canada, 1990.

15. Armstrong, D.K.; Carter, T.R. An updated guide to the subsurface Paleozoic stratigraphy of Southern Ontario. *Ont. Geol. Surv.* **2006**, *6191*, 214.

16. Ontario Geological Survey. *Bedrock geology of Ontario, southern sheet, Map 2544, scale 1:1 000 000*; Ontario Geological Survey: Sudbury, ON, Canada, 1991.

17. Ontario Power Generation's (OPG). *Deep Geologic Repository for Low and Intermediate Level Waste: Preliminary Safety Report*; OPG: Thunder Bay, ON, Canada, 2011; Program Document n 00216-SR-01320-00001.

18. Johnson, M.D.; Armstrong, D.K.; Sanford, V.V.; Telford, P.G.; Rutka, M.A. Paleozoic and Mesozoic geology of Ontario, in geology of Ontario. *Ont. Geol. Surv.* **1992**, *4*, 907–1008.

19. De Klein, G.V.; Hsui, A.T. Origin of cratonic basins. *Geology* **1987**, *15*, 1094–1098. [CrossRef]

20. Sleep, N.H. Thermal effects of the formation of Atlantic continental margins by continental break up. *Geophys. J. R. Astron. Soc.* **1971**, *24*, 325–350. [CrossRef]

21. Sleep, N.H.; Snell, N.S. Thermal contraction and flexure of mid-continent and Atlantic marginal basins. *Geophys. J. R. Astron. Soc.* **1976**, *45*, 125–154. [CrossRef]

22. Nunn, J.A.; Sleep, N.H.; Moore, W.E. Thermal subsidence and generation of hydrocarbon in Michigan Basin. *AAPG Bull.* **1984**, *68*, 296–315.

23. Howell, P.D.; van der Pluijm, B.A. Early history of the Michigan basin: Subsidence and Appalachian tectonics. *Geology* **1990**, *18*, 1195–1198. [CrossRef]

24. Howell, P.D.; van der Pluijm, B.A. Structural sequences and styles of subsidence in the Michigan basin. *Geol. Soc. Am. Bull.* **1999**, *111*, 974–991. [CrossRef]

25. AECOM. *Regional Geology—Southern Ontario; Report for the Nuclear Waste Management Organization NWMO DGR-TR-2011-15 R000*; AECOM Canada Ltd.: Calgary, AB, Canada; Itasca Consulting Canada Inc.: Sudbury, ON, Canada, 2011.

26. Coakley, B.J.; Nadon, C.; Wang, H.F. Spatial variations in tectonic subsidence during Tippecanoe I in the Michigan Basin. *Basin Res.* **1994**, *6*, 131–140. [CrossRef]

27. Coakley, B.; Gurnis, M. Far-field tilting of Laurentia during the Ordovician and constraints on the evolution of a slab under an ancient continent. *J. Geophys. Res. Solid Earth* **1995**, *100*, 6313–6327. [CrossRef]

28. Beaumont, C.; Quinlan, G.; Hamilton, J. Orogeny and stratigraphy: Numerical models of the Paleozoic in the eastern interior of North America. *Tectonics* **1988**, *7*, 389–416. [CrossRef]

29. Bethke, C.M.; Marshak, S. Brine migrations across North America—The plate tectonics of groundwater. *Annu. Rev. Earth Planet. Sci.* **1990**, *18*, 287–315. [CrossRef]

30. Wang, H.F.; Crowley, K.D.; Nadon, G.C. Thermal history of the Michigan basin from apatite fission-track analysis and vitrinite reflectance. In *Basin Compartments and Seals*; Ortoleva, P.J., Ed.; AAPG Memoir 61; American Association of Petroleum Geologists: Tusla, OK, USA, 1994.

31. Van der Voo, R. Paleozoic paleogeography of North America, Gondwana, and intervening displaced terranes: Comparison of paleomagnetism with paleoclimatology and biogeographical patterns. *Geol. Soc. Am. Bull.* **1988**, *100*, 311–324. [CrossRef]

32. Sonnenfeld, P.; Al-Aasm, I.S. The Salina evaporites in the Michigan basin. In *Early Sedimentary Evolution of the Michigan Basin*; Catacosinos, P.A., Daniels, P.A., Jr., Eds.; Geological Society of America: Boulder, CO, USA, 1991; Volume 256, pp. 139–153.

33. Birchard, M.C.; Rutka, M.A.; Brunton, F.R. Lithofacies and geochemistry of the Lucas Formation in the subsurface of Southwestern Ontario: A high purity limestone and Potential high purity dolostone resource. *Ont. Geol. Surv.* **2004**, *6137*, 57.

34. Armstrong, D.K.; Goodman, W.R. *Stratigraphy and Depositional Environments of Niagaran Carbonates, Bruce Peninsula, Ontario*; Field Trip No. 4 Guidebook; American Association of Petroleum Geologists, Eastern Section Meeting; Ontario Petroleum Institute: London, ON, Canada, 1990.

35. Dickson, J.A. Carbonate identification and genesis as revealed by staining. *J. Sediment. Res.* **1966**, *36*, 491–505.

36. Evamy, B.D. The application of a chemical staining technique to a study of dolomitization. *J. Int. Assoc. Sedimentol.* **1963**, *2*, 164–170. [CrossRef]

37. Al-Aasm, I.S.; Taylor, B.E.; South, B. Stable isotope analysis of multiple carbonate samples using selective acid extraction. *Chem. Geol.* **1990**, *80*, 119–125. [CrossRef]

38. Hall, D.L.; Sterner, S.M.; Bodnar, R.J. Freezing point depression of NaCl-KCl-H_2O solutions. *Econ. Geol.* **1988**, *83*, 197–202. [CrossRef]

39. Bodnar, R.J. Revised equation and table for freezing point depressions of H_2O-salt solutions. *Econ. Geol.* **1992**, *83*, 197–202.

40. Roedder, E. Fluid inclusions. *Mineral. Soc. Am. Rev. Mineral. Sediment. Petrol.* **1984**, *57*, 967–975.

41. Lawler, J.P.; Crawford, M.L. Stretching of fluid inclusions resulting from a low temperature microthermometric technique. *Econ. Geol.* **1983**, *78*, 527–529. [CrossRef]

42. Pourmand, A.; Dauphas, N.; Ireland, T.J. A novel extraction chromatography and MC-ICP-MS technique for rapid analysis of REE, Sc and Y: Revising CI-chondrite and Post-Archean Australian Shale (PAAS) abundances. *Chem. Geol.* **2012**, *291*, 38–54. [CrossRef]

43. Bau, M.; Dulski, P. Distribution of yttrium and rare-earth elements in the Penge and Kuruman iron-formations, Transvaal Supergroup, South Africa. *Precambr. Res.* **1996**, *79*, 37–55. [CrossRef]

44. Kucera, J.; Cempirek, J.; Dolnicek, Z.; Muchez, P.; Prochaska, W. Rare earth elements and yttrium geochemistry of dolomite from post-Variscan vein-type mineralization of the Nizky Jesenik and Upper Silesian Basins, Czech Republic. *J. Geochem. Explor.* **2009**, *103*, 69–79. [CrossRef]

45. Felitsyn, S.; Morad, S. REE patterns in latest Neoproterozoic–early Cambrian phosphate concretions and associated organic matter. *Geochem. Geol.* **2002**, *187*, 257–265. [CrossRef]

46. Tortola, M. Petrographic and Geochemical Attributes of Silurian and Devonian Dolomitized Formations in the Huron Domain, Michigan Basin. Electronic Theses and Dissertation. 2019. Available online: https://scholar.uwindsor.ca/etd/7740 (accessed on 13 September 2019).

47. Sibley, D.F.; Gregg, J.M. Classification of dolomite rock textures. *J. Sediment. Res.* **1987**, *57*, 967–975.

48. Middleton, K.; Coniglio, M.; Sherlock, R.; Frape, S.K. Dolomitization of Middle Ordovician carbonate reservoirs, southwestern Ontario. *Bull. Can. Pet. Geol.* **1993**, *41*, 150–163.

49. Hurley, N.F.; Lohmann, K.C. Diagenesis of Devonian reefal carbonates in the Oscar Range, Canning Basin, Western Australia. *J. Sediment. Res.* **1989**, *59*, 127–146.

50. Land, L.S. The origin of massive dolomite. *J. Geol. Educ.* **1985**, *33*, 112–125. [CrossRef]

51. Veizer, J.; Ala, D.; Azmy, K.; Bruckschen, P.; Buhl, D.; Bruhn, F.; Carden, G.A.F.; Diener, A.; Ebneth, S.; Godderis, Y.; et al. 87Sr/86Sr and δ13C and δ18O evolution of Phanerozoic seawater. *Chem. Geol.* **1999**, *161*, 59–88. [CrossRef]

52. Qing, H.; Barnes, C.R.; Buhul, D.; Veizer, J. The strontium isotopic composition of Ordovician and Silurian brachiopods and conodont: Relationships to geological events and implications for coeval seawater. *Geochim. Cosmochim. Acta* **1998**, *62*, 1721–1733. [CrossRef]

53. Boggs, S. *Petrology of Sedimentary Rocks*; Macmillan Publishing Company: New York, NY, USA, 1992; p. 707.

54. Azmy, K.; Brand, U.; Sylvester, P.; Gleeson, S.A.; Logan, A.; Bitner, M.A. Biogenic and abiogenic low-Mg calcite (bLMC and aLMC): Evaluation of seawater-REE composition, water masses and carbonate diagenesis. *Chem. Geol.* **2011**, *280*, 180–190. [CrossRef]

55. Gao, G.; Land, L.S. Early ordovician cool creel dolomite, middle Arbuckle group, silick Hills, SW Oklahoma, USA: Origin and modification. *J. Sediment. Res.* **1991**, *61*, 161–173.

56. Haeri-Ardakani, O.; Al-Aasm, I.S.; Coniglio, M. Petrologic and geochemical attributes of fracture-related dolomitization in Ordovician carbonates and their spatial distribution in southwestern Ontario, Canada. *Mar. Pet. Geol.* **2013**, *3*, 409–422. [CrossRef]

57. Gregg, J.M.; Sibley, D.F. Epigenetic dolomitization and the origin of xenotopic dolomite texture. *J. Sediment. Res.* **1984**, *54*, 908–931.

58. Gregg, J.M.; Shelton, K.L. Dolomitization and dolomite Neomorphism in the back-reef facies of the Bonneterre and Davis formations (Cambrian), southwestern Missouri. *J. Sediment. Res.* **1990**, *60*, 549–562.

59. Hiatt, E.E.; Pufahl, P.K. Cathodoluminescence petrography of carbonate rocks: Application to understanding diagenesis, reservoir quality, and pore system evolution. In *Cathodoluminescence and Its Application to Geoscience*; Short Course Series; Coulson, I.M., Ed.; Mineralogical Association of Canada: Québec, QC, Canada, 2014; Volume 45, pp. 75–96.

60. Dawans, J.M.; Swart, P.K. Textural and geochemical alteration in the late Cenozoic Bahamian dolomites. *Sedimentology* **1988**, *35*, 385–403. [CrossRef]

61. Land, L.S. Dolomitization of the Hope Gate Formation (north Jamaica) by seawater: Reassessment of mixing-zone dolomite. In *Stable Isotope Geochemistry: A Tribute to Samuel*; Geochem. Society: San Antonio, TX, USA, 1991.

62. Kupecz, J.A.; Land, L.S. Progressive recrystallization and stabilization of early-stage dolomite: Lower Ordovician Ellenburger Group, west Texas. In *Dolomites. A Volume in Honor of Dolomieu*; Purser, B., Tucker, M., Zenger, D., Eds.; International Association of Sedimentologists: Gent, Belgium, 1994; Volume 21, pp. 155–166.

63. Sibley, D.F.; Gregg, G.M.; Brown, R.G.; Laudon, P.R. Dolomite crystal size distribution. In *Carbonate Microfabrics, Frontiers in Sedimentology*; Rezak, R., Lavoie, D.L., Eds.; Springer: New York, NY, USA, 1993; pp. 195–204.

64. Machel, H.G. Concepts and models of dolomitization: A critical reappraisal. In *The Geometry and Petrogenesis of Dolomite Hydrocarbon Reservoirs*; Braithwaite, C.J.R., Rizzi, G., Drake, G., Eds.; Geological Society Special Publication: London, UK, 2004; Volume 245, pp. 7–63.

65. Montanez, I.P.; Read, J.F. Fluid-rock interaction history during stabilization of early dolomites, Upper Knox Group (Lower Ordovician), U.S. Appalachian. *J. Sediment. Res.* **1992**, *62*, 753–778.

66. Machel, H.G. Recrystallization versus neomorphism, and the concept of "significant recrystallization" in dolomite research. *Sediment. Geol.* **1997**, *113*, 161–168. [CrossRef]

67. American Geological Institute (AGI). *Glossary of Geology*; AGI: Alexandria, VA, USA, 1987; p. 788.

68. Mazzullo, S.J. Geochemical and neomorphic alteration of dolomite: A review. *Carbonate Evaporites* **1992**, *7*, 21–37. [CrossRef]

69. Al-Aasm, I.S. Chemical and isotopic constraints for recrystallization of sedimentary dolomites from the Western Canada Sedimentary basin. *Aquat. Geochem.* **2000**, *6*, 229–250. [CrossRef]

70. Al-Aasm, I.S.; Packard, J.J. Stabilization of early-formed dolomite: A tale of divergence from two Mississippian dolomites. *Sediment. Geol.* **2000**, *131*, 97–108. [CrossRef]

71. Adam, J.; Al-Aasm, I.S. Petrologic and geochemical attributes of calcite cementation, dolomitization and dolomite recrystallization: An example from the Mississippian Pekisko Formation, west-central Alberta. *Bull. Can. Pet. Geol.* **2017**, *65*, 235–261. [CrossRef]

72. Al-Aasm, I.S.; Lu, F. Multistage dolomitization of the Mississippian turner valley formation, quirk Creek field, alberta: Chemical and petrologic evidence. *Can. Soc. Pet. Geol.* **1994**, *17*, 657–675.

73. Qing, H.; Mountjoy, E.W. Rare element geochemistry of dolomite in the Middle Devonian Presqu'ile barrier, Western Canada Sedimentary Basin: Implication fluid-rock ratios during dolomitization. *Sedimentology* **1994**, *41*, 787–804. [CrossRef]

74. Walker, K.R.; Jernigan, D.G.; Weber, L.J. Petrographic criteria for the recognition of marine, syntaxial overgrowths, and their distribution in geologic time. *Carbonates Evaporites* **1990**, *5*, 141–161. [CrossRef]

75. Choquette, P.W.; James, N.P. Diagenesis in limestones. The deep burial environment. *Geosci. Can.* **1987**, *14*, 3–35.

76. Flügel, E. *Microfacies of Carbonate Rocks, Analysis, Interpretation and Application*; Springer: Berlin, Germany, 2010; p. 976.

77. Al-Aasm, I.S.; Crowe, R. Fluid compartmentalization and dolomitization in the Cambrian and Ordovician successions of the Huron Domain, Michigan Basin. *Mar. Pet. Geol.* **2018**, *92*, 160–178. [CrossRef]

78. Reinhold, C. Dog-tooth cements; indicators of different diagenetic environments. *Zbl. Geol. Paläont.* **1999**, *10*, 1221–1235.

79. Land, L.S. The isotopic and trace element geochemistry of dolomite: The state of the art. In *Concepts and Models of Dolomitization*; Zenger, D.H., Dunham, J.B., Ethington, R.L., Eds.; SEPM Special Publication: Broken Arrow, UK, USA, 1980; Volume 28, pp. 87–110.

80. Allan, J.R.; Wiggins, W.D. *Dolomite Reservoirs: Geochemical Techniques for Evaluating Origin and Distribution*; Continuing Education Course Note Series #36; American Association of Petroleum Geologists: Tulsa, OK, USA, 1993.

81. Al-Aasm, I.S. Origin and characterization of hydrothermal dolomite in the Western Canada Sedimentary Basin: Jour. *Geochem. Explor.* **2003**, *78–79*, 9–15. [CrossRef]

82. Land, L.S. The application of stable isotopes to studies of the origin of dolomite and to problems of diagenesis of clastic sediments. In *Stable Isotopes in Sedimentary Geology*; Society of Economic Paleontologists and Mineralogists: New York, NY, USA, 1983; p. 129.

83. Azmy, K.; Veizer, J.; Bassett, M.G.; Copper, P. Oxygen and carbon isotopic composition of Silurian brachiopods: Implications for coeval seawater and glaciations. *Geol. Soc. Am. Bull.* **1998**, *110*, 1499–1512. [CrossRef]

84. Baldermann, A.; Deditius, A.P.; Dietzel, M.; Fichtner, V.; Fischer, C.; Hippler, D.A.; Leis, A.; Baldermann, C.; Mavromatis, V.; Stickler, C.P.; et al. The role of bacterial sulfate reduction during dolomite precipitation: Implications from Upper Jurassic platform carbonates. *Chem. Geol.* **2015**, *412*, 1–14. [CrossRef]

85. Warren, J. Dolomite: Occurrence, evolution and economically important associations. *Earth Sci. Rev.* **2000**, *1–3*, 1–81. [CrossRef]

86. Dorobek, S.L.; Smith, T.M.; Whitsit, P.M. Microfabrics and geochemistry of meteorically altered dolomite in Devonian and Mississippian carbonates, Montana and Idaho. In *Carbonate Microfabrics*; Rezak, R., Lavoie, D.L., Eds.; Springer: Berlin, Germany, 1993; Volume 16, pp. 205–224.

87. Durocher, S.; Al-Aasm, I.S. Dolomitization and neomorphism of Mississippian (Visean) upper Debolt Formation, Blueberry Field, NE British Columbia, Canada: Geologic, petrologic and chemical evidence. *Am. Assoc. Pet. Geol. Bull.* **1997**, *81*, 954–977.

88. Geldern, R.A.; Joachimski, M.M.; Day, J.; Jansen, U.; Alvarez, F.; Yolkin, E.A.; Ma, X.P. Carbon, oxygen and strontium isotope records of Devonian brachiopod shell calcite. *Palaeogeogr. Palaeoclimatol. Palaeoecol.* **2006**, *240*, 47–67. [CrossRef]

89. Worthington, S.R.H. *Karst Assessment*; NWMO DGR-TR-2011-22; Canadian Nuclear Safety Commission: Toronto, ON, Canada, 2011.

90. Barnes, D.A.; Parris, T.M.; Grammer, G.M. *Hydrothermal Dolomitization of Fluid Reservoirs in the Michigan Basin, USA*; Search and Discovery Article #50087; AAPG/Datapages, Inc.: San Antonio, TX, USA, 2008.

91. Legall, F.D.; Barnes, C.R.; Macqueen, R.W. Thermal maturation, burial history and hotspot development, Paleozoic strata of southern Ontario-Quebec, from conodont and acritach color alteration index. *Bull. Can. Pet. Geol.* **1981**, *29*, 492–539.

92. Sibley, D.F. Unstable to stable transformations during dolomitization. *J. Geol.* **1990**, *98*, 967–975. [CrossRef]

93. Gregg, J.M.; Howard, S.A.; Mazzullo, S.J. Early diagenetic recrystallization of Holocene (<3000 years old) peritidal dolomites, Ambergris Cay, Belize. *Sedimentology* **1992**, *39*, 143–160. [CrossRef]

94. White, T.; Al-Aasm, I.S. Hydrothermal dolomitization of the Mississippian Upper Debolt Formation, Sikanni gas field, northeastern British Columbia, Canada. *Bull. Can. Pet. Geol.* **1997**, *45*, 297–316.

95. Malone, M.J.; Baker, P.A.; Burns, S.J. Recrystallization of dolomite: An experimental study from 50–200 °C. *Geochim. Cosmochim. Acta* **1996**, *60*, 2189–2207. [CrossRef]

96. Swart, P.K. The geochemistry of carbonate diagenesis: The past, present and future. *Int. Assoc. Sedimentol.* **2015**, *62*, 1233–1304. [CrossRef]

97. Veizer, J. Chemical diagenesis of carbonates: Theory and application of trace element technique. In *Stable Isotopes in Sedimentary Petrology*; Arthur, M.A., Anderson, T.F., Kaplan, I.R., Veizer, J., Land, L.S., Eds.; Society of Economic Paleontologists and Mineralogists: New York, NY, USA, 1983; Volume 10, pp. 3–100.

98. Gao, G. Geochemical and isotopic constraints on the diagenetic history of a massive, late Cambrian (Royer) dolomite, Lower Arbukle Group, Silick Hills, SW Oklahoma, USA. *Geochim. Cosmochim. Acta* **1990**, *54*, 1979–1989. [CrossRef]

99. Frank, J.R. Dolomitization in the Taum Sauk Limestone (Upper Cambrian), southeast Missouri. *J. Sediment. Res.* **1981**, *51*, 7–18.

100. Lorens, R.B.; Mn, S.C. Distribution coefficients in calcite as a function of calcite precipitation rate. *Geochim. Cosmochim. Acta* **1981**, *45*, 553–561. [CrossRef]

101. Machel, H.G.; Anderson, J.H. Pervasive subsurface dolomitization of the Nisku Formation in central Alberta. *J. Sediment. Res.* **1989**, *59*, 891–911.

102. Banner, J.L. Application of the trace element and isotope geochemistry of strontium to studies of carbonate diagenesis. *Sedimentology* **1995**, *42*, 805–824. [CrossRef]

103. Dunham, J.B.; Olson, E.R. Shallow subsurface dolomitizaion of subtidally deposited carbonate sediments in the Hanson Creek Formation (Ordovician-Silurian) of Central Nevada. In *Concepts and Models of Dolomitization*; Zenger, D.H., Dunham, J.B., Ethington, R.L., Eds.; Society of Economic Paleontologists and Mineralogists: New York, NY, USA, 1980; Volume 28, pp. 139–161.

104. Webb, G.E.; Nothdurft, L.D.; Kamber, B.S.; Kloprogge, J.T.; Zhao, J.X. Rare earth element geochemistry of scleractinian coral skeleton during meteoric diagenesis: A before and-after sequence through neomorphism of aragonite to calcite. *Sedimentology* **2009**, *56*, 1433–1463. [CrossRef]

105. Allwood, A.C.; Kamber, B.S.; Walter, M.R.; Burch, I.W.; Kanik, I. Trace elements record depositional history of an Early Archean stromatolitic carbonate platform. *Chem. Geol.* **2010**, *270*, 148–163. [CrossRef]

106. Banner, J.L.; Hanson, G.N.; Meyers, W.J. Rare earth element and Nd isotopic variations in regionally extensive dolomites from the Burlington-keokuk formation (Mississippian): Implications for REE mobility during carbonate diagenesis. *J. Sediment. Res.* **1988**, *58*, 415–432.

107. Frimmel, H.E. Trace element distribution in Neoproterozoic carbonates as palaeoenvironmental indicator. *Chem. Geol.* **2009**, *258*, 338–353. [CrossRef]

108. Zhao, Y.; Zheng, Y.F.; Chen, F. Trace element and strontium isotope constraints on sedimentary environment of Ediacaran carbonates in southern Anhui, South China. *Chem. Geol.* **2009**, *265*, 345–362. [CrossRef]
109. Chaudhuri, S.; Clauer, N. (Eds.) *Isotopic Signatures and Sedimentary Records*; Lecture Notes in Earth Sciences; Springer: Berlin, Germany, 1992; Volume 43, pp. 497–529.
110. Murray, R.W.; Buchholtz ten Brink, M.R.; Gerlach, D.C.; Russ, P.G., III; Jones, D.L. Interoceanic variation in the rare earth, major, and trace element depositional chemistry of chert: Perspectives gained from the DSDP and ODP record. *Geochim. Cosmochim. Acta* **1992**, *56*, 1897–1913. [CrossRef]
111. Madhavaraju, J.; Ramasamy, S. Rare earth elements in limestones of Kallankurichchi formation of Ariyalur Group, Tiruchirapalli Cretaceous, Tamil Nadu. *J. Geol. Soc. India* **1999**, *54*, 291–301.
112. De Baar, H.J.W. On cerium anomalies in the Sargasso Sea. *Geochim. Cosmochim. Acta* **1991**, *55*, 2981–2983. [CrossRef]
113. Nothdurft, L.D.; Webb, G.E.; Kamber, B.S. Rare earth element geochemistry of Late Devonian reefal carbonates, Canning Basin, Western Australia: Confirmation of a seawater REE proxy in ancient limestones. *Geochim. Cosmochim. Acta* **2004**, *68*, 263–283. [CrossRef]
114. Tostevin, R.; Shields, G.A.; Tarbuck, G.M.; He, T.; Clarkson, M.O.; Wood, R.A. Effective use of cerium anomalies as a redox proxy in carbonate-dominated marine settings. *Chem. Geol.* **2016**, *438*, 146–162. [CrossRef]
115. Vugrinovich, R. Shale compaction in the Michigan Basin: Estimates of former depth of burial and implications for paleogeothermal gradients. *Bull. Can. Pet. Geol.* **1988**, *36*, 1–8.
116. Cercone, K.R.; Pollack, H.N. Thermal maturity of the Michigan Basin. In *Early Sedimentary Evolution of the Michigan Basin*; Catacosinos, P., Daniels, P., Eds.; Geological Society of America: Boulder, CO, USA, 1991; Special Paper 256; pp. 1–11.
117. Friedman, I.; O'Neil, J.R. Compilation of stable isotope fractionation factors of geochemical interest. In *Data of Geochemistry*, 6th ed.; Chapter, K.K., Ed.; U.S. Government Printing Office: Washington, DC, USA, 1977.
118. Graven, G. Continental-scale groundwater flow and geological processes. *Annu. Rev. Earth Planet. Sci.* **1995**, *23*, 89–118. [CrossRef]
119. Bjorlykke, K. Fluid-flow processes and diagenesis in sedimentary basins. In *Geofluids: Origin, Migration and Evolution of Fluids in Sedimentary Basins*; Parnell, J., Ed.; Geological Society: London, UK, 1994; Volume 87, pp. 127–140.
120. Girard, J.P.; Barnes, D.A. Illitization and paleothermal regimes in the middle Ordovician St. Peter sandstone, central Michigan Basin: K–Ar, oxygen isotope, and fluid inclusion data. *Am. Assoc. Pet. Geol. Bull.* **1995**, *79*, 49–69.
121. Ma, L.; Castro, M.C.; Hall, C.M. Atmospheric noble gas signatures in deep Michigan Basin brines as indicators of a past thermal event. *Earth Planet. Sci. Lett.* **2009**, *277*, 137–147. [CrossRef]
122. Sutcliffe, C.N.; Thibodeau, A.M.; Davis, D.W.; Al-Aasm, I.S.; Parmenter, A.; Zajacz, Z.; Jensen, M. Hydrochronology of a proposed deep geological repository for Low and Intermediate nuclear waste in southern Ontario from U-Pb dating of secondary minerals: Response to Alleghanian events. *Can. J. Earth Sci.* **2019**. [CrossRef]

Article

Geochemical and Dynamic Model of Repeated Hydrothermal Injections in Two Mesozoic Successions, Provençal Domain, Maritime Alps, SE-France

Namam Salih [1,2,*], Howri Mansurbeg [3,4] and Alain Préat [5]

1   Faculty of Science, Department of Petroleum Geosciences, Soran University, Soran, Erbil 44008, The Kurdistan Region of Iraq, Iraq
2   Scientific Research Center (SRC), Soran University, Soran, Erbil 44008, The Kurdistan Region of Iraq, Iraq
3   General Directorate of Scientific Research Center, Salahaddin University-Erbil, Erbil 44002, The Kurdistan Region of Iraq, Iraq; Howri.Mansurbeg@uwindsor.ca
4   School of the Environment, University of Windsor, Windsor, ON N9B 3P4, Canada
5   Res. Grp.—Biogeochemistry & Modelling of the Earth System, Université Libre de Bruxelles, 1050 Brussels, Belgium; apreat@ulb.ac.be
*   Correspondence: namam.muhammed@gmail.com

Received: 9 August 2020; Accepted: 30 August 2020; Published: 31 August 2020

Abstract: A field, petrographic and geochemical study of two Triassic–Jurassic carbonate successions from the Maritime Alps, SE France, indicates that dolomitization is related to episodic fracturing and the flow of hydrothermal fluids. The mechanism governing hydrothermal fluids has been documented with the best possible spatio-temporal resolutions specifying the migration and trapping of hydrothermal fluids as a function of depth. This is rarely reported in the literature, as it requires a very wide range of disciplines from facies analysis (petrography) to very diverse and advanced chemical methods (elemental analysis, isotope geochemistry, microthermometry). In most cases, our different recognized diagenetic phases were mechanically separated on a centimetric scale and analyzed separately. The wide range of the $\delta^{18}O_{VPDB}$ and $^{87}Sr/^{86}Sr$ values of diagenetic carbonates reflect three main diagenetic realms, including: (1) the formation of replacive dolomites (Type I) in the eogenetic realm, (2) formation of coarse to very coarse crystalline saddle dolomites (Types II and Type III) in the shallow to deep burial mesogenetic realm, respectively, and (3) telogenetic formation of a late calcite cement (C1) in the telogenetic realm due to the uplift incursion of meteoric waters. The Triassic dolomites show a lower $^{87}Sr/^{86}Sr$ ratio (mean = 0.709125) compared to the Jurassic dolomites (mean = 0.710065). The Jurassic calcite ($C1_J$) shows lower Sr isotopic ratios than the Triassic $C1_T$ calcite. These are probably linked to the pulses of the seafloor's hydrothermal activity and to an increase in the continental riverine input during Late Cretaceous and Early Cenozoic times. This study adds a new insight into the burial diagenetic conditions during multiple hydrothermal flow events.

Keywords: hydrothermal dolomite; diagenetic settings; optical petrography; geochemical; Triassic-Jurassic successions; Provençal Domain

---

1. Introduction

The formation of low- versus high-temperature dolomites in carbonate successions is attributed to separate diagenetic conditions and fluid origins [1,2]. Low temperature dolomitizing fluids are associated with near surface diagenetic fluids [3–5]. Hot dolomitizing fluids are either the result of high burial temperatures or the flux of hot hydrothermal (HT) fluids into colder, shallower buried carbonate succession environments [6–9]. The HT fluids are commonly formed by the upward flow of basinal brines [10], or deeply percolating meteoric waters [11] that interact with crystalline basement rocks [12]. In this paper, the petrographical characteristics and geochemical and isotopic compositions of dolomite and calcite in the Triassic-Jurassic successions are used to constrain the diagenetic conditions encountered within the framework of a constructed paragenetic sequence of dolomites in the Provençal Domain in SE France.

2. Regional Geology

The study area is located in the Provençal Domain close to the northeastern part of the Maritime Alps, in the southeast of France, bordered by NW Italy (Figure 1). The area, which is considered as the interior-most part of the European Proximal Margin [13], is underlain by the Argentera Massif from the southern part of the Dauphinois Domain. The Provençal Domain consists of Mesozoic rocks in shallow marine environmental settings [14] and Upper Carboniferous–Permian continental sediments overlain by Triassic rocks consisting of siliciclastics (Lower Triassic), tidal carbonates (Middle Triassic) and evaporites (Upper Triassic) [15].

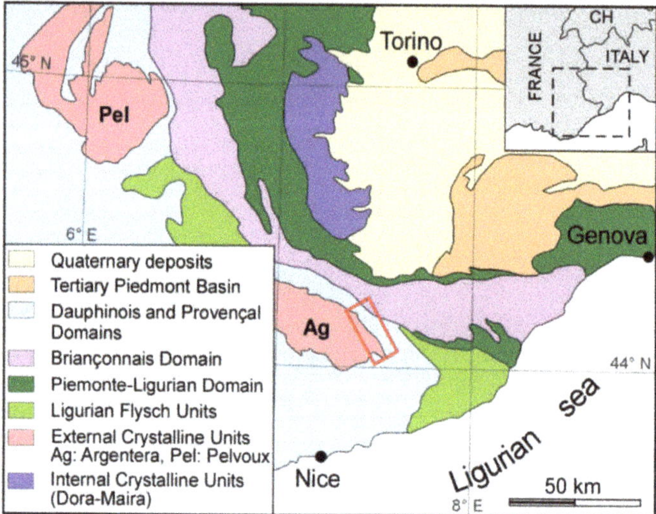

Figure 1. General geological map of the south-west Alps. The red rectangle shows a part of the study area (Provençal Domain) (after [14]).

The European paleo-margin experienced extensional and transtensional tectonics during the opening of the Alpine Tethys during the Jurassic and Cretaceous [16]. A subsequent regional unconformity followed the Triassic succession and encompasses the Late Triassic–Early Jurassic succession [14]. The emersion of the Provençal Domain during Early Jurassic times led to the formation of a carbonate platform followed by a continuous decrease in the sea level and the consequent establishment of a peritidal environment during the Berriasian [17,18]. The deposition of shallow-water marine carbonates continued during the Cretaceous [14]. The Mesozoic succession is separated from the Cenozoic deposits by a regional unconformity that is equivalent to the late Cretaceous-Middle Eocene interval [14]. A rapid lateral facies variation is observed in both Provençal and Dauphinois Domains and the Mesozoic succession is thinner toward the Provençal area [14].

3. Methods and Materials

Conventional and cathodoluminescence (CL) petrography was carried on 50 thin sections (out of total 98 samples) stained with Alizarin Red. The CL examinations were carried out on representative polished thin sections using CL model 8200 at operating conditions of 9–12 kV and a 300–400 µA gun current.

A total of 1.5 mg of the 43 dolomite and calcite samples were dissolved in 3 M HNO_3 for the purpose of elemental analysis. The solution was diluted with 2 mL of deionized H_2O (N18.2 MΩ cm^{-1}). The element concentrations are reported in ppm. The analytical errors are given as $\pm$ %RSD. The 1σ-reproducibility for the major and trace elements of the two standard materials are $\pm$0.18 wt. % for Ca, $\pm$0.081 wt. % for Mg, $\pm$22 ppm for Sr, $\pm$17 ppm for Fe, $\pm$1 ppm for Mn (CRM-512, $n = 111$), $\pm$0.36 wt. % for Ca, $\pm$0.002 wt. % for Mg, $\pm$1 ppm for Sr, $\pm$12 ppm for Fe and $\pm$1 ppm for Mn (CRM-513, $n = 111$).

The oxygen and carbon stable isotopic analyses were performed by the reaction of powdered, micro-sampled calcite and dolomite with 100% phosphoric acid at 70 °C using a Gasbench II connected to a ThermoFisher Delta V Plus mass spectrometer. All values are reported in per mil relative to V-PDB. The reproducibility and accuracy were monitored by a replicate analysis of laboratory standards calibrated to international standards NBS19, NBS18 and LSVEC. Isotopic values are referred to the University of Erlangen PDB standard with reproducibilities of $\pm$0.2‰ or better.

The $^{87}Sr/^{86}Sr$ analyses were performed at the Institute for Geology, Mineralogy and Geophysics, Ruhr-Universität Bochum (Germany). According to the ICP-OES, the Sr concentration measurements need a carbonate sample with 250 to 350 ng of Sr digested with 6 M supra-pure HCl for about 24 h at room temperature in PFA beakers and evaporated until dry. Subsequently, the dried samples were re-dissolved in 0.4 mL HNO_3 3M. The Sr fraction was recovered using an ion exchange resin (Sr-resin TRISEM) with 0.05 M and 3 M HNO_3 applying 2 mL of distilled water. After evaporation, the dried sample material was treated with 1 mL of a 1:1 mixture of concentrated HNO_3:H_2O_2 removing organic remains and subsequently dried again. Later, the samples were converted to chloride-form by applying 0.4 mL 6 M HCl. The mean of 279 analyses of NIST standard NBS 987 was 0.710242 with a mean standard error of 0.000002 ($\pm$2 se) and mean standard deviation of 0.000032 ($\pm$2 sd), whereas the mean of 253 analyses of the USGS EN-1 standard was 0.709162 with a mean standard error of 0.000002 ($\pm$2 se) and mean standard deviation of 0.000026 ($\pm$2 sd).

4. Results

4.1. Field Observation

The studied carbonates display a complex fracture system with carbonate cemented breccias and zebra fabrics (Figure 2A,B and Figure 3). The brecciated fragments show angular edges cemented by coarse-crystalline dolomite. Brecciation is most extensive in the lower part of the Triassic section (Figure 2B). The development of stratabound zebra fabric in the Triassic series is associated with fracture networks with predominantly conjugated orientations extending for decimeters (Figure 2A). Dolomitization is the most common diagenetic alteration in the host carbonates in the Maritime Alps [18].

4.2. Petrography of the Triassic–Jurassic Succession

Petrographic characteristics are used to document the diagenetic fluids that influenced the Mesozoic carbonate successions in the Maritime Alps (Provençal Domain—SE France), their relative timing and burial history. The diagenesis of the carbonate is complex and includes several dolomite generations.

Figure 2. Triassic outcrops characterized by fractures arranged in horizontal and sub-vertical bands to the bedding planes, resulting in zebra-like structures (**A**), open fractures zones, and breccias (**B**).

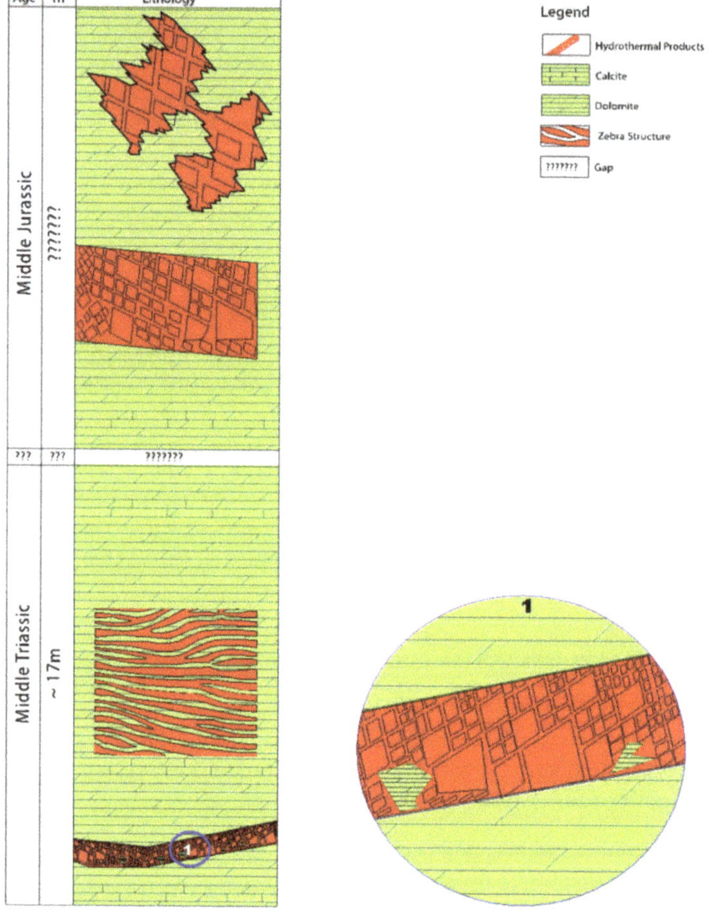

Figure 3. Basic stratigraphic litho-log of the Triassic and Jurassic carbonate rocks in Provençal Domain. The lower most part of Triassic rock (1) is extensively brecciated and the floated fragments have been cemented by saddle dolomite (see the enlarged part of this breccia on the right part "1"). The breccia is followed by another structure, i.e., zebra dolomites.

4.2.1. Middle Triassic Section

The micrite and replacive microdolomite ($D1_T$) composition is more common in breccia and zebra fabrics of the Triassic than in the Jurassic carbonates. Despite strong dolomitization, the depositional features are still preserved including laminated algal/microbial mats and slightly compacted micritized ooids with darker micritic envelopes in grain-supported fabrics (Figure 4A,B). Dolomite 2 ($D2_T$) is composed of medium to coarse grained anhedral–subhedral crystals (70–200 µm) in a nonplanar to planar-s fabric. $D2_T$ replaced the $D1_T$ dolomite and display relict ooids (Figure 4A,B).

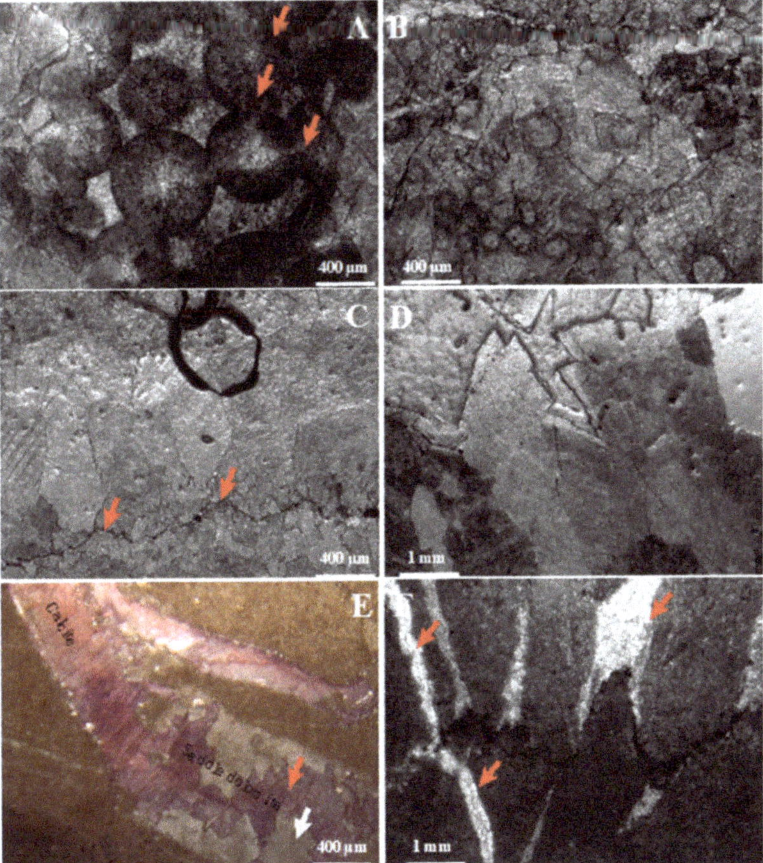

Figure 4. (**A**) The depositional texture is grain-supported with ooids (grainstones). The intergranular contacts between ooids are sutured and convexo–concave due to overburden load (arrows). (**B**) Transition from a mostly anhedral dolomite $D2_T$ to a euhedral saddle dolomite $SD2_T$ separated by an irregular stylolite indicating that these dolomites were formed at different time. (**C**) Early stylolite (arrows) cross-cuts $SD1_T$ with precipitation of sulfide minerals. (**D**) Light brown saddle dolomite $SD3_T$, showing coarse crystal size and pressure dissolution. (**E**) Macroscopic fracture showing Fe-rich $SD3_T$ (red arrow) and Fe-poor $SD2_T$ (white arrow). A late-stage Fe-poor calcite cement appears at the top left (staining). (**F**) Arrows point to sets of tension gashes filled by Fe-rich $SD3_T$ fracture. The fractures and columns of stylolite (red arrows) are in the same direction and this could be related to the mechanical overburden in the deep burial diagenetic setting.

Saddle dolomite (SD1$_T$), which occurs as irregular compacted crystals (0.3 to 0.6 mm) with nonplanar-s fabric, replaced D2$_T$. SD1$_T$ commonly occurs in the dolostones and, less commonly, as vug-lining cement in fractures. Coarser saddle dolomite SD2$_T$ crystals (0.5 up to 4 mm) consist of homogeneous euhedral slightly curved, sometimes planar, crystals. In contrast to SD2$_T$, SD1$_T$ appears as a cement in irregular fenestrae and breccias and as zebra dolomite fabric. The brecciated dolostones D2$_T$ (clast sizes up to a few cm) are cemented by various dolomitic types. SD1$_T$ is distinguished from SD2$_T$ crystals by its fabric retentive, inclusion-free rims and inclusion-rich cores. SD1$_T$ and SD2$_T$ saddle dolomites are characterized by a similar growth zoning with bright red and dark red luminescence. The gray bands of D2$_T$ alternate with the white bands of the saddle dolomites (SD1, SD2), forming the zebra fabric and are thus genetically related to the fracturing of the dolostones. SD1$_T$ and SD2$_T$ are crosscut by low amplitude stylolites (Figure 3C) that were aligned parallel to fracture-filling saddle dolomite.

Light brown colored, coarse-grained, euhedral, slightly planar, Fe-rich saddle dolomite crystals (SD3$_T$) are found in the breccias (Figure 4D–F). The breccias vary from clast-supported (jigsaw puzzle) to cement-supported textures (Figure 2B). SD3$_T$ is associated with stylolites with sets of fractures or tension gashes. The tension gashes are parallel to the columns of stylolites and terminated at the wider end of the stylolite seam. SD3$_T$ exhibited the typical saddle characteristics with a wavy extinction and curved crystal faces under a cross-polarized light, and has a uniform bright red luminescence under CL.

Euhedral, Fe-poor SD4$_T$ saddle dolomite (Figure 5A) displayed alternating inclusion-rich cores and -free rims. The margins of SD4$_T$ crystals were corroded and replaced by calcite (C1$_T$; Figure 5B,C). Similar calcite filled the vugs and fractures and replaced D2$_T$, SD3$_T$ and SD4$_T$ too.

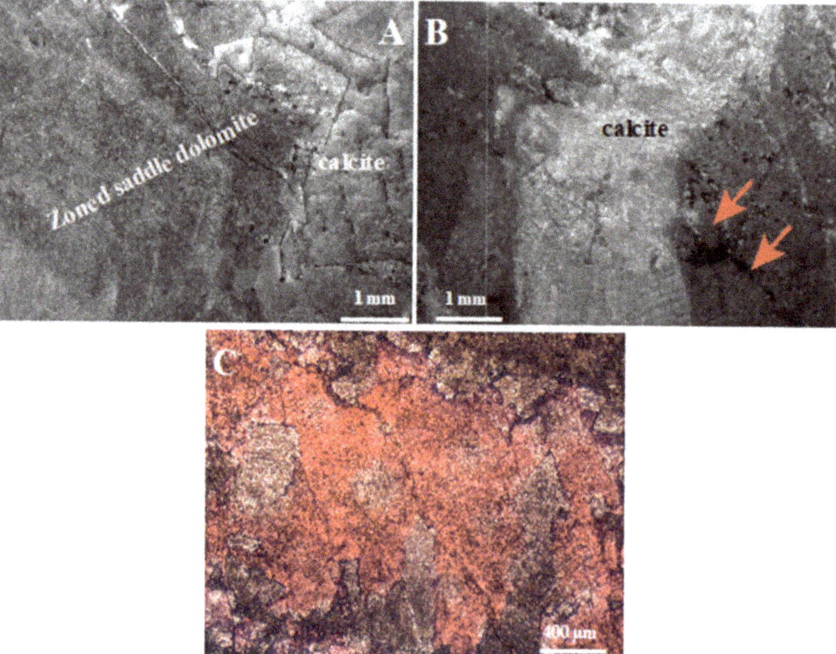

Figure 5. (**A**) The former saddle dolomite SD3$_T$ leaving traces along the core and cortex of zoned SD4$_T$, reveals that SD4$_T$ was formed under different diagenetic conditions. C1$_T$ filled the remaining secondary pores, and postdating the latter saddle dolomite. (**B,C**) Millimeter-sized fracture-filling Fe-poor, late stage calcite cement (C1$_T$), postdating late stylolite (arrows).

4.2.2. Middle-Upper Jurassic

The peritidal Middle–Upper Jurassic carbonates in the Provençal region include ooidal packstones and grainstones, which are replaced by microcrystalline dolomite (10–300 μm) ($D1_J$) (Figure 6A). $SD1_J$ replaced D_J (Figure 6B). A fabric destructive, second saddle dolomite generation ($SD2_J$) with subhedral–euhedral, nonplanar crystals (>1 mm) fill fractures and vugs (Figure 6C).

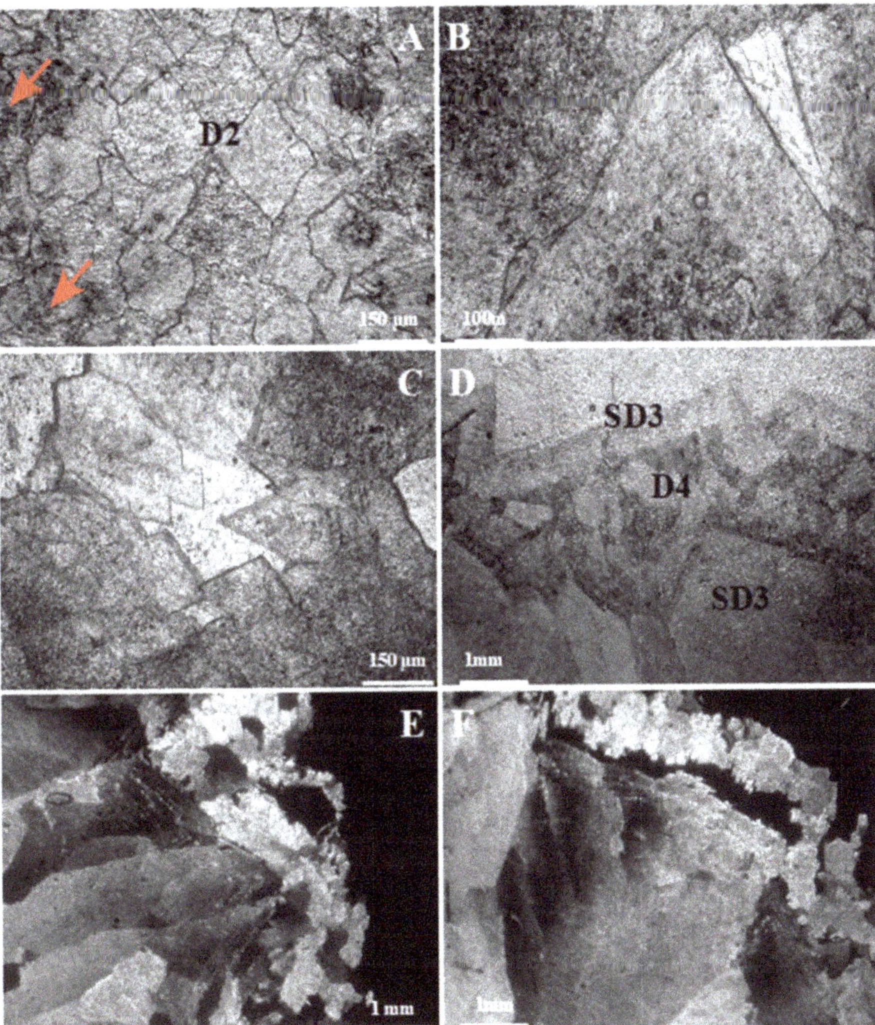

Figure 6. (**A**) Mostly planar-s dolomite $D2_J$ with remaining of a former replacement dolomite D_J "lithorelic" (arrows) showing dissolution and irregular compacted grain boundaries. Micrite of precursor facies on the top left corner. (**B**) Saddle dolomite-type $SD1_J$ matrix. The core is rich in inclusions. Thus, this dolomite facies is a key for the replacement of a former dolomite. (**C**) Secondary porosity filled by saddle dolomite $SD2_J$ cement. (**D**) Secondary porosity occluded by late dolomite D4 of different grain sizes, within $SD3_J$. (**E,F**) Saddle dolomite $SD5_J$ is characterized by strongly curved crystal boundaries and a sweeping extinction under crossed nicols. D, Dolomite; SD, Saddle dolomite.

Yet another phase of coarse-crystalline (several mm across), euhedral, nonplanar to planar, beige colored saddle dolomite (SD3$_J$) (Figure 6D), which fills the fractures or replaces the dolomitic matrix, was recognized. SD3$_J$ is postdated by a dark-grey dolomite with different sizes and shapes (D4) of filling space between SD3$_J$ crystals (Figure 6D).

Another coarse-crystalline, euhedral, nonplanar, multizoned, red luminescent fracture and vug-filling saddle dolomite (SD5$_J$) (Figure 6E,F) was observed. SD5$_J$ crystals are characterized by alternating inclusion-rich and inclusion-free zones. High-amplitude stylolites cross cut the fracture-filling saddle dolomites; no tension gashes were observed. The SD6$_J$ dolomite is distinguished from the previous saddle dolomites by its elongated crystal shapes (Figure 7A–C). This SD6$_J$ dolomite is densely packed and systematically associated with D4 (Figure 6D). CL revealed an alteration of thick red luminescence (SD5$_J$) and thick nonluminescent SD6$_J$ (Figure 8B,C). In addition, thin bright red luminescence is associated with fractures that postdated SD6$_J$.

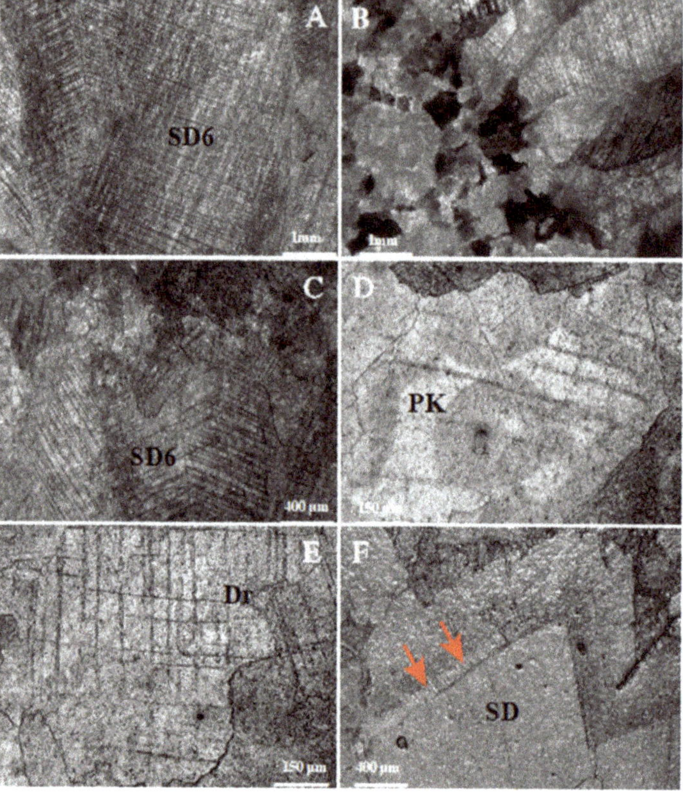

Figure 7. (**A–C**) More than 1mm sized-dolomite crystals SD6$_J$ with two sets of cleavages, precipitated in fractures in contact with recrystallized dolomite. SD6$_J$ shows etching and overgrowth of coarse dolomite. (**D,E**) Saddle dolomite rhomb with an irregular, selectively dissolved outline which predates the poikilotopic texture within the recrystallized dolomite matrix. Two sets of curvature twin laminae, and dissolution of high amplitude stylolite intersecting coarse recrystallized dolomite crystals (the upper side part of Dr and lower side part of E). (**F**) Bladed calcite cement (arrows) precipitated on saddle dolomite crystal: note the trace of calcite across saddle dolomite; such cement strongly refers to late meteoric alteration at near-surface condition. SD, Saddle dolomite; Dr, Dolomite recrystallization.

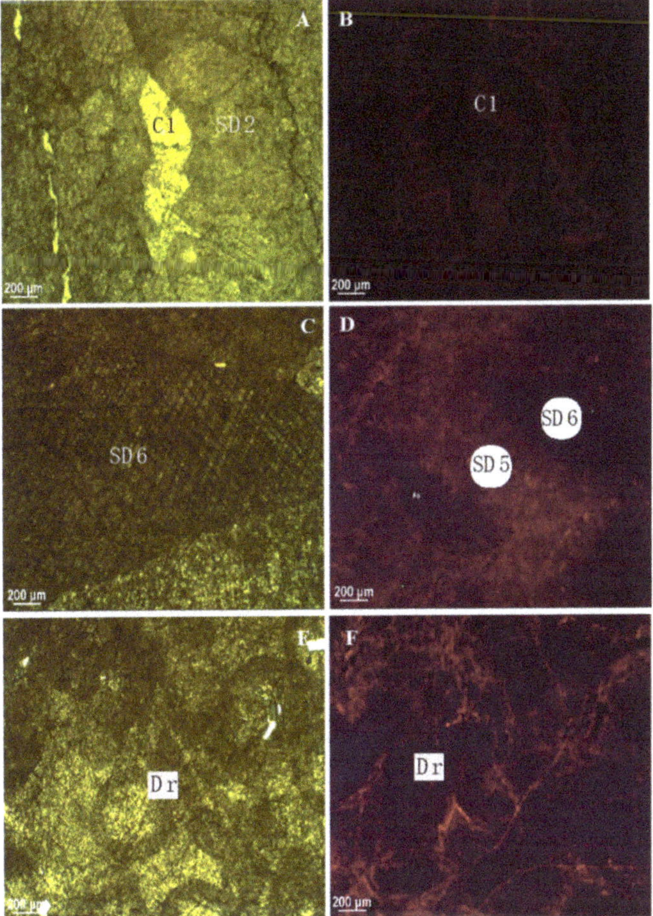

Figure 8. Photomicrograph of (**A**) saddle dolomite (SD2) and calcite (C1) in PPL, calcite overgrowth on the saddle dolomite is clearly highlighted under cathodoluminescence in (**B**). (**C**) Saddle dolomite (SD6) in PPL and (**D**) cathodoluminescence (CL) photomicrograph, illustrating the general pattern in luminescence, including thicker zones of red and nonluminescence with shades of thin bright red luminescence, reflecting formation in pore waters that chemically changed. (**E**) Recrystallized dolomite (Dr) in PPL, (**F**) the same as in (**E**) CL photomicrograph, the recrystallized dolomite shows thick nonluminescent zones with thin bright red luminescent bands.

A Fe-poor calcite cement (C1$_J$) fills the central parts of the fractures and vugs, hence the latest cement (Figure 8A,B). The calcite sizes are ranged from fine to coarse crystals (>400 μm), precipitated and postdated saddle dolomite formation, and had crystals that were transparent with whitish and yellowish colors. CL revealed an alteration of thick red luminescence (SD5$_J$) and thick nonluminescent SD6$_J$ (Figure 8B,C). In addition, thin bright red luminescence was associated with small veins and fractures that postdated SD6$_J$.

Dolomite recrystallization (Dr) occurred after all saddle dolomitic phases, where the ghosts of the pristine oolitic facies still resisted diagenesis, and in places the coarse sized poikilotopic cement fabric (>700 μm) developed. Dr displays the same characteristics as those observed in SD6$_J$ under CL, i.e., nonluminescent thick with thin bright red luminescent bands (Figure 8E,F).

Because of the small sizes of fractures and cavities, and also the fact that sometimes two dolomite phases occur together in a fracture during mixing the carbonate phases are categorized into four groups to obtain precise data: type 1 with D_J, $D1_T$ and $D2_T$; type II with saddle dolomites from $SD1_{T/J}$ and $SD2_{T/J}$; type III with $SD3_{T/J}$, $SD4_T$, $D4_J$, $SD5_{T/J}$, Dr, and $SD6_J$ and type IV with $C1_{T/J}$. This phase succession will also make it easier to interpret the geochemical composition, especially when two carbonate phases are mixed.

4.3. Early and Late Diagenetic Stylolites

Most stylolites are used as a key of the burial depth, especially those aligned parallel to the bedding planes. The Triassic–Jurassic carbonate succession is characterized by two types of stylolites: (i) low amplitude stylolites (no more than 1 cm in length) crosscut Type I dolomites. The low amplitude stylolites (early stylolites) were the first ones produced during the Triassic–Jurassic diagenesis and are associated with a jump-up in McArthur's graph [19]; (ii) early stylolites are followed by larger stylolites (several centimeters in length with high amplitudes) associated with a set of fractures and tension gashes. Stylolite-related fractures (tension gashes) are rare diagenetic features formed by increasing and progressive mechanical loading during burial, evidenced by the parallel orientation with the bedding plane [20,21].

4.4. Chemical Composition (Major and Trace Elements)

Nineteen samples were selected for trace element analyses from both outcrops (Table 1).

Table 1. Elemental analysis of diagenetic phases from Triassic and Jurassic samples ($n = 19$). Type I = $D1_{T/J}$ and $D2_{T/J}$. Type II = $SD1_{T/J}$ and $SD2_{T/J}$. Type III = $SD3_{T/J}$, $SD4_T$, $D4_J$, $SD5_{T/J}$, Dr_J, and $SD6_J$. Type IV = $CT1_{T/J}$. T = Triassic, J = Jurassic, and Lst = Host limestone. SE France sections.

Sample No.	Type	Mn (ppm)	Fe (ppm)	Ba (ppm)	Sr (ppm)	Mg (wt. %)	Ca (wt. %)	Ca (wt. %)/ Mg (wt. %)
NM4	Type III	147	433	3	125	12.7	21.8	1.7
NM8	Type IV	13	27	4	291	0.5	39.4	86.0
NM10a	Type II	34	152	3	60	12.8	21.9	1.7
NM12	Type II	37	78	6	197	11.4	23.7	2.1
NM16	Type II	108	317	4	134	12.1	22.7	1.9
NM23	Type I	110	1991	2	73	12.4	21.5	1.7
NM23	Type III	66	1076	2	106	11.9	22.7	1.9
NM28	Type I	107	718	4	83	12.7	21.8	1.7
NM30	Type IV	146	204	3	368	0.6	38.9	68.0
NM34	Type III	91	282.3	2	86	11.5	23.4	2.0
NM4	Type III	140	441	3	61	12.2	20.6	1.7
NM21	Type III	226	730	4	180	11.9	21.8	1.8
NM22	Type III	40	168	3	103	12.6	21.8	1.7
NM11	Type I	51	463	2	83	11.4	22.1	1.9
NM34	Type III	61	291	1	79	11.4	21.9	1.9
NM32	Type II	60	337	2	47	12.3	21.2	1.7
NM23	Type III	149	1225	8	257	10.8	23.1	2.1
NM44	Lst-Triassic	113	1309	2	79	12.2	20.4	1.7
NM39	Lst-Jurassic	10	154	3	127	0.3	37.8	135.5

Generally, Ca (mean: 22.1 wt. %; $n = 15$) and Mg (mean: 12 wt. %; $n = 15$), Ca (mean: 39.2 wt. %; $n = 2$) and Mg (mean: 0.6 wt. %; $n = 2$) are the major elements in dolomite and calcite phases, respectively. Other elements like Sr (mean: 112 ppm; $n = 15$ for dolomite; mean: 330 ppm; $n = 2$ for calcite), Fe (mean: 580 ppm; $n = 15$ for dolomite; mean: 116 ppm; $n = 2$ for calcite), Mn (mean: 95 ppm; $n = 15$ for dolomite; mean: 180 ppm; $n = 2$ for calcite), Ba (mean: 3 ppm; $n = 15$ for dolomite; mean: 4 ppm; $n = 2$ for calcite), represent crucial "trace elements" in the carbonate diagenetic setting (Figure 9). The relative abundance of major elements and scarcity of trace elements in dolomite and calcite, are likely the recorded evidence of diagenetic processes (e.g., [22]).

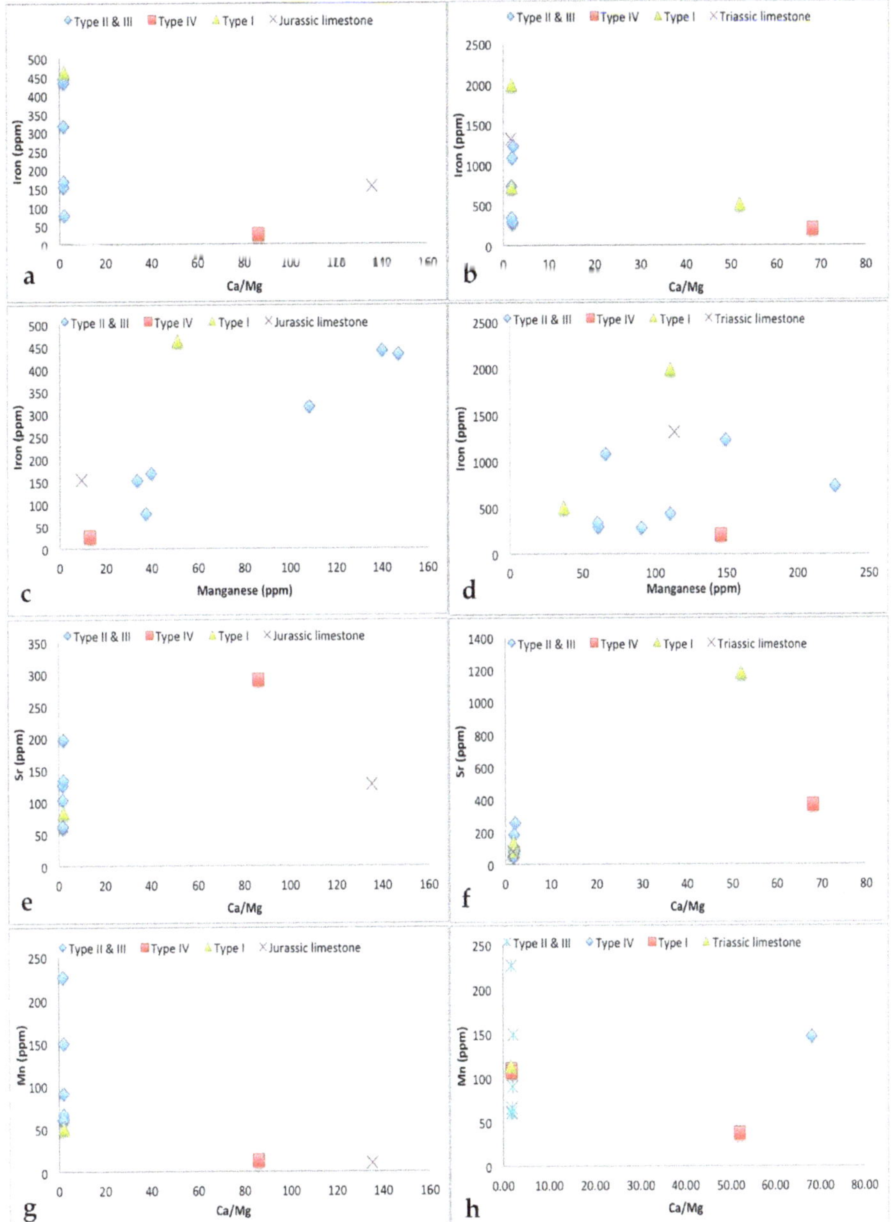

Figure 9. Diagram of the main chemical elements involved in the diagenesis and of the host limestones from the study area (*n* = 20). For more detail about Type I, II, III, IV, see text. (**a**) Ca/Mg vs. Fe for Jurassic carbonate samples, (**b**) Ca/Mg vs. Fe for Triassic carbonate samples, (**c**) Mn vs. Fe for Jurassic carbonate samples, (**d**) Mn vs. Fe for Triassic carbonate samples, (**e**) Ca/Mg vs. Sr for Jurassic carbonate samples, (**f**) Ca/Mg vs. Sr for Triassic carbonate samples, (**g**) Ca/Mg vs. Mn for Jurassic carbonate samples, and (**h**) Ca/Mg vs. Mn for Triassic carbonate samples.

The Ca/Mg ratio of the saddle dolomites (Types II and III) ranges between 1.72 and 2.14 from the Triassic samples, and 1.69 and 2.08 from the Jurassic samples. The Ca/Mg ratio of the late calcite phase in the Triassic is 68 and in the Jurassic is 85 (Figure 9). The Jurassic dolomites show a gradual increase in Fe and Mn, starting from 34 going up to 226 ppm for Mn, and 78 to 730 ppm for Fe. The Triassic dolomites exhibit high Fe and Mn values (mean: 846 and 92 ppm, respectively), with a wide range. Rarely do the Mn contents of the early dolomites in both sections overlap the Mn contents of the later dolomites. The early dolomites display the highest Fe contents. The Fe and Mn contents in the Triassic and Jurassic limestones share the same values with a few late dolomite samples. In contrast, late calcite-Type IV ($C1_J$ and $C1_T$) contents exhibit the lowest Fe contents ($C1_T$; Fe = 204 ppm; Mn = 116 ppm) ($C1_J$: Fe = 27 ppm; Mn = 13 ppm). Higher Fe and Mn contents in $C1_T$ could be related to $SD3_T$, which was overlain by the calcite cement, i.e., due to mixing of $SD3_T$ and $C1_T$ during preparation.

4.5. $\delta^{18}O_{VPDB}$ and $\delta^{13}C_{VPDB}$ Isotopes

The stable isotopic compositions of ten limestone samples (Table 2) shows that the $\delta^{18}O_{VPDB}$ and $\delta^{13}C_{VPDB}$ values vary between −0.2‰ to −4.2‰ and −0.5‰ to +3.1‰, respectively. The isotopic composition of the four dolomite types (I, II, III and IV) shows a range in $\delta^{13}C_{VPDB}$ values from −0.6‰ to +2.6‰ and $\delta^{18}O_{VPDB}$ values from −14.1‰ to −4.9‰. In more detail, the $\delta^{18}O_{VPDB}$ values of Type I ($D1_{T/J}$ and $D2_{T/J}$) display a narrow range between −7.1‰ and −4.8‰ in the Jurassic and from −7.4‰ to −5.7‰ in the Triassic. The $\delta^{13}C_{VPDB}$ values for both sections display similar values (average +1.1‰ and +1.9‰, respectively). All $\delta^{18}O_{VPDB}$ values are depleted in comparison with the original isotopic compositions of the marine dolomite signature during Triassic and Jurassic times [22–25].

Table 2. $\delta^{13}C$ (% VPDB), $\delta^{18}O$ (% VPDB), (Sr^{87}/Sr^{86}) values of selected samples from Triassic–Jurassic succession ($n = 100$). Type I = $D1_{T/J}$ and $D2_{T/J}$. Type II = $SD1_{T/J}$ and $SD2_{T/J}$. Type III = $SD3_{T/J}$, $SD4_T$, $D4_J$, $SD5_{T/J}$, Dr_J, and $SD6_J$. Type IV = $CT1_{T/J}$. D is for dolomite, SD = saddle dolomite, C= calcite, Dr = recrystallized dolomite, T = Triassic, J = Jurassic, and Lst = Host limestone. SE France sections.

Age	Sample No.	Type	$\delta^{13}C$ (‰ VPDB)	$\delta^{18}O$ (‰ VPDB)	Sr^{87}/Sr^{86}
	NM1	Type III	1.72	−9.09	
	NM2	Type II	1.78	−8.25	
	NM3	Type III	1.46	−9.35	
	NM3	Type III	1.78	−9.94	
	NM3	Type II	2.06	−8.03	
	NM4	Type I	1.81	−6.44	
	NM5	Type I	1.74	−7.07	
	NM5	Type II	1.76	−7.85	
	NM5	Type III	1.83	−9.99	
	NM6	Type I	2.04	−4.87	
	NM7	Type III	1.69	−9.82	
Upper Jurassic	NM7	Type III	1.72	−9.43	
	NM7	Type III	1.72	−9.83	
	NM8	Type III	1.81	−8.75	
	NM8	Type III	1.63	−9.61	
	NM8	Type IV	0.38	−11.83	0.70868
	NM9	Type III	1.81	−10.86	
	NM10a	Type II	1.55	−8.02	0.71030
	NM10b	Type II	1.44	−8.18	0.710552
	NM11	Type III	1.48	−10.78	
	NM11	Type III	2.07	−10.56	
	NM12	Type III	1.78	−9.14	
	NM13	Type III	1.76	−9.28	

Table 2. *Cont.*

Age	Sample No.	Type	$\delta^{13}C$ (‰ VPDB)	$\delta^{18}O$ (‰ VPDB)	Sr^{87}/Sr^{86}
Upper Jurassic	NM14	Type IV	−0.59	−11.12	
	NM14	Type IV	−0.02	−9.98	0.710355
	NM15	Type III	1.85	−9.42	
	NM15	Type III	1.86	−9.71	0.710154
	NM16	Type I	1.76	−6.31	0.708757
	NM16	Type III	1.42	−9.65	
	NM17	Type III	1.79	−9.31	
	NM17	Type III	1.8	−9.47	
	NM18	Type IV	0.92	−10.42	
	NM18	Type III	1.59	−10.26	
	NM18	Type III	1.57	−9.66	
	NM19	Type III	1.57	−9.72	0.711203
	NM19	Type III	1.48	−9.23	
	NM20	Type IV	0.98	−9.62	
	NM20	Type IV	−0.15	−10.52	
	NM20	Type III	1.29	−9.31	
	NM21	Type III	1.61	−11.63	0.710609
	NM21	Type I	1.98	−6.82	
	NM22	Type III	1.82	−9.05	0.708879
	NM22	Type III	1.84	−9.45	
	NM22	Type III	1.83	−9.37	
Triassic	NM23	Type II	2.3	−6.2	
	NM23	Type II	2.3	−6.8	
	NM24	Type III	2.3	−9.98	
	NM24	Type III	1.97	−9.09	
	NM25	Type III	2.56	−9.86	
	NM27	Type II	2.31	−7.58	
	NM27	Type III	2.21	−8.28	0.708299
	NM27	Type II	2.21	−7.48	
	NM26	Type I	1.98	−5.7	
	NM26	Type I	1.31	−6.44	0.707735
	NM28	Type III	2.54	−8.51	
	NM28	Type III	1.46	−9.35	
	NM29	Type II	2.2	−6.67	0.708671
	NM30	Type IV	0.65	−14.14	0.712023
	NM31	Type III	2.11	−8.44	
	NM32	Type III	2.2	−8.57	
	NM32	Type III	2.02	−9.91	
	NM33	Type III	2.49	−8.74	
	NM33	Type III	2.03	−9.44	0.70942
	NM34	Type III	1.81	−9.63	
	NM34	Type II	1.97	−7.93	
	NM34	Type III	2	−9.16	0.708184
	NM35	Type III	2.26	−9.07	
	NM35	Type IV	2.09	−12.08	
	NM36	Type I	0.21	−7.35	0.708595
	NM36	Type I	0.93	−7.35	0.711628
	NM37	Type III	1.16	−10.35	0.710469
Upper Jurassic	NM38	Limestone	2.07	−2.33	
	NM39	Limestone	−0.53	−4.21	0.707862

Table 2. *Cont.*

Age	Sample No.	Type	$\delta^{13}C$	$\delta^{18}O$	Sr^{87}/Sr^{86}
			(‰ VPDB)	(‰ VPDB)	
Triassic	NM40	Limestone	1.54	−2.31	
	NM41	Limestone	1.73	−2.5	
	NM42	Limestone	1.88	−3.36	
	NM43	Limestone	1.98	−1.67	
	NM44	Limestone	2.53	−2.37	0.708766
	NM45	Limestone	2.31	−3.36	
	NM46	Limestone	3.05	−0.23	
	NM47	Limestone	2.71	−3.25	

The $\delta^{18}O_{VPDB}$ values of Type II dolomite show a signature with depleted values varying between −8.3‰ and −7.9% in the Jurassic samples, and from −7.9‰ to −6.2‰ in the Triassic samples. These $\delta^{18}O_{VPDB}$ values of saddle dolomites decrease with the increase in crystal size. The $\delta^{13}C$ values of Type II and Type III exhibit a narrow range in the Triassic (average +2.0‰ and +2.4‰) and in the Jurassic (average +1.7‰ and +1.9‰). Type IV late-stage calcite shows significantly depleted $\delta^{18}O_{VPDB}$ values in the Triassic (−14.1‰ and −12.1 ‰) and in the Jurassic (between −11.8 and −9.6‰).

4.6. $^{87}Sr/^{86}Sr$ Data

The $^{87}Sr/^{86}Sr$ data of dolomite and calcite (Table 2) in the Triassic section show a wide range (0.707735 and 0.712023), while in the Jurassic section the range is between 0.708680 and 0.71120, so the ratios are lower than those of the host limestone sample.

The two unaltered limestone samples yielded an $^{87}Sr/^{86}Sr$ ratio of 0.708766 and 0.707862 for the Triassic and Jurassic limestones, respectively. The Triassic signal displays a value different from the seawater signal when compared to the data of Korte et al. [26], which reported the $^{87}Sr/^{86}Sr$ values between 0.707562 and 0.70804. Jurassic samples have been documented by Jones et al. [27] with an Sr^{87}/Sr^{86} ratio in the range of 0.707060–0.707280.

5. Interpretation and Discussion

5.1. Paragenetic Sequence Related to Dolomitizing Fluids

The petrographic and geochemical data obtained suggest that dolomitization has been accomplished by various episodes of hydrothermal flow events under various burial diagenetic conditions. The earliest dolomitization, which affected mainly the micrite mud matrix, probably occurred at near-surface conditions. This is evidenced by the anhedral, non-saddle dolomite fabric [28] and the uncompacted ooids. The nonplanar crystal shape of dolomite Type I D2$_T$ could be linked to a diagenesis at a higher temperature (e.g., ≥ 50 °C, [28]). However, these nonplanar dolomites are probably related to impurities in the growth media, which stabilize the crystal faces [29], and not to high temperature fluids. This is supported by petrographic evidence—for example, D2$_T$ crystals do not show any characteristics of hot fluid involvement "i.e., lack of curved faces and sweeping extinction".

It is well reported in the literature that saddle dolomite forms under high temperatures typically by flow of hydrothermal fluids [30]. Type II is dissolved by low-amplitude, early stylolites, in high temperature conditions (*senso*, [31]) in the shallow–intermediate depth without necessarily implying deep burial conditions "i.e., eogenesis".

Brecciation and concomitant dolomitization (Type II dolomites) suggest a flow of hydrothermal fluids along fractures (Figure 10). Brecciation was accompanied by the dissolution of the host carbonates, which resulted in the creation of new pathways for fluid of hydrothermal fluids and cementation by producing the Type III saddle dolomite. Further cementation by the latter dolomite was probably related to hydrofracturing and resetting of the zebra and breccia fabrics. Stylolite-related fractures are parallel to the amplitude of stylolite (Figure 4F), and according to Nelson [20] the tension gashes are not true tension fractures, but are extensional fractures derived from the same compressive state of stress as the fractures of stylolites [32]. Filling the tension gash fractures with SD3 confirmed that the late stylolites are coeval with Type III saddle dolomite formation (Figures 11 and 12). Stylolite-related fractures (tension gashes), which were produced during deep burial mesogenesis, were cemented by Type III saddle dolomite (Figure 4F). Hence, Type III saddle dolomite formation was presumably formed by the flux of hot basinal fluids along stylolites during compressional basin tectonics [33].

Dol 3
Dol 4
Number of measurements
1
2
3
4
5
6
7
8
9
10
11
12
170
180
190
200
210
220
230
240
250
260
Homogenization temperature (°C)

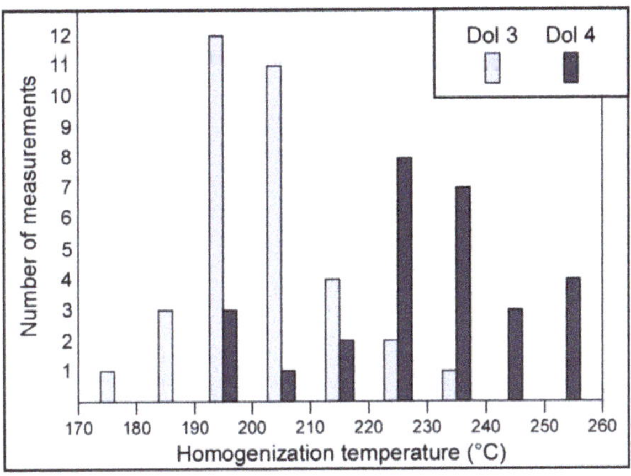

Figure 10. Histogram of homogenization temperatures obtained for saddle dolomite and fibrous dolomite cements [14].

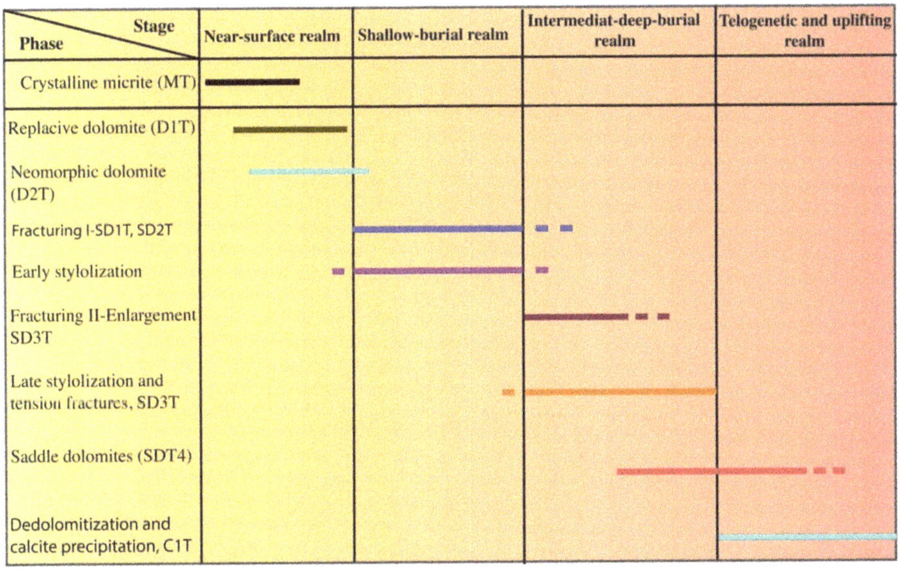

Stage
Phase
Near-surface realm
Shallow-burial realm
Intermediat-deep-burial realm
Telogenetic and uplifting realm
Crystalline micrite (MT)
Replacive dolomite (D1T)
Neomorphic dolomite (D2T)
Fracturing I-SD1T, SD2T
Early stylolization
Fracturing II-Enlargement SD3T
Late stylolization and tension fractures, SD3T
Saddle dolomites (SDT4)
Dedolomitization and calcite precipitation, C1T

Figure 11. Paragenetic succession illustrating the diagenetic phases of Triassic section. Provençal Domain, SE France (see text for explanation).

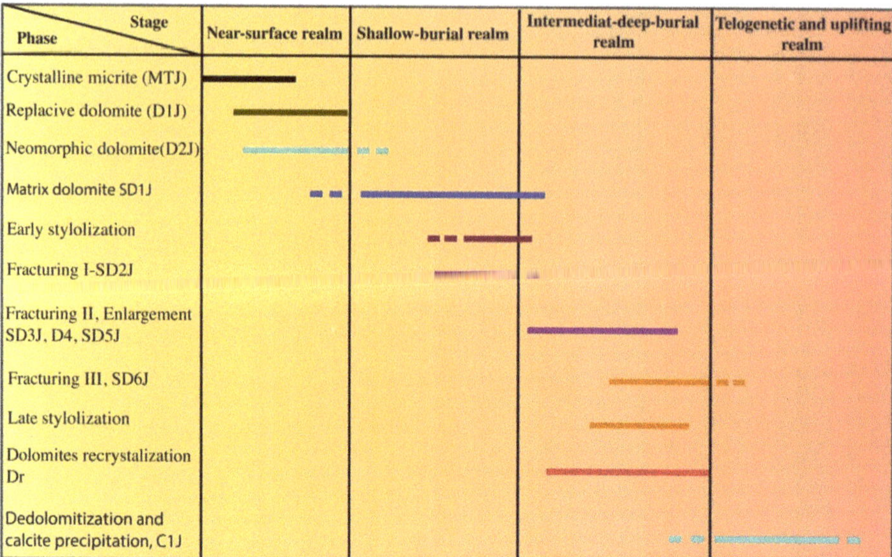

Figure 12. Paragenetic succession illustrating the diagenetic phases of Jurassic section. Provençal Domain, SE France (see text for explanation).

Subsequent openings of the veins or fractures in Type II facies led the hot secondary brine fluid to percolate along these pathways, which resulted in the precipitation of Type III saddle dolomites. The hydrothermal activity associated with hydrofracturing affecting the studied Mesozoic series has also been reported by [34–36].

The homogenization temperature of saddle dolomites from the Dauphinois–Provençal Domains varied between 170 and 260 °C ([14]; Figure 10), which rules out a low temperature origin for dolomite. The final melting temperature of fluid inclusions have been reported as highly saline fluids, and the salinity was approximately equivalent to 20% to 23% $CaCl_2$ [14], this confirms the interaction of hot circulating fluids with Argentera basement. The high temperature dolomitizing fluid is further supported by low $\delta^{18}O_{VPDB}$ values. The depletion of the $\delta^{18}O_{VPDB}$ values follows the increase in the size of saddle crystal from Type II to Type III (up to −11.6‰ in Type III dolomites). Therefore, the plausible explanation that can adequately explain the range of homogenization temperatures and considerable range of oxygen isotope values is that the precipitation of dolomite phases occurred at different diagenetic settings during which a repeated injection of hydrothermal fluid occurred (Figures 13 and 14). This interpretation is further supported by the following lines of evidence (Figures 4 and 6): (1) suture intergranular contacts between the ooids are engulfed by the saddle dolomite cement; (2) occurrence of sulfides and dissolution of stylolite-related fractures in type III saddle dolomites; (3) co-occurrence of Type III saddle dolomites and late stylolite-related fractures, later the cements inside the fractures were dissolved and etched; (4) coarse-crystalline poikilotopic cement fabric (>700μm) etched and corroded the Dr boundaries; (5) dissolution, compacted saddle dolomite and association with sulfides; (6) the presence of high-amplitude stylolites parallel to the overburden force.

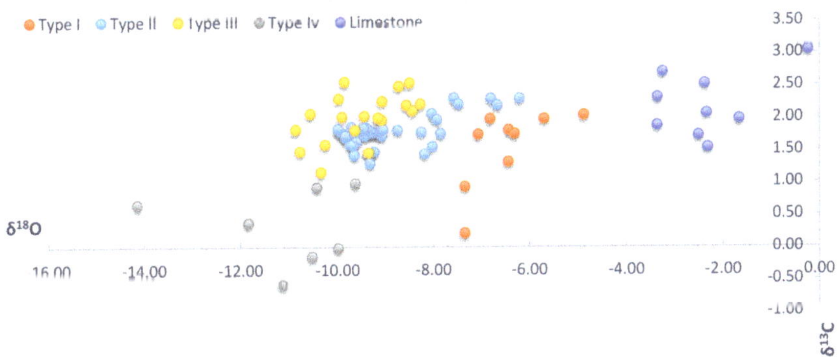

Figure 13. Cross-plot of oxygen and carbon isotopic compositions (VPDB) of selected dolomite, calcite phases (Type I to IV: see text) and pristine limestone from Triassic–Jurassic succession ($n = 80$).

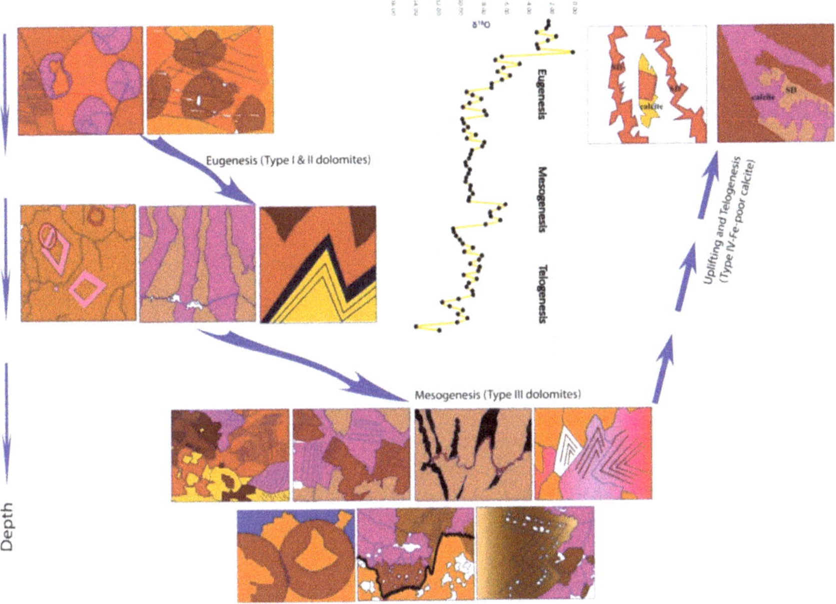

Figure 14. Petro-physical and geochemical model tracing the Triassic–Jurassic succession in Provençal Domain, SE France. The detailed microfacies in each diagenetic setting is described by digitized microphotographs under optical microscope. The initial precipitation of saddle dolomite is started in a shallow burial system (see the key evidence in digitized microphotographs). The mechanisms of hydrothermal (HT) fluid movement under deep burial conditions (mesogenesis) created another active pathway for hot fluid with different phase of saddle dolomite (see the zoned saddle dolomites, suture contact boundaries between the ooid grains, and stylolite-related fractures) from that of shallow burial conditions (see the low amplitude stylolite). The diagenesis in the Triassic–Jurassic succession ended by precipitation of Fe-poor calcite during uplifting (close up the large fracture that postdates all saddle phases on the right side of digitized microphotograph, while the one on the left side showing the digitized thin section under CL which result in a band of luminescence due to a change in fluid chemistry.

Therefore, according to the abovementioned points, our interpretation agrees with the one of Wierzbicki et al. [37], who suggested that the partial dissolution of dolomite crystal outlines fits the deep burial mesogenesis. This late evolution could be linked to the subsidence and recrystallization processes in the Provençal Domain during Late Cretaceous–Early Eocene times, where the main extensional and transtensional tectonics affected the European paleomargin of the Alpine Tethys [38].

The calcitization of saddle dolomite and the precipitation of the fracture-filling Fe-poor calcite cement (C1$_{T/J}$, Type IV; Figure 14), which cross cut the high-amplitude stylolites, are the last diagenetic event affecting the Triassic series by the incursion of meteoric waters. Furthermore, the precipitation of bladed calcite cement around saddle crystals implies that a new activity led to the uplift of the studied series and could reactivated the Alpine transcurrent shear zone during Cenozoic [39].

5.2. Geochemical Evolution of Diagenetic Fluids

A crucial question in any study of carbonate chemical composition is whether a primary marine signal is preserved or not. Carbon and oxygen isotopic compositions of the regional limestones span a considerable range with δ^{13}C values between −0.5‰ and +3.1‰ and δ^{18}O$_{VPDB}$ values between −4.2‰ and 0.2‰. Their isotopic compositions are similar to the values expected for Triassic and Jurassic marine carbonates [23–27].

Type I dolomite and the last event of calcite precipitation (Type IV) have considerably lighter δ^{18}O$_{VPDB}$ compared to those expected for the carbonate precipitated in equilibrium with ambient seawater (Figure 13). The δ^{18}O and δ^{13}C values in both types of dolomite and calcite fit the inverse "J" Lohmann curve [40] pointing to a diagenetic alteration near surface conditions. The initial nucleation of dolomite in the muddy facies was therefore driven by dolomite supersaturation caused by the mixing of fresh and marine groundwater [40–42]. Mixed groundwater filled the pore spaces in the oolite facies and led to undersaturation of calcite, and the dissolution and precipitation of dolomite in the case of Type I dolomite. The Triassic and Jurassic rocks developed on a shelf in which the water depth was extremely shallow, probably a few meters deep, as suggested by the oolites and algae. In such a shallow depositional setting these carbonates were probably altered and dolomitized by mixed water, as no evaporitic minerals were observed in this study. In the Triassic and Jurassic sections, the δ^{18}O$_{VPDB}$ shifts toward a lower value in comparison with the original depositional facies, are quite abrupt and vary around their respective average (−6.7‰ and −6.4‰) in the range of the "meteoric" [3]. The strong deviation in δ^{13}C$_{VPDB}$ values from this line is related to the rock–water interaction in the Type I Triassic samples. Very weak variations or near constant δ^{13}C$_{VPDB}$ values in the Type I Jurassic samples suggest decreased rock–water interactions in a distal setting from the meteoric recharge with a consequent buffering by the rocks.

Types II and III have depleted δ^{18}O values (up to −10.4‰ for Type III in the Triassic and up to −11.6‰ for Type III in the Jurassic) with more or less constant δ^{13}C$_{VPDB}$ values in the Jurassic (Types II and III) and a wider range of δ^{13}C$_{VPDB}$ values in Type III in the Triassic. The δ^{13}C values are similar to the ones of the early dolomitic facies suggesting that the saddle dolomitizing fluid was mainly buffered by interactions with early dolomite rocks (Type I). The same buffered system has been inferred from our trace element analyses and showed that positive and gradual variation in Fe and Mn contents observed between Types II and III saddle dolomites due to water–rock interactions (Figure 13).

The significant shift to lower δ^{18}O$_{VPDB}$ values from Type II to Type III in both sections is probably due to the interaction of the dolomitized rocks (Type I) with hydrothermal fluids (e.g., [12,43]). Similar carbon and oxygen isotopic values (−9.0‰ to −7.3‰ and +1.8‰ to +2.4‰, respectively) were reported by Nader et al. [42] in the Jurassic rocks at Lebanon as the result of hydrothermal hot fluids under deep conditions. The tension-gashed stylolitization phase during the saddle emplacement between Type II and Type III saddle dolomites suggests that the chemical compaction or burial was coeval with the late phase of hot fluids (Type III dolomite). This phase is associated the most widespread depleted δ^{18}O$_{VPDB}$ values (Figure 14). Triassic Type II facies show some similarities and overlapping δ^{18}O$_{VPDB}$ values, such as those of Type I, suggesting that saddle precipitation started in the same burial conditions of the latter, but after early stylolitization. These complex scenarios of paragenetic models show that Types II and III correspond to complex polyphased dolomitizing fluids and Type III saddle dolomite was precipitated under deep burial–mesogenetic diagenesis. Their geometries (irregular, strata-discordant, zebras, even faults on the field) may suggest that the faults constituted a hydrothermal fluid conduit operating during two major steps with increasingly $^{18}O_{VPDB}$ depleted values in the saddles as evidenced by two different isotopic domains in both sections. Further evidence for deep burial setting and mesogenesis includes the dissolution of saddle dolomite along stylolites (e.g., [44]) and sutured intergranular contacts between ooids.

Type IV Fe-poor calcite records a dedolomitization and cementation by similar calcite due to the incursion of meteoric waters [40,45]. The wide variations of the $\delta^{18}O_{VPDB}$ and $\delta^{13}C_{VPDB}$ values reveals an inverse "J" Lohmann curve [40,45], at least in the Jurassic and probably in the Triassic, supporting a telogenetic meteoric origin. The $\delta^{18}O_{VPDB}$ and $\delta^{13}C_{VPDB}$ values from calcite samples overlap with the values of the HT dolomite; a similar situation has been reported from NW Spain by Shah et al. [46] and Nader et al. [42] in Lebanon on hydrothermal-hosted Jurassic rocks. These authors concluded that the wide range of depleted isotopic oxygen values with a slight shift in the isotopic carbon values of pristine facies indicate different temperatures of the dolomitizing fluids, while the extra-shifted carbon isotope values and highly depleted isotopic oxygen values in the calcite facies indicate surface-derived fluids. The abrupt Fe and Mn drops, Sr rise and weak Mn variations only reported in the Triassic samples that could be related to the SD3 overgrowth by $C1_T$ suggest a significant lack in carbonate-buffering system. In addition, the saddle dolomitic contribution to calcite oxygen contents is very low compared to the contribution of the meteoric fluids [46] due to high $\delta^{18}O_{water}/\delta^{18}O_{rock}$ ratios. These arguments support that the calcite cement marks the end-member of the diagenetic sequence as a result of a late uplift of the carbonate platform, which initiated this late telogenetic phase (Figure 14).

5.3. Impact of Basement–Fluid Interactions, Continental Riverine and Oceanic Fluxes on Strontium Isotope Signatures

Multiple parameters controlling the radiogenic strontium isotope in carbonate rocks, including: (i) the isotopic composition of seawater from which they precipitated and terrigenous (i.e., siliciclastics) sedimentary rocks; (ii) the $^{87}Sr/^{86}Sr$ ratio is mainly related to the nature of the parent rocks and their continental weathering rates (riverine) and oceanic fluxes of nonradiogenic strontium. However, diagenesis will be the main factor controlling the significant variation of the $^{87}Sr/^{86}Sr$ ratio in marine carbonates as it will be affected by hydrothermal or cold fluids that control the strontium ratio.

$^{87}Sr/^{86}Sr$ in dolomites (Types I, II, III) vary from 0.707735 and 0.712023 in the Triassic and from 0.708680 up to 0.711203 in the Jurassic (Figure 15). These ratios are associated to the early and late stylolites, prior to the latest deformation during the uplift (Figure 14). The $^{87}Sr/^{86}Sr$ ratios are higher than those of ambient Triassic and Jurassic seawater [26,27]. A few of these ratios in the Triassic dolomites are lower than those of the marine carbonate obtained from host limestone samples (0.708766; Figure 15), while in the Jurassic dolomites the ratios are higher than those in the marine carbonate samples (0.707862; Figure 15).

When comparing our radiogenic values with the ones of the Triassic $^{87}Sr/^{86}Sr$ signals of the Lombardic Alps, the strontium values of the Lombardic Alps increased from 0.70805 in the Anisian to 0.708148 in the Ladinian as a result of the various rates of weathering and erosion of young volcanic rocks and old granitic rocks [47], are lower than our values in Provençal Domain (0.708766; Figure 15). Thus, it is more likely that an increase in the riverine flux was the dominant factor causing the $^{87}Sr/^{86}Sr$ seawater signature rise during the Triassic times in Provençal Domain (cf. [26]), as well as the involved Triaasic carbonates by meteoric waters.

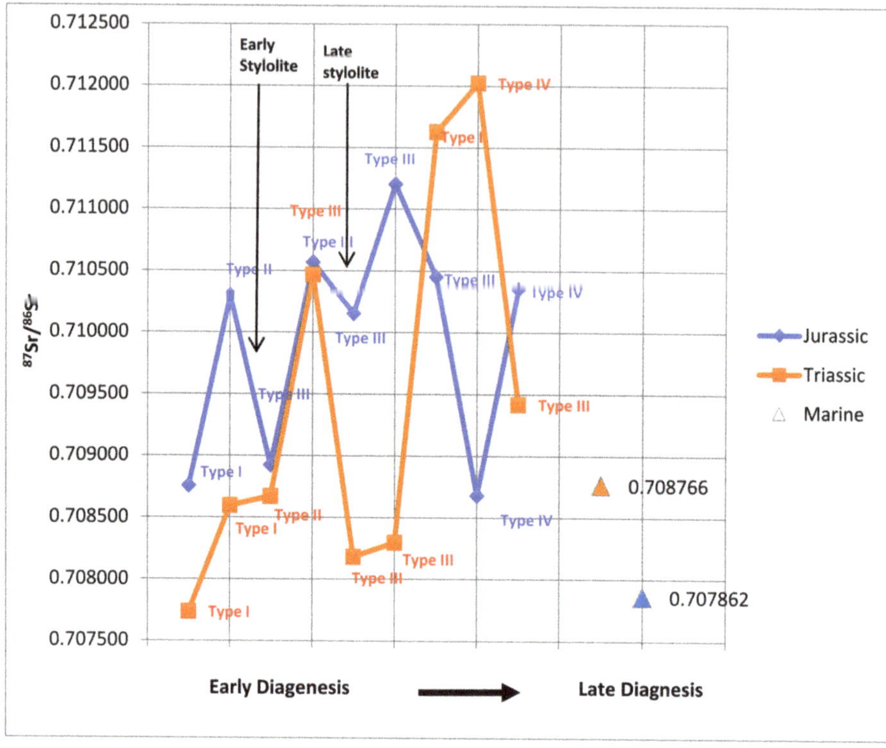

Figure 15. The rate of change of $^{87}Sr/^{86}Sr$ according to time in order to illustrate the number of jumps between the different diagenetic conditions in both sections separately. The number of jumps (down and up) can calculate the number of fault reactivation ($n = 20$).

The repeated injection of hydrothermal fluids into the Mesozoic succession in the Provençal Domain and formation of many phases of saddle dolomites might have been expelled by from overpressured lithologies and fluxed through deep-seated faults (Figures 4, 6, 10 and 14; e.g., [30]). The high Sr isotopic ratios of saddle dolomite suggest that the hydrothermal fluids interacted with the crystalline basement and/or siliciclastic deposits [12,30].

The radiogenic enrichment of the Jurassic samples could be deviated from open ocean evaporite, however, the dolomitizing fluids are not in agreement with the involvement of the evaporites which are missing here. If a high strontium ratio represents the involvement of evaporites, then it is expected that a high strontium ratio would correlate with the depletion of oxygen values. Mostly, the dolomitizing fluids that have high strontium ratios would have been enriched with heavier oxygen isotope values, confirming the involvement of evaporites (e.g., [48]). However, this scenario is not in agreement with our Jurassic dolomite samples and the regional geology, therefore the interaction of hot dolomitizing fluid mostly interacted with the Argentera crystalline basement before flowing up to the Jurassic carbonate rocks.

In contrast to Jurassic dolomites, some Triassic dolomites have significantly lower $^{87}Sr/^{86}Sr$ values with respect to the open marine facies values. The depletion of $^{87}Sr/^{86}Sr$ ratios started from the Late Triassic up to the Turonian recording pulses of mid-oceanic hydrothermal fluxes that related to the opening and closure of the North Atlantic Ocean [27]. There were several short-term excursions of radiogenic depletion of $^{87}Sr/^{86}Sr$ ratios highlighting these pulses. The authors of [49] reported excursions of very low $^{87}Sr/^{86}Sr$ ratios during three short periods (Early Jurassic, Early Cretaceous and Late Cretaceous) that are linked to the rate of hydrothermal activity as a consequence of increased ocean-crust production at the mid-oceanic ridges. Therefore, some of our data in the Triassic could be partly influenced by the short-term excursions of strontium depletion after the deposition of Triassic sediments. In addition, following McArthur's Graph, our data display that Type II saddle dolomites are associated with a jump-down of the $^{87}Sr/^{86}Sr$ isotope values. The repetition of the jump-down and jump-up graph of each type of dolomite follows the same trends in the Triassic and Jurassic samples (Figure 15). The graph suggests that the Triassic and Jurassic sections had the same stratigraphic position (not separated) prior to the precipitation of the calcite phase (Type IV).

The ultimate diagenetic phase (Type IV) records the highest enrichment radiogenic composition (0.712023) from the Triassic and the lowest $^{87}Sr/^{86}Sr$ ratio (0.708680) from the Jurassic samples. This phase attributed to the late stage of calcite cementation and was consistent with the very depleted $\delta^{18}O_{VPDB}$ values and the lowest $\delta^{13}C_{VPDB}$ values (cf. [50]). The meteoric water has the lowest $\delta^{18}O_{VPDB}$ values and highest $^{7}Sr/^{86}Sr$ ratios (0.709630 to 0.70948, [51]). The increasing recharge from meteoric water could apply to the late stage of Fe-poor calcite precipitation and is probably related to an uplift of the buried Triassic–Jurassic rocks (Figures 11–14). The late calcite cement C1$_J$ in the Jurassic section shifted toward decreasing Sr isotope values, and appears to be more affected by other sources besides meteoric water (in contrast to Triassic calcite value). On the basis of the tectonic graph, C1$_T$ and C1$_J$ are attributed to two different fluids, which were stratigraphically separated during the uplifting of these units (outcrops). Consequently, the lowering of Sr isotope compositions (C1$_J$) in this study, especially in the Jurassic section, could be partly linked to the pulses of the seafloor hydrothermal activity that lowered the $^{87}Sr/^{86}Sr$ ratios. During the late calcite event, the former dolomitized rocks were strongly brecciated with an in situ accumulation of crushed fragments. This event reveals that an intense tectonic activity was ongoing during the late fluid migration [30,52,53], which could be associated to Late Cretaceous–Eocene times, where general strike-slip fault systems occurred in the European paleomargin [52,53].

The Triassic–Jurassic carbonate seawater in Provençal Domain could partly have been influenced by the input of various rates of weathering and erosion of young volcanic rocks and/or old granitic rocks. These inputs probably persisted after the deposition of Mesozoic succession in Provencal Domain. Finally, fluctuations in the rising and falling of $^{87}Sr/^{86}Sr$ seawater signatures have already been reported [26,27], therefore, the values of the $^{87}Sr/^{86}Sr$ fluctuations are the important factors and may have influence the diagenesis products beside the hydrothermal fluids and meteoric waters in Provencal Domain (Figure 15).

6. Conclusions

The main conclusions derived from this integrated field, petrographic and geochemical study of dolomitization in the Mesozoic carbonate succession, Provençal Domain, Maritime Alps, SE France show that:

1. The succession is extensively dolomitized with abundant breccia and zebra textures, and complex fracturing paths due to the influence of repeatedly injections of hydrothermal fluids.
2. Three diagenetic settings are recognized:

    (i) An eogenetic realm with weak diagenetic imprints. This setting is confirmed by $\delta^{18}O_{VPDB}$ and $\delta^{13}C_{VPDB}$ values that fit the inverse "J" Lohmann curve and meteoric line models.

    (ii) A second mesogenetic setting in two phases, the first one started with the generation of wispy stylolites ("early dolomite"), and various networks of fractures as well as zebra and breccia fabrics. The fabrics are associated with a first generation of medium- to coarse-grained dominantly euhedral saddle dolomites (Type II) formed through the migration of hot dolomitizing fluids. This occured during early Cretaceous times and was related to the extensional and transtensional tectonics of the European paleomargin of the Alpine Tethys. In the second phase (our Type III dolomite), further physical brecciation and cataclastic fractures cross cut the Type II dolomite under deeper burial conditions with late stylolite-related fractures (tension gashes). The $\delta^{18}O_{VPDB}$ values become more depleted with subsequent increases in the saddle crystal sizes. The $\delta^{18}O_{VPDB}$ saddle dolomite values (up to $-8.3‰$ in Type II and up to $-11.6‰$ in Type III) suggest that at least two pulses of hydrothermal fluids during the precipitation of saddle dolomites occurred. The chemical composition of the successive saddle dolomites was progressively modified from nonferroan to ferroan-rich in an open system and the positive co-variant trends of $^{87}Sr/^{86}Sr$ vs. $\delta^{18}O_{VPDB}$ strongly support an origin from from hot fluids in a deep burial-reducing mesogenetic environment.

(iii)   The ultimate diagenesis is related to the late phase of calcite cement precipitation, synchronous with Fe and Mn depletion and ultra negative $\delta^{18}O_{VPDB}$-$\delta^{13}C_{VPDB}$ values. A meteoric water under near-surface conditions during telogenesis and uplifting of the Triassic–Jurassic succession is inferred.

3.   The depletion and enrichment of radiogenic strontium signals could also be linked to the pulses of mid-oceanic hydrothermal fluxes and an input of riverine fluxes during the opening and closure of the North Atlantic Ocean, beside the involvement of hydrothermal fluids and meteoric waters on Triassic–Jurassic successions.

Author Contributions: Conceptualization, N.S.; methodology, N.S.; software, N.S.; validation, N.S. and A.P.; formal analysis, N.S. and A.P.; investigation, N.S. and A.P.; resources, N.S.; data curation, N.S.; writing—original draft preparation, N.S.; writing—review and editing, H.M. and A.P.; visualization, N.S.; supervision, A.P. and H.M.; project administration, N.S., A.P. and H.M. All authors have read and agreed to the published version of the manuscript.

Funding: This research received no external funding.

Acknowledgments: The authors kindly thank Luca Martire, Luca Barale, and Carlo Bertok from Dipartimento di Scienze della Terra, Università di Torino, Italy for collecting samples at the field "Provençal Domain". We acknowledge also the support of Philippe Muchez from Department of Earth and Environmental Sciences, KUL-university-Belgium for helpful revision of the manuscript.

Conflicts of Interest: The authors declare no conflict of interest.

References

1. Machel, H.G. Concepts and models of dolomitization: A critical reappraisal. In *The Geometry and Petrogenesis of Dolomite Hydrocarbon Reservoirs*; Braithwaite, C.J.R., Rizzi, G., Darke, G., Eds.; Geological Society of London: London, UK, 2004; Volume 235, pp. 7–63.

2. Fu, Q.; Quing, H. Medium and coarsely crystalline dolomites in the Middle Devonian Ratner Formation, southern Saskatchewan, Canada: Origin and pore evolution. *Carbonates Evaporites* **2011**, *26*, 111–125. [CrossRef]

3. Land, S. The origin of massive dolomite. *J. Geol. Educ.* **1985**, *33*, 87–110. [CrossRef]

4. Machel, G.; Mountjoy, W. Chemistry and environment of dolomitization: A reappraisal. *Earth Sci. Rev.* **1986**, *23*, 175–222. [CrossRef]

5. Hardie, A. Dolomitization: A critical view of some current views. *J. Sediment. Petrol.* **1987**, *57*, 166–183. [CrossRef]

6. Qing, H.; Mounjoy, W. Formation of coarsely crystalline, hydrothermal dolomite reservoir in the Presqu'ile Barrier, western Canada Sedimentary Basin. *Am. Assoc. Pet. Geol.* **1994**, *78*, 55–77.

7. Al-Aasm, I.S. Origin and characterization of hydrothermal dolomite in the Western Canada Sedimentary Basin. *J. Geochem. Explor.* **2003**, *78–79*, 9–15. [CrossRef]

8. Smith, B. Origin and reservoir characteristics of Upper Ordovician Trenton-Black River hydrothermal dolomite reservoirs in New York. *Bull. Am. Assoc. Pet. Geol.* **2006**, *90*, 1691–1718. [CrossRef]

9. Lavoie, D.; Chi, G. Lower Paleozoic foreland basins in eastern Canada: Tectono-thermal events recorded by faults, fluids and hydrothermal dolomites. *Bull. Can. Pet. Geol.* **2010**, *58*, 17–35. [CrossRef]

10. Hardie, A. On the significance of evaporites. *Annu. Rev. Earth Planet. Sci.* **1991**, *19*, 131–168. [CrossRef]

11. Iannace, A.; Gasparrini, M.; Gabellone, T.; Mazzoli, S. Late dolomitization in basinal limestones of the southern Apennines fold and thrust belt (Italy). *Oil Gas Sci. Technol. Rev. d'IFP Energ. Nouv.* **2012**, *67*, 59–75. [CrossRef]

12. Davies, G.; Smith, L. Structurally controlled hydrothermal dolomite reservoir facies: An overview. *Am. Assoc. Pet. Geol. Bull.* **2006**, *90*, 1641–1690. [CrossRef]

13. Mohn, G.; Manatschal, G.; Muntener, O.; Beltrando, M.; Masini, E. Unravelling the interaction between tectonic and sedimentary processes during lithospheric thinning in the Alpine Tethys margins. *Int. J. Earth Sci.* **2010**, *99*, 75–101. [CrossRef]

14. Barale, L.; Bertok, C.; Talabani, S.N.; d'Atri, A.; Martire, L.; Piana, F.; Préat, A. Very hot, very shallow hydrothermal dolomitization: An example from the Maritime Alps (north-west Italy–south-east France). *Sedimentology* **2016**, *63*, 2037–2065. [CrossRef]

15. Durand, M. Permian to Triassic continental successions in southern Provence (France): An overview. *Geol. Soc. Italy* **2008**, *127*, 697–716.

16. De Graciansky, C.; Lemoine, M. Early Cretaceous extensional tectonics in the southwestern French Alps: A consequence of North-Atlantic rifting during Tethyan spreading. *Bull. Société Géologique Fr.* **1988**, *8*, 733–737. [CrossRef]

17. Dardeau, G.; Pascal, A. La régression fin-Jurassique dans les Alpes-Maritimes: Stratigraphie, faciès, environnements sédimentaires et influence du bâti structural dans l'arc de Nice. *Bull. Bur. Rech. Géologiques Minières* **1982**, *3*, 193–204.

18. Debrand-Passard, S.; Courbouleix, S.; Lienhart, J. Geological synthesis of south-east France. *Bur. Geol. Min. Res. Orléans* **1984**, *125*, 615.

19. McArthur, M.; Howarth, J.; Shields, A. Strontium isotope stratigraphy. In *The Geologic Time Scale 2012*; Gradstein, F.M., Ogg, J.G., Schmitz, M., Ogg, G.M., Eds.; Elsevier: Boston, MA, USA, 2012; Volume 2, pp. 127–144.

20. Nelson, A. Significance of fracture sets associated with stylolite zones. *Am. Assoc. Pet. Geol. Bull.* **1981**, *65*, 2417–2425.

21. Eren, M. Origin of stylolite related fractures in Atoka Bank Carbonates, Eddy County, New Mexico, U.S.A. *Carbonates Evaporites* **2005**, *20*, 42–49. [CrossRef]

22. Jones, E.; Tarney, J.; Baker, H.; Gerouki, F. Tertiary granitoids of Rhodope, Northern Greece: Magmatism related to extensional collapse of the Hellenic orogen? *Tectonophysics* **1992**, *210*, 295–314. [CrossRef]

23. Veizer, J.; Hoefs, J. The nature of O^{18}/O^{16} and C^{13}/C^{12} secular trends in sedimentary carbonate rocks. *Geochim. Cosmochim. Acta* **1976**, *40*, 1387–1395. [CrossRef]

24. Veizer, J.; Ala, D.; Azmy, K.; Bruckschen, P.; Buhl, D.; Bruhn, F.; Carden, G.A.F.; Diener, A.; Ebneth, S.; Goddéris, Y.; et al. $^{87}Sr/^{86}Sr$, $\delta^{13}C$ and $\delta^{18}O$ evolution of Phanerozoic seawater. *Chem. Geol.* **1999**, *161*, 59–88. [CrossRef]

25. Katz, M.E.; Wright, J.D.; Miller, K.G.; Cramer, B.S.; Fennel, K.; Falkowski, P.G. Biological overprint of the geological carbon cycle. *Mar. Geol.* **2005**, *217*, 323–338. [CrossRef]

26. Korte, C.; Kozur, W.; Bruckschen, P.; Veizer, J. Strontium isotope evolution of Late Permian and Triassic seawater. *Geochim. Cosmochim. Acta* **2003**, *67*, 47–62. [CrossRef]

27. Jones, E.; Jenkyns, C.; Coe, L.; Hesselbo, P. Strontium isotopic variations in Jurassic and Cretaceous seawater. *Geochim. Cosmochim. Acta* **1994**, *58*, 3061–3074. [CrossRef]

28. Gregg, J.M.; Sibley, D.F. Epigenetic dolomitization and origin of xenotopic dolomite texture. *J. Sediment. Res.* **1984**, *54*, 908–931.

29. Sibley, F.; Gregg, M. Classification of dolomite rock textures. *J. Sediment. Res.* **1987**, *57*, 967–975.

30. Salih, N.; Mansurbeg, H.; Kolo, K.; Gerdes, A.; Préat, A. In situ U-Pb dating of hydrothermal diagenesis in tectonically controlled fracturing in the Upper Cretaceous Bekhme Formation, Kurdistan Region-Iraq. *Int. Geol. Rev.* **2019**, *0020–6814*, 1938–2839. [CrossRef]

31. Kretz, R. Carousel model for the crystallization of saddle dolomite. *J. Sediment. Petrol.* **1992**, *62*, 551–574.

32. Nelson, A. Natural fracture systems: Description and classification. *Am. Assoc. Pet. Geol. Bull.* **1979**, *63*, 2214–2232.

33. Morad, S.; Al Suwaidi, M.; Mansurbeg, H.; Morad, D.; Ceriani, A.; Papagoni, M.; Al-Aasm, I. Diagenesis of a limestone reservoir (Lower Cretaceous), Abu Dhabi, United Arab Emirates: Comparison between the anticline crest and flanks. *Sediment. Geol.* **2019**, *380*, 127–142. [CrossRef]

34. Carraro, F.; Dal Piaz, V.; Franceschetti, B.; Malaroda, R.; Sturani, C.; Zanella, E. Note Illustrative della Carta Geologica del Massiccio dell'Argentera alla scala 1: 50.000. *Mem. Soc. Geol. Italy* **1970**, *9*, 557–663.

35. Malaroda, R. L'Argentera meridionale–Memoria illustrativa della "Geological Map of Southern Argentera Massif (Maritime Alps) 1:25,000". *Sci. Géologiques–Mémoires* **1999**, *51–52*, 241–331.

36. Bigot-Cormier, F.; Poupeau, G.; Sosson, M. Dénudations différentielles du massif cristallin externe alpin de l'Argentera (Sud-Est de la France) révélées par thermochronologie traces de fission (apatites, zircons). *Comptes Rendus l'Académie Sci. Paris* **2000**, *330*, 363–370. [CrossRef]

37. Wierzbicki, R.; Dravis, J.; Al-Aasm, I.; Harland, N. Burial dolomitization and dissolution of Upper Jurassic Abenaki platform carbonates, Deep Panuke reservoir, Nova Scotia, Canada. *Am. Assoc. Pet. Geol.* **2006**, *90*, 1843–1861. [CrossRef]

38. Crampton, I.; Allen, A. Recognition of forebulge unconformities associated with early-stage Foreland Basin development—Example from the North Alpine Foreland Basin. *Am. Assoc. Pet. Geol. Bull.* **1995**, *79*, 1495–1514.

39. D'Arti, A.; Piana, F.; Barale, L.; Bertok, C.; Martire, L. Geological setting of the southern termination of Western Alps. *Int. J. Earth Sci.* **2016**, *105*, 1831–1858.

40. Lohmann, C. Geochemical patterns of meteoric diagenetic systems and their application to studies of paleokarst. In *Paleokarst*; James, N.P., Choquette, P.W., Eds.; Springer: New York, NY, USA, 1988; pp. 58–80.

41. Allan, R.; Mathhews, K. Carbon and oxygen isotopes as diagenetic and stratigraphic tools: Data from surface and subsurface Barbados, West Indies. *Geology* **1977**, *5*, 16–20. [CrossRef]

42. Nader, H.; Swennen, R.; Ottenburgs, R. Karst-meteoric dedolomitisation in Jurassic carbonates, Lebanon. *Geol. Belg.* **2003**, *6*, 3–23.

43. Moore, H. Carbonate Diagenesis and Porosity. *Dev. Sedimentol.* **1989**, *46*, 338.

44. Wright, P.; Harris, P. Carbonate Dissolution and Porosity Development in the Burial (mesogenetic) Environment. In Proceedings of the Oral presentation at American Association of Petroleum Geologists, Annual Convention and Exhibition, Pittsburgh, PA, USA, 19–22 May 2013.

45. Salih, N.; Kamal, M.; Préat, A. Classical observations and stable isotopes for identification the diagenesis of Jeribe formation at Jambour oil Fields-Kurdistan Region-Iraq. *Pet. Sci. Technol.* **2019**, *37*, 1548–1556. [CrossRef]

46. Shah, M.; Nader, H.; Garcia, D.; Swennen, E.R. Hydrothermal dolomites in the Early Albian (Cretaceous) platform carbonates (NW Spain): Nature and origin of dolomites and dolomitizing fluids. *Oil Gas Sci. Technol.* **2012**, *67*, 97–122. [CrossRef]

47. Faure, G.; Assereto, R.; Tremba, L. Strontium isotope composition of marine carbonates of Middle Triassic to Early Jur assic age, Lombardic Alps, Italy. *Sedimentology* **1978**, *25*, 523–543. [CrossRef]

48. Li, S.; Levin, E.; Chesson, A. Continental scale variation in [17]O-excess of meteoric waters in the United States. *Geochim. Cosmochim. Acta* **2015**, *164*, 110–126. [CrossRef]

49. Jones, E.; Jenkyns, C. Seawater strontium isotopes, oceanic anoxic events, and seafloor hydrothermal activity in the Jurassic and Cretaceous. *Am. J. Sci.* **2001**, *301*, 112–149. [CrossRef]

50. Salih, N.; Mansurbeg, H.; Kolo, K.; Préat, A. Hydrothermal Carbonate Mineralization, Calcretization, and Microbial Diagenesis Associated with Multiple Sedimentary Phases in the Upper Cretaceous Bekhme Formation, Kurdistan Region-Iraq. *Geosciences* **2019**, *9*, 459. [CrossRef]

51. Dong, S.; Chen, D.; Qing, H.; Zhou, X.; Wang, D.; Guo, Z.; Jiang, M.; Qian, Y. Hydrothermal alteration of dolostones in the Lower Ordovician, Tarim Basin, NW China: Multiple constraints from petrology, isotope geochemistry and fluid inclusion microthermometry. *Mar. Pet. Geol.* **2013**, *46*, 270–286. [CrossRef]

52. Sanchez, G.; Rolland, Y.; Jolivet, M.; Brichau, S.; Corsini, M.; Carter, A. Exhumation controlled by transcurrent tectonics: The Argentera-Mercantour massif (SW Alps): Exhumation controlled by transcurrent tectonics in SW Alps. *Terra Nova* **2011**, *23*, 116–126. [CrossRef]

53. Dumont, T.; Schwartz, S.; Guillot, S.; Simon-Labric, T.; Tricart, P.; Jourdan, S. Structural and sedimentary records of the Oligocene revolution in the Western Alpine arc. *J. Geodyn.* **2012**, *56–57*, 18–38. [CrossRef]

Article

REE Characteristics of Lower Cretaceous Limestone Succession in Gümüşhane, NE Turkey: Implications for Ocean Paleoredox Conditions and Diagenetic Alteration

Merve Özyurt [1,2,*], M. Ziya Kırmacı [1], Ihsan Al-Aasm [3], Cathy Hollis [2], Kemal Taslı [4] and Raif Kandemir [5]

[1] Department of Geological Engineering, Karadeniz Technical University, 61080 Trabzon, Turkey; kirmaci@ktu.edu.tr

[2] School of Earth and Environmental Science, University of Manchester, Manchester M13 9PL, UK; Cathy.Hollis@manchester.ac.uk

[3] School of the Environment, University of Windsor, Windsor, ON N9B 3P4, Canada; alaasm@uwindsor.ca

[4] Department of Geological Engineering, Mersin University, 33343 Mersin, Turkey; ktasli@mersin.edu.tr

[5] Department of Geological Engineering, Recep Tayyip Erdoğan University, 53100 Rize, Turkey; raif.kandemir@erdogan.edu.tr

[*] Correspondence: merveyildiz@ktu.edu.tr

Received: 2 July 2020; Accepted: 22 July 2020; Published: 30 July 2020

Abstract: Trace and rare earth elements (REEs) are considered to be reliable indicators of chemical processes for the evolution of carbonate systems. One of the best examples of ancient carbonate successions (Berdiga Formation) is widely exposed in NE Turkey. The Lower Cretaceous limestone succession of Berdiga Formation may provide a case study that reveals the effect of ocean paleoredox conditions on diagenetic alteration. Measurement of major, trace and REEs was carried out on the Lower Cretaceous limestones of the Berdiga Formation, to reveal proxies for paleoredox conditions and early diagenetic controls on their geochemistry. Studied micritic limestone microfacies (MF-1 to MF-3) indicate deposition in the inner platform to a deep shelf or continental slope paleoenvironment during the Hauterivian-Albian. The studied limestone samples mainly exhibit low Mg-calcite characteristics with the general chemical formula of $Ca_{98.35-99.34}Mg_{0.66-1.65}(CO_3)$. They are mostly represented by a diagnostic REE seawater signature including (1) slight LREE depletion relative to the HREEs (ave. 0.72 of Nd/Yb_N and ave. 0.73 of Pr/Yb_N), (2) negative Ce anomalies ($Ce/Ce^* = 0.38-0.81$; ave. 0.57), (3) positive La anomaly ($La/La^* = 0.21-3.02$; ave. 1.75) and (4) superchondritic Y/Ho (ave. 46.26). Studied micritic limestones have predominantly low Hf (bdl to 0.5 µg/g), Sc (bdl to 2 µg/g) Th (bdl to 0.9 µg/g) contents suggesting negligible to minor shale contamination. These findings imply that micritic limestones faithfully record chemical signals of their parental and diagenetic fluids. The succession also exhibits high ratios of Eu/Eu^* (1.01–1.65; ave. 1.29 corresponding to the positive Eu anomalies), Sm/Yb (1.26–2.74; ave. 1.68) and La/Yb ratios (0.68–1.35; ave. 0.9) compared to modern seawater and wide range of Y/Ho ratios (29.33–70.00; ave. 46.26) which are between seawater and hydrogenetic Fe-Mn crusts. Several lines of geochemical evidence suggest water-rock interaction between parental seawater and basaltic rocks at elevated temperatures triggered by hydrothermal activity associated with Early Cretaceous basaltic magma generation. The range of Ce/Ce^* values is suggestive of mostly oxic to dysoxic paleoceanographic conditions, with a sudden change to dysoxic conditions ($Ce/Ce^* = 0.71-0.81$), in the uppermost part of the MF-1. This is followed by an abrupt deepening paleoenvironment with a relative increase in the oxic state of the seawater and deposition of deeper water sediments (MF-2 and MF-3) above a sharp transition. The differences in microfacies characteristics and foraminifera assemblage between MF-1 and overlying facies (MF-2 and MF-3) may also confirm the change in paleoceanographic conditions. Therefore, REEs data obtained from studied limestones have the potential to contribute important information as to regional paleoceanographic conditions of Tethys during an important period in Earth history.

Keywords: platform carbonates; REE + Y chemistry; paleoceanographic proxies; diagenetic proxies; NE Turkey

---

1. Introduction

The rare earth element and Y (REE + Y) signature of carbonates have been widely used to reconstruct the paleoenvironmental history of seawater in deep time [1–6]. REE + Y have similar coherent chemical properties in the marine system and exhibit systematic changes in the chemical properties across the REE (La to Lu) series due to the progressive filling of the f-electron shell [7–13]. This causes different fractionations of Lantonides in natural systems; heavier REEs (HREE, from Ho to Lu) tend to be preferentially complexed, while lighter REEs (LREE, from La to Nd) are preferentially scavenged by particles [13]. Further, the rare earth elements (REEs) occur in the trivalent state in seawater with the exception of multiple oxidation states for Ce and Eu [13]. Ce (Ce^{3+}, Ce^{4+}) and Eu (Eu^{2+}, Eu^{3+}) are redox-sensitive elements, and they show distinct geochemical behaviour compared to other Lantonides [13,14]. Ce (Ce^{3+}, Ce^{4+}) and Eu (Eu^{2+}, Eu^{3+}) are generally considered a natural proxy for revealing interaction processes between particles and solutions, and redox reactions [15–18]. These geochemical features allow carbonate sedimentologists to reconstruct ancient environmental conditions [19–22].

The Upper Jurassic-Lower Cretaceous Berdiga Formation, which consists dominantly of platform carbonates are widely exposed in the Eastern Pontides, NE Turkey. This carbonate succession is one of the best-preserved examples of ancient shallow marine environmental systems in the passive continental margin of Laurasia [23]. The carbonate succession is well defined in terms of its stratigraphical and sedimentological characteristics [23–37]. Sediments are platform carbonates which are represented by varying lithofacies ranging from supratidal to continental slope [23,27–33,38]. The platform carbonates have undergone multiple stages of diagenesis, including dolomitization and recrystallisation [23,31–34,39]. Dolomitization was interpreted to have been initiated by seawater and/or partly modified seawater prior to compaction during shallow-moderate burial, and massive dolomite bodies were then continuously recrystallised by the circulation of episodic diagenetic fluids with elevated temperatures during deeper burial [31]. The recrystallisation processes have been related to widespread Late Cretaceous to Cenozoic magma generation [23,31–33].

Although the diagenetic history seems complex, the mechanisms and origins of dolomitization have been well described in recent studies [31]. However, there has not been an attempt to investigate the micritic limestone strata that are least diagenetically altered and should, therefore, provide proxies for paleoredox conditions and early diagenetic controls on their geochemistry. Consequently, this study focuses on the limestone strata outcropping in the Mescitli area in the southern part of Eastern Pontide (NE Turkey). Here, we present a petrographic analysis coupled with REE+Y characterisation of the limestone succession. The main purpose of the current study is (1) to determine ancient seawater composition using marine carbonates of NE Turkey, from inner platform to slope paleoenvironments; (2) to define the redox conditions and discuss the paleoceanographic proxies and (3) to discuss the role of potential diagenetic influence on REE signals of ancient carbonates.

2. Geological Setting

Turkey comprises a series of tectonic units, which are separated by ophiolitic suture zones [40,41]. The Sakarya Zone consists of east-west trending Alpine continental fragments along with northern Turkey and is generally known as the Eastern Pontides, constituting a significant part of the Alpine-Himalayan system. Magmatic, metamorphic and sedimentary Mesozoic to Cenozoic rocks are widely exposed in Eastern Pontide (e.g., [40,41]), and a well-constrained tectono-magmatic and stratigraphic framework has been established (Figure 1). Eastern Pontide can roughly be separated into two main subzones based on its dominant lithological differences from north to south [40–42] (Figure 1). The northern subzone is mainly characterized by Late Mesozoic and Early Cenozoic magmatic rocks [42–49], while the southern subzone is predominantly characterized by Jurassic carbonate and clastic sedimentary rocks [50–54].

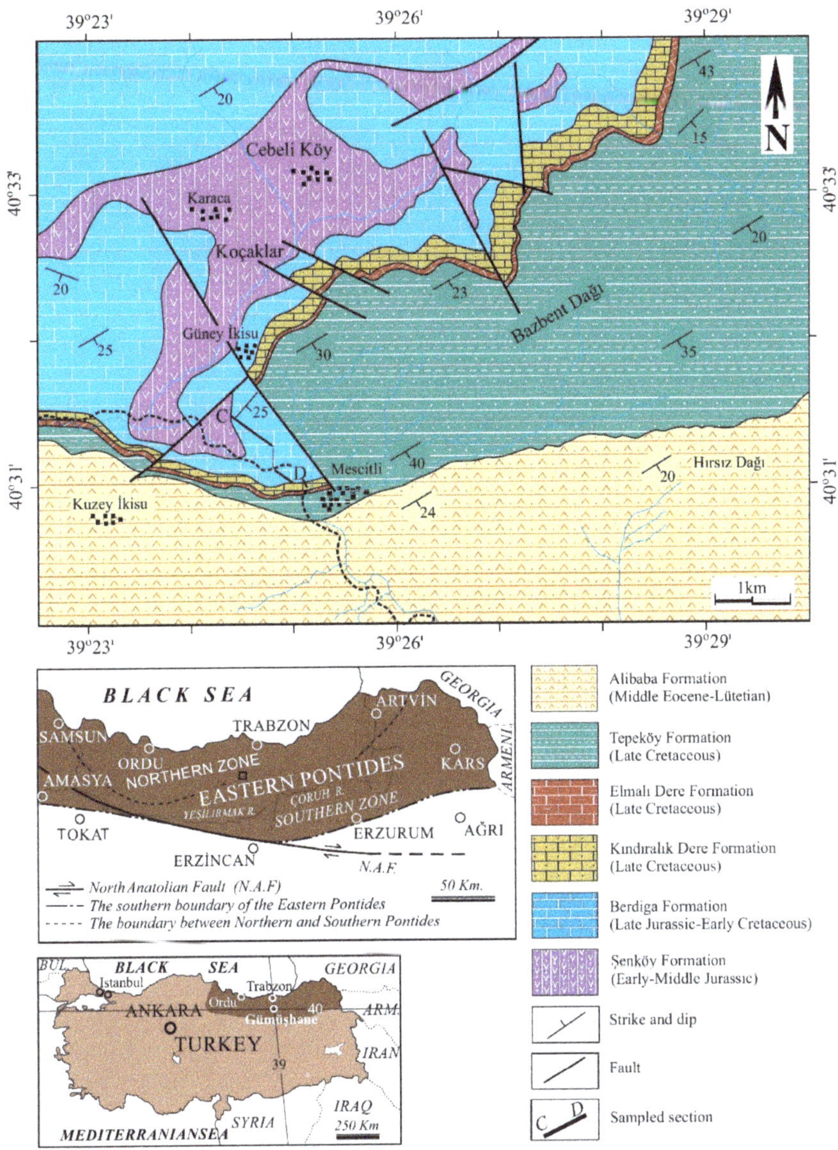

Figure 1. Location and simplified geological map of the Mescitli area [54].

The Mescitli Area (Gümüşhane, NE Turkey) is located in the southern part of the Eastern Pontide, where Upper Jurassic-Lower Cretaceous carbonates are well exposed (Figure 1). In the southern part, the Hercynian basement mainly constitutes a metamorphic unit dated as 320.3 ± 1.7 Ma [55] and Upper Carboniferous Gümüşhane granitoid [56,57]. These basement rocks are generally overlain by the Lower to Middle Jurassic volcano-sedimentary successions consisting mainly of siliciclastics, basalt–andesite and associated volcanoclastic, deposited within an extensional (rift) basin, and rarely Ammonitico-Rosso limestone facies [58,59]. Upper Jurassic to Lower Cretaceous carbonate successions (Berdiga Formation) [35] lie conformably over the volcano-sedimentary successions. Based on benthic foraminiferal assemblages, the platform carbonates in the Eastern Pontides are of Oxfordian to Albian age [27–29,35–37]. In the Mescitli Area, the lower part of the Berdiga Formation is completely dolomitized [31] while the upper part comprises a well-preserved limestone succession which is the focus of this study. The succession is overlain by an Upper Cretaceous unit which comprises three different sedimentary assemblages: (1) yellowish coloured, sandy limestones, (2) globotruncanid-bearing—red pelagic limestones and (3) siliciclastics locally with interbedded felsic tuff [40,53,54,60,61]. Both Hercynian basement and post-Hercynian volcano-sedimentary associations are cut by Eocene granitic intrusions and unconformably overlain by the Early Cenozoic volcano-sedimentary sequence [44,46,62,63].

3. Studied Section

The studied Upper Jurassic Lower Cretaceous carbonate successions are widely exposed in the southern part of Eastern Pontides [23,27–39]. The succession is generally composed of platform carbonates, and their lithofacies reflects lateral and vertical changes in the environment from a supratidal to a continental slope facies [23,27–39]. This study is based upon the carbonate succession located in the Mescitli area of the Eastern Pontides (Figures 1, 2 and 3a–c). Here, the succession is estimated to be up to 400 m thick [31,38] and is subdivided into two informal lithological units/intervals based on the vertical change in macro-facies and microfacies characteristics [31] (Figure 2).

At the base, a 150 m thick-dolomite lithofacies overlies the basaltic rocks of Early to Middle Jurassic volcano-sedimentary successions (Figure 3a). The dolostones are light grey to grey coloured, thick to medium bedded (30–50 cm) and locally massive (Figure 3a). The middle and upper parts comprise approximately 230 m thick limestone succession which does not display significant diagenetic alteration petrographically, and hence primary microfacies textures are well preserved (Figures 2, 3a–c and 4a–i), namely: (1) Benthic foraminiferal packstone microfacies (MF-1), (2) Reworked skeletal grainstone/packstone microfacies (MF-2) and (3) Sponge spicule wackestone/mudstone microfacies (MF-3) [31,38]. Of these, the benthic foraminiferal packstone microfacies (M-1) forms the basal part of the limestone succession. It is nearly 70 m thick, grey to dark grey coloured, thick to medium bedded (0.5–1.5 m) and locally massive (Figure 2). It includes *Pseudolituonella gavonensis* Foury, 1968, *Arenobulimina* spp., *Praechrysalidina* sp., Miliolidae and Textularidae. The presence of *Pseudolituonella gavonensis*, *Arenobulimina* spp., *Praechrysalidina* sp. suggest a Hauterivian-Aptian age [31,38]. The high abundance of benthic foraminifera assemblages and a relative absence of deep-water bioclasts implies deposition in an inner platform paleoenvironment.

Reworked skeletal grainstone/packstone microfacies (MF-2) conformably overly the benthic foraminiferal packstone microfacies (MF-1) and comprise approximately 80 m thick, light grey to grey coloured, thick to medium bedded (25–75 cm) limestones (Figure 3b). The predominant carbonate components are diverse, fragmented skeletal assemblage (echinoids, pelecypoda, brachiopoda (punctate shells), thin-shelled ostracoda, bryozoa) and peloids. Most of the biotic components are broken and reworked, except for small benthic foraminifera [64]. Rare plagioclase and basaltic extraclasts are observed in the base of the microfacies. The absence of a diverse and abundant benthic foraminiferal community and the fragmented assemblage of heterotrophic skeletal fragments implies that the deposition of MF-2 took place in a deeper water environment than MF-1 [31,38].

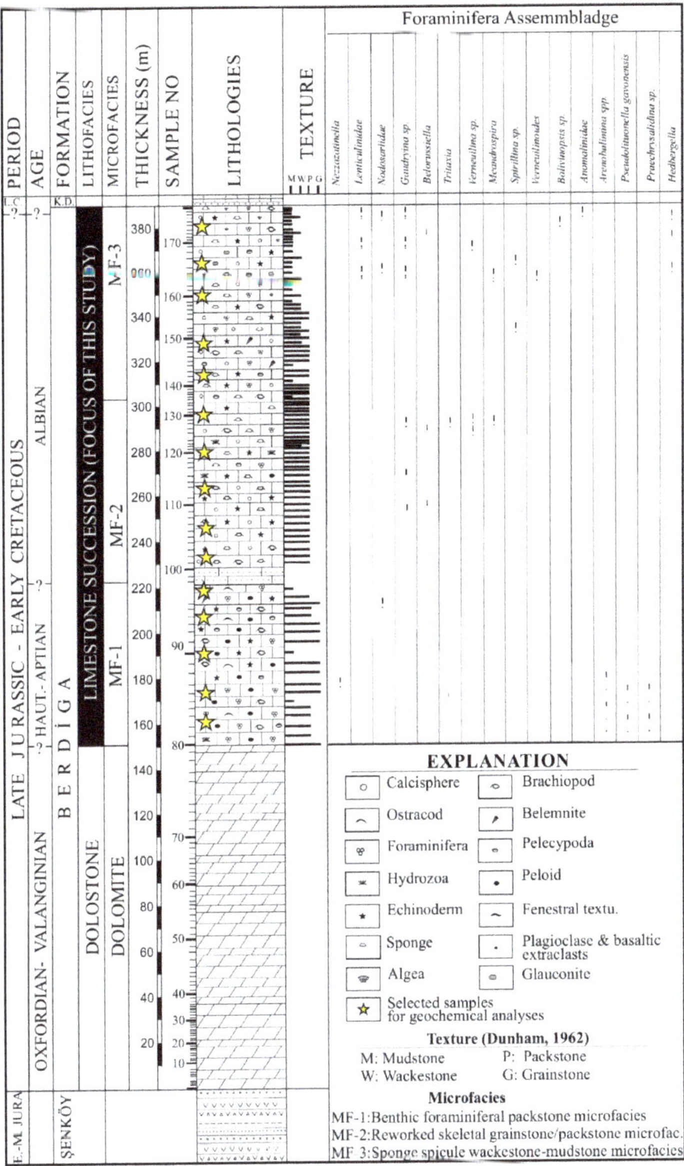

Figure 2. Stratigraphy, lithofacies and foraminifera distribution in the Berdiga Formation of the Mescitli (Gümüşhane, NE Turkey) [31,38].

Sponge spicule wackestone/mudstone microfacies (MF-3) constitute most of the upper part of the studied succession (Berdiga Formation). The contact between microfacies MF-2 and MF-3 is a gradual facies change. MF-3 microfacies comprises approximately 180 m thick, grey to dark grey mudstone and wackestone that is mostly characterized by sponge spicules and glauconite. The other carbonate components are allochthonous benthic foraminifera, intraclasts and rarely peloids. Besides, rare small benthic foraminifera assemblage including Lenticulinidae, Nodosariidae, *Gaudryina* sp., *Verneuilina* sp., *Bolivinopsis* sp., *Spirillina* sp. are also observed in the lower part of the lithofacies. The abundance and size of allochthonous skeletal fragments gradually decrease throughout MF-3 whilst the proportion of micrite increases upward. Planktonic foraminifera, including *Microhedbergella,* are also observed in the most upper part of the MF-3. The mud-rich texture and decrease in abundance/size of allochthonous skeletal fragments indicate that the depositional environment was deeper than the MF-2. Overall, the assemblage and the presence of the planktonic foraminifera may suggest that this was the deepest part of the succession, corresponding to a deep shelf or slope. This is consistent with coeval carbonate succession, which is reported in the Başoba Yayla area (Trabzon, NE Turkey) [34,37]. It has also been noted that the age of the most upper part of the formation is the Albian owing to the presence of *Microhedbergella rischi* Moullade, 1974 [65]. Therefore, it can be inferred that the age of MF-2 and MF-3 is likely Albian [31,38].

Figure 3. General lithological and macro-micro sedimentological features of the inner platform to slope environment in the interval of the Berdiga Formation (Hauterivian-Albian). (**a**). Field view of the Mescitli stratigraphic section; (**b**) thick to medium bedded Limestone (MF-1); (**c**) limestone with chert (MF-2 and MF-3).

4. Material and Methods

Upper Jurassic-Lower Cretaceous carbonates are well exposed along the Mescitli section (Figures 1 and 3a–c). The study area is situated in Mescitli-İkisu (40.528656, 39.380240), nearly 10 km north-west of Gümüşhane (NE Turkey). The 230 m of Upper Jurassic to Lower Cretaceous neritic micritic limestone of the Berdiga Formation was examined and sampled, and 100 rock samples were collected with a 20 to 50 cm spacing. A total of 80 samples of the Mescitli section were petrographically analyzed. Microfacies types were described based on fundamental principles of limestone classification and microfacies concept [66,67].

Representative samples from the different microfacies (Figures 2, 3a–c and 4a–i) were selected for geochemical analysis (Table 1). Selected samples are all micritic limestones (MF-1 to MF-3) not exhibiting petrographic evidence of significant post-depositional alteration such as recrystallization. Representative thin sections and a mirror-image slab of each thin section were polished to evaluate (i) weathering (ii) presence of clay minerals or basalt extraclasts (iii) micro-fissures filled with calcites or clastic components and (iv) fracturing, all of which were avoided during geochemical sampling.

The micritic orthochemical carbonate particles corresponding to the matrix micrite of each polished slab were sampled by micro-drilling using a hand drill.

Major, minor and trace elements, including REE of selected matrix micrite of each sample, were carried out by ACME Analytical Laboratories, Ltd. (Vancouver, BC, Canada). The major elements, trace elements, and REEs in the carbonates were determined by inductively coupled plasma–mass spectrometry (ICP–MS). Analyses used ~0.2 g of powdered sample digested in 10 mL 8 N HNO_3, of which 1 mL was diluted with 8.8 mL deionised water, and 0.1 mL HNO_3. To monitor the precision and accuracy, 1 mL of an internal standard (including Bi, Sc, and In) was added to the solution. For more details of these methods, please see the website of http://acmelab.com. Detection limit (dtl) of $CaCO_3$, $MgCO_3$, SiO_2, Al_2O_3, Na_2O, K_2O, TiO_2, P_2O_5, MnO is 0.01 wt % with exception of Fe_2O_3 (dtl = 0.04 wt %) and Cr_2O_3 (dtl = 0.002 wt %). Detection limits for Ba, Ni and Sc, are 1, 20 and 1 µg/g, respectively. Detection limits for Hf, Zr, Y, La, Ce are 0.1 µg/g. Detection limits for Th and Nd are 0.2 and 0.3 µg/g, respectively. Detection limits for Tm, Tb, Lu are 0.01 µg/g; for Pr, Eu and Hu are 0.02 µg/g; for Er is 0.03 µg/g; for Sm, Gd, Dy and Yb are 0.05 µg/g.

The measured REEs data of all micritic limestone samples were normalized to those of post-Archean Australian Shale (PAAS) which are previously reported (e.g., [68]). The equations of (i) Eu/Eu* ratio = $Eu_N/(Sm_N + Gd_N)0.5$, (ii) Pr anomaly = $Pr/Pr* = Pr_N/(0.5Ce_N + 0.5Nd_N)$, (iii) La anomaly = $La/La* = La_N/(3Pr_N - 2Nd_N)$, and (iv) Ce anomaly = $Ce/Ce* = 3Ce_N/(2La_N + Nd_N)$ are used to express Eu, Pr, La and Ce anomalies in the studied limestones [69,70].

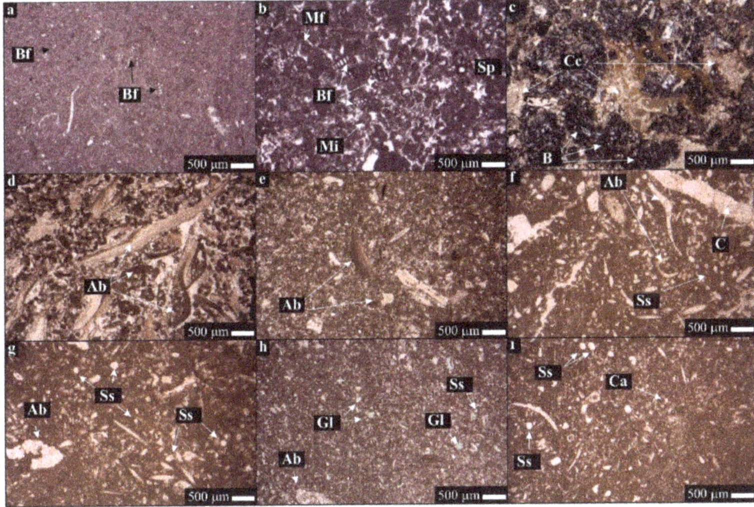

Figure 4. Benthic foraminiferal packstone microfacies (MF-1) (**a,b**). The transition zone between MF-1 and MF-2, showing basalt extraclasts and carbonate components (**c**). Reworked skeletal grainstone/packstone microfacies (MF-2) (**d–f**). Sponge spicule wackestone-mudstone microfacies (MF-3) (**g–i**). Bf: benthic foraminifera, Mf: micritized foraminifera, Mi: miliolid, Sp: sparite, Cc: carbonate component, B: basalt extraclast, Ab: allochthonous bioclasts, Ss: Sponge spicule, Gl: Glauconite, Ca: Calcisphere.

Table 1. Major, trace and rare earth elements of the limestone succession from the Mescitli area (NE Turkey).

	1	2	3	4	5	6	7	8	9	10	11	12	13	14	15
Sample No	M-82	M-83	M-84	M-85	M-89	M-91	M-92	M-94	M-106	M-110	M-112	M-116	M-137	M-141	M-146
Analyte	Rock Pulp	Rock Pulp	Rock Pulp	Rock Pulp	Rock Pulp	Rock Pulp	Rock Pulp	Rock Pulp	Rock Pulp	Rock Pulp	Rock Pulp	Rock Pulp	Rock Pulp	Rock Pulp	Rock Pulp
Microfacies	MF-1	MF-1	MF-1	MF-1	MF-1	MF-2	MF-2	MF-2	MF-2	MF-2	MF-3	MF-3	MF-3	MF-3	MF-3
$CaCO_3$	99.20	99.22	99.24	99.22	99.14	99.18	99.16	99.34	99.33	99.18	99.03	98.35	99.20	99.26	99.29
$MgCO_3$	0.80	0.78	0.7	0.78	0.86	0.82	0.84	0.66	0.67	0.82	0.97	1.65	0.80	0.74	0.71
SiO_2	0.67	12.22	0.75	0.57	3.51	3.79	7.99	3.42	1.78	3.88	4.12	21.90	4.86	2.84	1.96
Al_2O_3	0.24	0.76	0.19	0.14	1.10	0.72	2.48	0.88	0.27	0.82	0.44	3.57	0.20	0.34	0.35
Fe_2O_3	0.11	0.27	0.09	0.08	0.18	0.27	0.19	0.20	0.20	0.21	0.25	0.46	0.12	0.13	0.15
Na_2O	<0.01	0.01	<0.01	<0.01	<0.01	<0.01	0.02	<0.01	<0.01	<0.01	<0.01	0.03	<0.01	0.01	<0.01
K_2O	0.07	0.22	0.06	0.04	0.57	0.23	1.79	0.62	0.09	0.38	0.14	0.90	0.06	0.11	0.11
TiO_2	<0.01	0.03	<0.01	<0.01	0.02	0.02	0.03	0.02	<0.01	0.02	0.01	0.08	<0.01	0.01	0.01
P_2O_5	0.02	0.13	0.07	0.02	0.02	0.04	0.04	0.02	0.03	0.03	0.02	0.09	0.03	0.35	0.32
MnO	<0.01	0.01	0.01	<0.01	<0.01	0.01	<0.01	<0.01	0.01	<0.01	0.02	0.01	<0.01	0.01	0.01
Cr_2O_3	<0.002	<0.002	<0.002	<0.002	<0.002	<0.002	0.002	<0.002	<0.002	<0.002	<0.002	0.003	<0.002	<0.002	<0.002
Ba	3	3	7	4	7	5	27	9	2	6	3	21	4	3	3
Ni	<20	<20	<20	<20	<20	<20	<20	<20	<20	<20	<20	<20	<20	<20	<20
Sc	<1	<1	<1	<1	<1	<1	<1	<1	<1	<1	<1	2	<1	<1	<1
LOI	43.6	38.2	43.5	43.7	41.6	41.9	38.8	41.9	43.1	41.6	41.8	32.6	41.7	42.3	42.6
Hf	<0.1	0.1	<0.1	<0.1	0.1	0.1	0.2	0.2	<0.1	0.2	<0.1	0.5	<0.1	<0.1	0.1
Th	<0.2	0.4	<0.2	<0.2	0.2	0.3	0.3	<0.2	<0.2	<0.2	<0.2	0.9	<0.2	<0.2	<0.2
Zr	2.2	5.9	2.4	2.1	4.9	4.9	10.5	5.5	1.9	6.1	3.2	20.4	2.7	2.9	3.5
Y	1.4	5.3	1.1	0.7	2.2	5.0	3.2	2.7	4.0	2.7	3.6	8.8	2.9	4.8	4.0
La	1.2	4.4	0.7	0.6	2.5	3.8	3.5	2.7	2.7	2.6	2.9	10.8	2.1	3.0	2.9
Ce	0.9	4.2	0.8	0.7	3.4	3.0	3.8	3.1	2.2	3.0	2.4	10.8	2.0	2.5	2.5
Pr	0.17	0.79	0.33	0.09	0.40	0.56	0.53	0.47	0.42	0.41	0.42	2.06	0.31	0.42	0.44
Nd	0.7	3.1	0.4	0.3	1.5	2.5	2.0	1.7	1.7	1.6	1.7	8.5	1.3	1.9	1.8
Sm	0.11	0.59	0.11	0.05	0.25	0.50	0.31	0.31	0.38	0.27	0.31	1.67	0.25	0.29	0.43
Eu	0.02	0.16	<0.02	<0.02	0.09	0.13	0.12	0.08	0.09	0.09	0.09	0.45	0.05	0.09	0.10
Gd	0.10	0.79	0.06	0.07	0.33	0.54	0.45	0.39	0.39	0.38	0.42	1.68	0.27	0.47	0.50
Tb	0.02	0.11	0.02	<0.01	0.05	0.09	0.05	0.06	0.07	0.05	0.06	0.25	0.04	0.08	0.07
Dy	0.08	0.73	0.10	0.05	0.32	0.50	0.38	0.32	0.43	0.30	0.38	1.33	0.35	0.39	0.48
Ho	0.02	0.15	0.02	<0.02	0.06	0.11	0.06	0.07	0.08	0.07	0.08	0.30	0.05	0.10	0.09
Er	0.09	0.43	0.08	0.04	0.18	0.36	0.25	0.20	0.27	0.17	0.29	0.78	0.24	0.30	0.28
Tm	0.01	0.06	0.01	<0.01	0.02	0.05	0.04	0.03	0.04	0.03	0.04	0.11	0.02	0.05	0.04
Yb	0.06	0.37	0.07	<0.05	0.15	0.30	0.24	0.16	0.25	0.20	0.19	0.61	0.18	0.23	0.21
Lu	0.01	0.06	<0.01	<0.01	0.02	0.04	0.03	0.02	0.03	0.03	0.03	0.09	0.03	0.04	0.04
$\Sigma REEs$	4.89	21.24	3.80	2.60	11.47	17.48	14.96	12.31	13.05	11.90	12.92	48.23	10.09	14.66	13.88
Y/Ho	70.00	35.33	55.00	n.c.	36.67	45.45	53.33	38.57	50.00	38.57	45.00	29.33	58.00	48.00	44.44
Eu/Eu*	1.01	1.22	n.c.	n.c.	1.63	1.32	1.65	1.20	1.24	1.45	1.29	1.42	1.02	1.24	1.13
Ce/Ce*	0.47	0.54	0.38	0.71	0.81	0.48	0.66	0.66	0.49	0.69	0.51	0.55	0.58	0.52	0.52
Pr/Pr*	1.16	1.19	3.29	1.11	1.00	1.09	1.08	1.15	1.17	1.05	1.14	1.16	1.06	1.04	1.13
La/La*	2.11	1.47	0.21	1.28	1.48	2.70	1.59	1.27	1.83	1.65	1.97	1.58	2.15	3.02	1.94

5. Results

5.1. Petrography

The petrographical analyses of the Lower Cretaceous carbonate succession, corroborated by previously published sedimentological and paleontological data [31,38], allows us to determine microfacies characteristics (Figure 2 and Figure 10) for the studied stratigraphic section in the Mescitli area. Three microfacies were identified by petrographic analysis based on their depositional textures and fauna. These are 1) Benthic foraminiferal packstone microfacies (MF-1), (2) Reworked skeletal grainstone/packstone microfacies (MF-2) and (3) Sponge spicule wackestone/mudstone microfacies (MF-3). Based on the benthic foraminiferal assemblages (Figure 2), the studied limestone succession was deposited during the Hauterivian-Albian interval (Figure 2 and Figure 10).

Benthic foraminiferal packstone microfacies (MF-1) is dominated by shallow water benthic foraminiferal packstone with rare wackestone layers (Figure 4a). It has a mud-supported texture (micritic matrix) (10–60%), with frequent occurrence of small benthic foraminifera (30–50%) such as miliolidae (Figure 4a,b) that are strongly micritized (Figure 4b). Rare thin-shelled ostracods are observed along with peloids (15–25%) and fine-grain micritic intraclasts (5–10%). The fine-grain intraclasts are commonly carbonate mud (30–40%). In some parts of the thin section, sparite and micrite coexist. Besides, MF-1 show low biotic diversity and deep water bioclasts such as calcisphere and planktonic foraminifera are not observed. At the transition zone between MF-1 and MF-2 locally rare sand to silt-sized- plagioclase grains and basaltic extraclasts occur, that slightly decrease toward the top of the facies (Figure 4c).

Reworked skeletal grainstone/packstone microfacies (MF-2) comprise a high abundance skeletal component (30–50%), peloids (5–10%) and intraclasts (5–10%) (Figure 4d–f). The bioclastic components are represented by upward decreasing allochthonous skeletal fragments including, echinoids, molluscs, very rare small benthic foraminifera and undifferentiated shell fragments. The most frequent and common bioclasts through the MF-2 are echinoderms and bivalves. Other bioclasts, in order of abundance, are thin-shelled ostracods, sponge and sponge-spicules (Figure 4d–f). Most of the biotic components vary in size from 200 to 600 μm long and are broken and reworked, except for small benthic foraminifera. Benthic foraminiferas are rare and infrequent. Relatively high biotic diversity of bioclasts is observed compared to the MF-1 (Figure 4–f). Sponge spicule wackestone/mudstone microfacies (MF-3) is differentiated from MF-2 mainly by its mud-dominated texture, the predominance of the sponge spicules and lower abundance of reworked bioclasts (Figure 4g–i). MF-3 mostly consists of wackestone to mudstone beds, as well as sponge spicules, rare reworked skeletal components, pseudo-peloids, and rare intraclasts occur. The intraclasts are composed of rounded micritic grains. Reworked skeletal components become thinner and sparser towards the top of MF-3 (Figure 4g), and the most upper part of the microfacies consists of rare reworked skeletal components (Figure 4h), but with a high abundance of planktonic forams, including Calcisphaerulidae (Figure 4i). MF-3 also contains locally common authigenic components (glauconitic grains). Deep-water bioclasts such as calcispheres and sponge spicule are observed at the top of MF-3.

5.2. Major and Trace Elements

The chemical analyses of the micritic limestone samples are represented in Table 1. Studied limestone samples exhibit low Mg-calcite characteristics with the main chemical formula of $Ca_{98.35-99.34}Mg_{0.66-1.65}(CO_3)$. Major element contents in limestone (Mescitli Section) exhibit high $CaCO_3$ varying from 98.35 to 99.35 (mole %) and low $MgCO_3$ ranging between 0.66 and 1.65 (mole %). They have SiO_2 contents ranging from 0.57 to 21.90 wt % (average: 4.95 wt %), Al_2O_3 contents ranging from 0.14 to 3.57 wt % (average: 0.83 wt %), Fe_2O_3 contents ranging from 0.08 to 0.46 wt % (average: 0.19 wt %), K_2O contents varying 0.04 to 1.79 wt % (average: 0.36 wt %) and P_2O_5 varying from 0.02 to 0.35 wt % (average: 0.08 wt %). Na_2O (below detection limit (bdl), <0,01 to 0.03 wt %), MnO (<0.01 to 0.02 wt %), TiO_2 (<0.01 to 0.08 wt %; average: 0.03 wt %) and Cr_2O_3 (<0.002 to 0.003 wt

%) contents are all low (Table 1). The analytical results also show low contents of Hf (<0.1 to 0.50 μg/g), Sc (<1 to 2.0 μg/g), Th (<0.2–0.9 μg/g), Zr (1.9–20.4 μg/g) elements.

The studied limestones have variable ΣREEs (1.90–39.43 μg/g and ave. 10.74 μg/g), Y (0.70–8.80 μg/g; ave. 3.49 μg/g) and Ho (<0.02 to 0.30; ave. 0.09 μg/g) contents. Their Y/Ho (29.33–70.00; ave. 46.26) and Eu/Sm (0.18–0.39; ave. 0.28) ratios are mostly similar to those of seawater while their Sm/Yb (1.26–2.74; ave. 1.68) ratios are slightly higher than those of modern seawater [59] (Table 1). Following normalization of REEs to post-Archean Australian Shale (PAAS; [68]) the limestone samples exhibit (i) depleted LREE relative to HREE $(Nd/Yb)_N$ (0.46–1.12; average 0.72) and $(Pr/Yb)_N$ (0.50–1.40; 0.73) (ii) negative Ce/Ce* (0.38–0.81; ave. 0.57) (iii) slight positive Eu/Eu* (1.01–1.65; ave. 1.29), (iv) slightly flat Pr/Pr* (1.00–3.29; ave. 1.26) anomalies and (v) positive La/La* anomaly (0.21–3.02; ave. 1.75). The samples mostly plot in area IIIb of the Pr/Pr* vs. Ce/Ce* ratios [68], which confirm that the Ce anomalies are not an artefact of La—interference (Figure 5a). Eu/Eu* ratios exhibit no correlation with Ba/Sm (Figure 5b). Ce/Ce* ratios show fluctuation (Table 1) with a steady increase throughout the MF-1 part of the studied section. The uppermost part of MF-1 exhibits slightly lower Ce/Ce* anomalies and is followed by a sharp decrease in the Ce/Ce* anomalies at the base of MF-2. MF-2 and MF-3 are represented by relatively low Ce/Ce* anomalies. There is no significant difference in Ce anomaly between MF-2 and MF3.

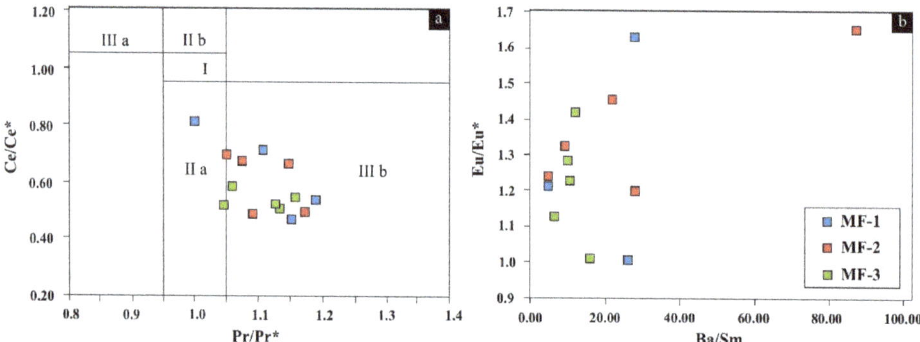

Figure 5. Plots of the Pr/Pr* vs. Ce/Ce* ratios diagram [69] (**a**), and Ba/Sm vs. Eu/Eu* ratios (**b**) of studied limestones. See the text for further details.

6. Discussion

6.1. Siliciclastic Impurities

To reveal the possible potential influence of terrigenous input on paleoredox indicators within the studied limestone, it is necessary to identify possible contamination by siliciclastics. This is important because contamination by terrestrial clays can control the primary REEs + Y contents of ancient marine carbonates [4,6,71,72]. A small quantity of terrestrial particulate matter (i.e., shale), which has high REE contents with distinctly non-seawater-like patterns [73], may dramatically modify original REE patterns and cause a decrease in both the extent of LREE depletion and Y/Ho ratios [3,4,74]. Several lines of evidence suggest that the studied limestone displays a diagnostic seawater signature: (1) negative Ce anomalies (0.38–0.81; average 0.57), (2) depleted LREE and (3) superchondritic Y/Ho (average 46.26) (Figure 6a) (e.g., [4]). Although the limestone exhibits a slight enrichment of LREE compared to modern seawater of South Pacific deep water [11] their PAAS normalized patterns exhibit a seawater-like signature compared to the North American Shale Composite (Figure 6a,b) [75]. Furthermore, their general PAAS normalized patterns (i.e., Ce* with an average of 0.57; Nd/Yb_N with an average of 0.72; Pr/Yb_N with an average of 0.73 values) are compatible with well-preserved brachiopods [8,75] (Figure 6b,c).

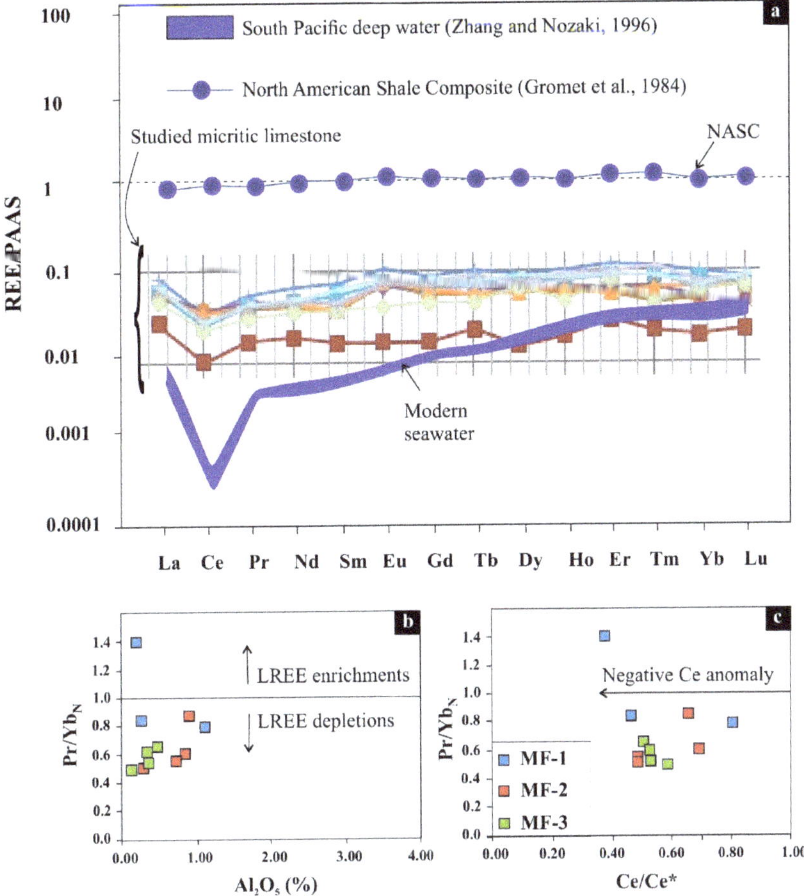

Figure 6. PAAS-normalized REE patterns of the studied limestone succession (**a**). Plot of the Pr/Yb$_N$ values vs Al$_2$O$_3$ (wt %) contents (**b**) and Pr/Yb$_N$ value vs. Ce/Ce* ratios (**c**). Modern seawater data from South Pacific deep water [11] and North American Shale Composite from [76].

The Y/Ho ratios can be used as a tracer to assess whether carbonate samples reflect primary marine signature or siliciclastic components [2,4,77–79]. Y and Ho are isovalent trace elements which have a similar charge and radius, and they are considered to display extremely coherent behaviour as twin pairs in a geochemical system characterized by charge-and-radius-controlled (CHARAC) [10]. According to the previous work [77,78,80], an aqueous marine system is characterised by non-CHARAC trace element behaviour, and electron structure must be considered as an important additional parameter which may cause trace element fractionation processes such as chemical complexation. The complexation behaviour of a trace element is additionally influenced by its electron configuration and by the character of chemical bonding between a central ion and a ligand [79,80]. For this reason, Ho is scavenged more readily from seawater than Y due to differences in geochemical behaviour in surface complexation [79–84]. According to the literature, high Y/Ho ratios (average of 61 ± 12) occur in the upper 200 m of seawater [7,11,12,64,85–91]. However, other studies have proposed that typical Y/Ho ratios in marine carbonates that are free of contamination from terrigenous material show a wider range of values (44–74) or a chondritic value of 28 [78,80]. The studied limestones display high Y/Ho ratios with an average of 46.26 (29.33–70.00), which is consistent with detritus-free marine sediments (average 45.50) [69]. Further, the lack of correlation between Y/Ho ratios, Ce/Ce*, and Al and Zr contents in the studied limestone implies little or negligible terrestrial contamination [74].

Finally, several trace elements including Hf, Sc, Th, Zr and Al_2O_3 are considered to be particularly sensitive to shale contamination, as they occur in detrital alumosilicate minerals in much greater volume in average shale than in marine carbonates [4–6,17–19,64,71,85,92,93]. The studied micritic limestones predominantly exhibit low Hf (<0.02 to 0.5 µg/g), Sc (<1 to 2 µg/g) Th (<0.2 to 0.9 µg/g) contents, which supports the observation that contamination by detrital minerals is minor to negligible (Table 1). Relatively low Zr (1.90–20.40 µg/g, an average of 5.27 µg/g) and Al_2O_3 (0.14–3.57 wt %, an average of 0.83 wt %), with a good correlation between Zr and Al_2O_3, suggest a minor amount of shale contamination (Figure 7a). However, total REE concentration is poorly correlated with Zr and Al_2O_3, implying that this small volume does not dramatically contaminate the REE concentration in these samples [2] (Figure 7b,c). Plotting Y/Ho, Eu/Eu* and Ce/Ce* versus and Zr contents in Figure 7d–f shows no positive correlation, as would be expected if there was low detrital contamination [3–5].

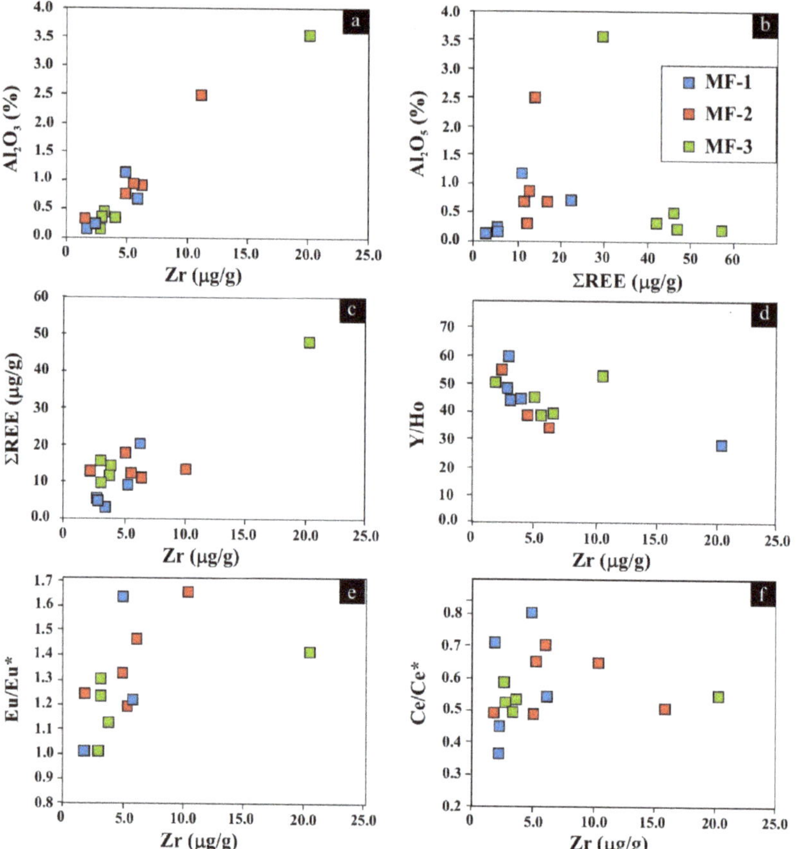

Figure 7. Plots of the Zr (µg/g) vs. Al_2O_3 (wt %) contents (**a**); ΣREE (µg/g) vs. Al_2O_3 (wt %) contents (**b**); Zr (µg/g) vs. ΣREE (µg/g) contents (**c**); Y/Ho ratios vs. Zr (µg/g) contents (**d**); Zr (µg/g) contents vs. Eu/Eu* ratios (**e**); Zr (µg/g) contents vs. Ce/Ce* ratios (**f**).

To sum up, the contamination of carbonate samples by terrestrial materials can be defined as minor or negligible based on these multiple and independent analyses including (1) their seawater signature; (2) similar Y/Ho ratios to those of the marine carbonates free of contamination and (3) lower Sc, Th, Zr and Al_2O_3 contents as discussed above.

6.2. Diagenetic Influence

REEs, in particular Ce and Eu, are widely used to reconstruct the redox evolution of the carbonate system [1,64,85,94]. Carbonate minerals, especially calcite, can be considered long-term repositories of lanthanides [8,95]. Therefore, carbonate sedimentologists have focused on a wide range of marine carbonate components such as reefal microbialites (e.g., [6,17]), reefal carbonate (e.g., [4]), laminae of ooids (e.g., [85]), belemnites (e.g., [8]) and micritic limestones (e.g., [86]) for the reconstruction of ancient paleoenvironmental conditions. Their REE + Y patterns do not always exhibit a typical seawater signature, even though the REE signature is thought to be inherited from their parental seawater [1–4,18,82,87–90] because original REE patterns can be significantly altered during diagenesis under high water-rock ratios [4,6,81,91]. However, complete mineral dissolution is necessary to lead to an effective loss of trace elements, including REEs, from the mineral lattice (e.g., [95]).

Many studies have focused on using REEs to provide a better understanding of the influence of thermodynamic and kinetic parameters on the partitioning of individual REEs during calcite precipitation [90,91,95,96]. Mineral dissolution is considered unlikely to destroy the geological record encoded by the distribution pattern of REEs in calcites based on experimental studies, though some have questioned this interpretation because of differences between the experimental system and natural settings (e.g., REE concentration in solution) [96]. Further, a large amount of information on REEs in natural carbonates have been compiled [95,96], and several parameters have been investigated to assess the potential diagenetic influence on the REE signature of carbonates [1–4]. It has been shown that the progressive REE scavenging during post-depositional modification may produce a positive correlation between the Ce anomaly and REE [70,71,91]. However, our samples do not display a notable positive correlation between Ce and REE content (Figure 8a). Furthermore, the studied carbonates show a negative correlation between Eu/Eu* and REE concentration (Figure 8b), implying that such a diagenetic influence is absent [91]. Finally, a negative correlation between Ce/Ce* and La/Sm_N ratios have been used as evidence of diagenetic influence on Ce [97–99], and this correlation is not apparent in our carbonates (Figure 8c). Similarly, an absence of correlation between Ce/Ce* and Eu/Eu* implies a negligible or absence of influence of post-depositional alteration on the measured Ce/Ce* [3,7,74,84,87,100]. Our samples display a weak correlation (Figure 8d–f), which implies negligible or no diagenetic impact on Ce anomalies. This conclusion is also supported by petrographic observations of the micritic limestone, which do not show significant recrystallization.

Given these conditions, the relationship between Ce/Ce* and Pr/Pr* can be used to characterize redox conditions using shale-normalized REE contents [69]. In assessing whether an apparent Ce anomaly is true, and not a function of La-enrichment, the data was analysed using the Ce/Ce* versus Pr/Pr* analysis [69] (Figure 5a), whereby area (I) no Ce anomaly; (IIa) positive La anomaly causes apparent negative Ce anomaly; (IIb) negative La anomaly causes apparent positive Ce anomaly; (IIIa) real positive Ce anomaly; (IIIb) real negative Ce anomaly; (IV) positive La anomaly disguises positive Ce anomaly [69,99]. The position of samples from this study in area IIIb indicates that they display real Ce anomalies (Figure 5a). In addition, Eu/Eu* can be contaminated by interference with Ba during ICP-MS analysis [101], but Eu/Eu* ratios exhibit no correlation with Ba/Sm, implying that calculated anomalies are real positive Eu anomalies (Figure 5b).

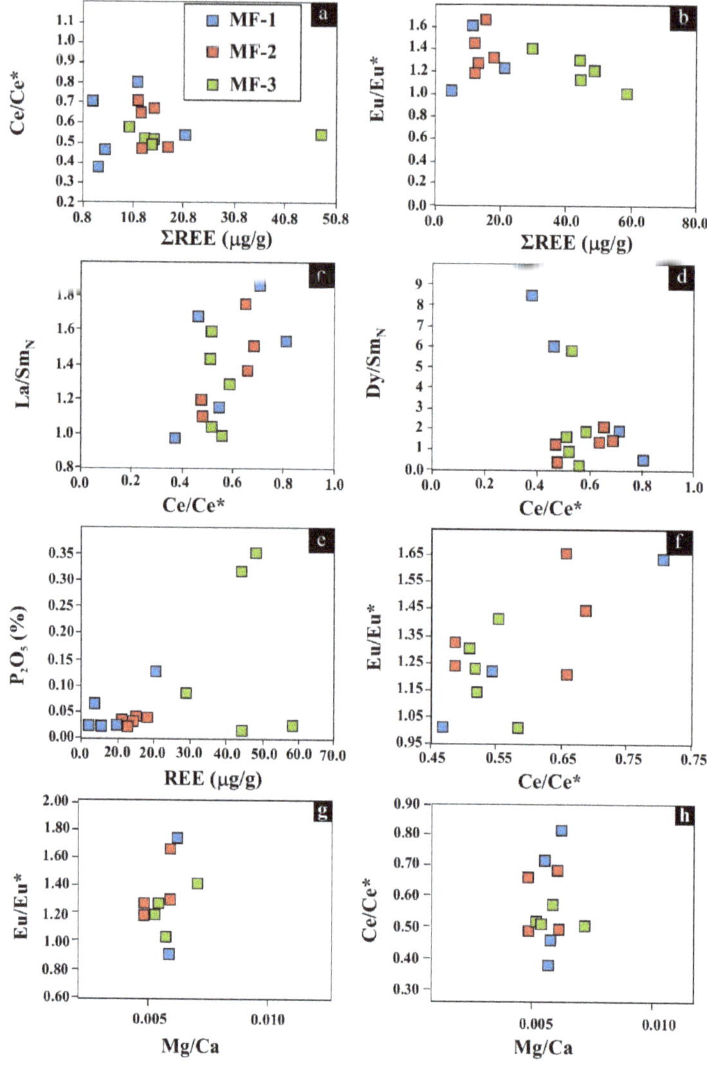

Figure 8. Plots of the ΣREE (μg/g) contents vs. Ce/Ce* ratios (**a**); ΣREE (μg/g) contents vs. Eu/Eu* ratios (**b**); Ce/Ce* vs. La/Sm$_N$ ratios (**c**); Ce/Ce* vs. Dy/Sm$_N$ ratios (**d**); ΣREE (μg/g) vs. P$_2$O$_5$ (wt %) contents (**e**); Eu/Eu* vs. Ce/Ce* ratios (**f**); Mg/Ca vs Eu/Eu* (**g**) and Mg/Ca vs. Ce/Ce* ratios (**h**).

It has been shown that the micritic limestones mainly show a seawater signature, with LREE depletion, positive La anomaly (La*) and a mostly negative Ce anomaly (Ce/Ce* = 0.38 to 0.81, average 0.57). However, they also exhibit a wide-range of Eu anomalies (Eu*/Eu = 1.01–1.65; 1.29). This pattern is missing in the general characteristics of normal marine carbonates [102] and is generally considered as an indicator of reducing, high-temperature fluids [77,83,103,104]. Similar patterns are recorded in many distinct marine systems affected by hydrothermal fluids associated with ferromagnesian rocks [71]. Such hydrothermal fluids are usually reported at mid-ocean ridges and back-arc spreading tectonic setting, where water-rock interactions with mafic rocks provide additional REE to ambient seawater [69,77,80]. Therefore, the Eu/Eu* ratios in this study may imply water-rock interaction with basalts. However, previous studies have shown that basaltic rocks are absent throughout the

Lower Cretaceous, and there is no evidence of Early Cretaceous magma generation in the study area (Eastern Pontide, NE Turkey). Although it is widely accepted that the Early Cretaceous was a stable tectonic regime due to a lack of igneous activity [50,61]. The eastern Pontide experienced multistage basaltic magma generation during the Late Cretaceous [43,49] and early Cenozoic acidic magma generation is also recorded [44–46,63]. This Upper Cretaceous to Cenozoic magma generation has led to recrystallisation and geochemical modification of the underlying dolomites at a late-stage of their diagenetic history [32–34]. However, the studied Lower Cretaceous limestones do not show petrographical evidence of such recrystallisation (Figure 4a,b). Furthermore, the studied limestone samples exhibit low Mg-calcite chemistries ($Ca_{98.81–99.52}Mg_{0.48–1.19}(CO_3)$) [105], indicating that they are chemically stable [106–113]. Further, there is neither petrographical evidence for recrystallization which associated with Late Cretaceous to Cenozoic magma generation nor geochemical pieces of evidence, for example, the lack of apparent correlation between Mg/Ca vs. Ce/Ce* and Eu/Eu* ratios which imply less influence of late diagenetic processes on their REE chemistry (Figure 8g,h). Therefore, Late Cretaceous to Cenozoic magma generation during this later diagenetic stage is probably not responsible for high the Eu/Eu* ratios of the studied limestone.

Nevertheless, Lower Cretaceous calc-alkaline lamprophyre and high-Nb alkaline basaltic dykes have been discovered in the Eastern Pontide [114], and this may have provided hydrothermal fluids to the depositional environment, which changed the temperature and/or redox conditions prior to diagenetic stabilization (e.g., [81]). To test this, Sm/Yb vs. Eu/Sm and La/Yb vs. La/Sm plots were constructed (Figure 9a,b). The data show an overlap with hydrogenetic-Fe-Mn crusts and relatively higher La/Yb_N values than seawater are observed, implying the presence of some early absorption processes [3,115] (Figure 9b). It has been proposed that hydrothermal fluid possess higher Sm/Yb and chondritic Y/Ho values (~28) than modern seawater [71]. Plotting Y/Ho versus Sm/Yb (Figure 9c), studied samples overlap with the samples of Pongola Iron Fm. which were affected by hydrothermal fluid during their sedimentation processes [71]. They also plot on the line of hydrothermal fluid (1–5%) suggesting a contribution of hydrothermal fluid Figure 9c. Although the studied samples have relatively lower Eu/Sm ratio compared to the high-T hydrothermal fluids. They possess similar Eu/Sm ratio to the hydrogenic Fe-Mn crusts (Figure 9d) and samples plot in an area between seawater and hydrogenetic- Fe-Mn crusts which may confirm the influence of the hydrothermal fluid. To sum up, even though the studied limestone formed in an open marine environment, (i) the positive Eu anomalies, (ii) high Sm/Yb and La/Yb_N ratios and (iii) the Y/Ho range between seawater and hydrogenetic Fe-Mn crusts suggest some modification by hydrothermal fluids which are most likely to be associated with Early Cretaceous magma generation (Hauterivian-Albian) [77,83,103,104,116,117].

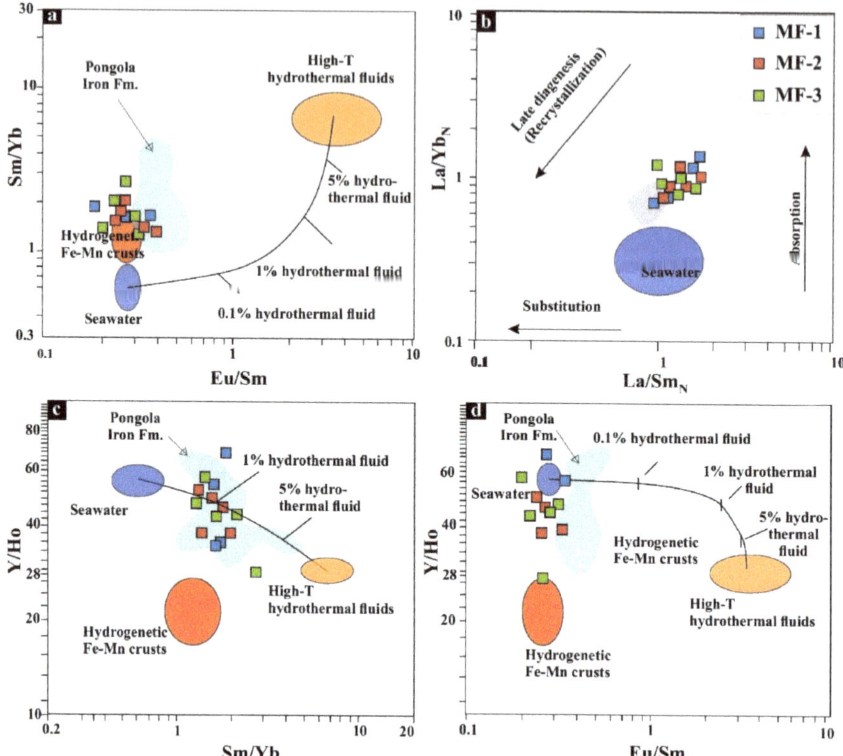

Figure 9. Plots of the Eu/Sm vs. Sm/Yb (**a**); (La/Sm)N vs. (La/Yb)N (**b**); Sm/Yb vs. Y/Ho (**c**); and Eu/Sm vs. Y/Ho (**d**) [21,71,115].

6.3. Early Cretaceous Paleoenvironmental Implications

It has been shown that the studied limestones mostly display a seawater signature including (1) negative Ce anomalies, (2) superchondritic Y/Ho (47 ± 4), and depleted LREE relative to HREEs ((Nd/Yb)$_N$ = 0.46–1.12; ave. 0.72 and (Pr/Yb)$_N$ = 0.50–1.40; 0.73) (Figures 6a–c and 10). The Ce* anomaly in carbonates is considered as a superior proxy for revealing the redox-state of the ancient seawater [3,8,74]. The studied samples exhibit negative Ce* anomalies, typical of modern oxygenated seawater [9,77,78,80,84,91,118]. However, anoxic conditions may weaken the Ce depletion as a result of redox reactions leading Ce^{3+} oxidation to insoluble Ce^{4+} because dissolved Ce^{3+} which is partially scavenged from seawater in the anoxic marine system [13,20,119,120]. The lack of significant negative-Ce anomalies in many marine sediments has been considered as evidence of the anoxic seawater [100,121]. Besides, it has been suggested that a pronounced negative Ce* anomaly can be divided into three: (a) smaller than <0.5; (b) ~0.6–0.9 and (c) ~0.9–1.0 which represents oxic, suboxic and anoxic marine water, respectively [19].

The studied limestone is interpreted to have experienced progressive modification through interaction with reducing, high-temperature diagenetic fluids, and this caused the slight enrichment of Eu. However, no notable positive correlation between Eu anomalies versus Ce are present, which may indicate that modification of Eu was not accompanied by the modification of their Ce* values. Consequently, the wide range of negative Ce distributions may point to a change in paleoceanographic conditions from slightly suboxic to oxic paleo-redox state of ancient seawater but still remaining <0.9 [3]. MF-1 located in the lower part of the studied limestone succession is interpreted to have been deposited in an inner platform paleoenvironment during the Hauterivian-Aptian based on the microfacies characteristics and foraminifera assemblages (e.g., the high abundance of the benthic characteristics of foraminifera assemblages, lack of deep-water bioclasts and low biotic diversity) [31,67] (Figure 1.) It exhibits a wide range of Ce anomalies (0.38–0.81), with a steadily increasing trend upward in the succession, coincident with interpreted increasing palaeo-water depth (Figure 10, Table 1). The uppermost part of MF-1 exhibits slightly weaker Ce/Ce* anomalies implying a relatively suboxic state of the ancient seawater. This is followed by a sharp decrease in the Ce/Ce* anomalies at the base of MF-2, suggesting a sudden change in the redox state of contemporaneous seawater. There is then a gradual transition to the overlying MF-3, indicating oxic seawater (Figure 10, Table 1) which remained relatively stable until the end of MF-3. This is despite the progressive deepening of the basin, which might have been caused a slight decrease in oxygenation as relative sea-level increased [15].

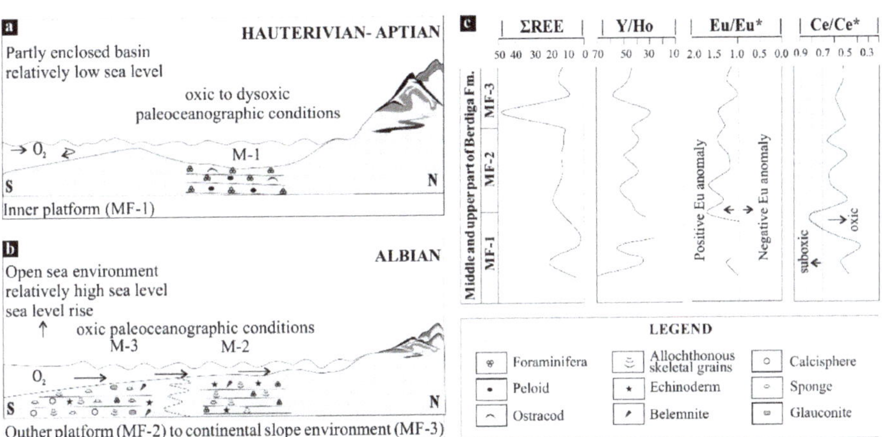

Figure 10. Diagrammatic sketch illustrating probable environment evolution of middle and upper part of the Berdiga Formation during the Hauterivian-Albian in Eastern Pontide. (**a**) During MF-1 deposition, the inner platform environment with relatively low sea level with oxic to dysoxic paleoceanographic conditions. (**b**) During MF-2 and MF-3 deposition, an outer platform to continental slope environment with relatively high relative sea level and a well-oxygenated open marine environment. (**c**) Inset showing the relative position of each phase with respect to ΣREE (µg/g) Y/Ho, Eu/Eu* and Ce/Ce* variations. Ce/Ce* shows dyspoxic (Ce* = 0.71–0.81) conditions in most upper part of the MF-1. MF-1: benthic foraminiferal packstone microfacies corresponding to inner platform environment, MF-2: reworked skeletal grainstone/packstone microfacies corresponding to other platform, and MF-3: Sponge spicule wackestone-mudstone microfacies continental corresponding to slope environment.

The differences in microfacies characteristics and foraminifera assemblage between MF-1 and overlying MF-2 may also confirm a change in paleoceanographic conditions because MF-2 is represented by the predominance of allochthonous bioclastic components which are broken and reworked, high biotic diversity and scarcity of small benthic foraminifera implying a deeper shelf environment relative to the MF-1. Furthermore, the presence of the rare plagioclase and basaltic extraclasts in the transition zone from MF-1 to MF-2 (Figure 4c) is likely associated with the basaltic magma generation, which has been recently reported [114]. This basaltic magma generation is probably caused by synsedimentary tectonic activity. The syn-sedimentary tectonic regime may also cause progressive deepening of the basin. Furthermore, the submarine basaltic magma generation may also influence on the sudden change in paleo-redox conditions of ancient depositional environment. This likely result in the oxygen-deficient conditions, which produce our relatively high Ce/Ce* data (0.71 to 0.81) in the uppermost part of MF-1.

The Ce/Ce* anomaly at the MF1 / MF2 boundary is marked, and coincides with a change in facies from shallow water, benthic foraminiferal packstone to reworked skeletal packstone that is dominated by fragmented, allochthonous, largely heterotrophic fauna. This could reflect a facies transition from shallow water, platform top (MF1) to deeper water slope (MF2) setting, consistent with the overall deepening of the succession. However, the presence of fragmented organisms suggests depositional energy remained moderate, and water depths were not sufficiently deep for oxygenation to be reduced; the overlying MF3 sediments are more likely to be deeper water but remain oxic. The decrease in the Ce/Ce* anomaly coincides with a decrease in ΣREE, Y/Ho and Eu/Eu* anomaly (Figure 10). Dating of benthic foraminifera indicates that a timing that is broadly consistent with the onset of OAE1a [122], where global temperatures rose, and ocean acidification occurred as a result of rising CO_2 concentrations during emplacement of Large Igneous Provinces. It is, therefore, possible that the MF1/MF2 boundary records this event, both in the increase in hydrothermal activity – which modified Eu/Eu*—and the decrease in seawater oxygenation that decreased Ce/Ce*. In this sense, the change in facies from MF1 to MF2 might not simply reflect a rise in relative sea level, and it might also indicate decreasing carbonate productivity and preservation.

Paleo-redox evolution can be controlled by the function of temperature, fluid chemistry of paleo-oceanic system triggered by complex tectono-sedimentary evolution of the Eastern Pontide. It is therefore recommended that paleoceanographic studies demonstrate more conclusively the occurrence of oxygen-deficient conditions in the Tethys ocean, within the study area. Our results have the potential to contribute new information to this discussion because a paleoceanographic reconstruction of the studied section can be established based on the faunal content, microfacies characteristics, and REE data. MF-1 represents an inner platform setting with mostly oxic conditions during the Hauterivian-Aptian. However, a sudden change in redox-state of ancient seawater, corresponding dysoxic conditions, is recorded in the most upper part of the MF-1. This is followed by an abrupt deepening paleoenvironment with a relative increase in the oxic state of the seawater and the deposition of the deeper water microfacies (MF-2 and MF-3) above a sharp transition. This represents a relative sea-level rise triggered by the evolution of the Tethys in the Eastern Pontide resulting deepening of the basin and is also recorded in age-equivalent carbonates in NE Turkey [37]. Previous studies have indicated that the extensional tectonics and rifting in the Cretaceous in NE Turkey [25,37,50,61] terminated in the middle Cretaceous, such that deeper paleoenvironmental conditions could be consistent with thermal subsidence, even though carbonate sedimentation continued until the end of Albian or Turonian [25,37,123]. Similarly, in this study, MF-1 represents an inner platform setting with mostly oxic conditions during the Hauterivian-Aptian. However, a sudden change in redox-state of ancient seawater, corresponding dysoxic conditions, is recorded in the most upper part of the MF-1 (Figure 10). This is followed by an abrupt deepening paleoenvironment with a relative increase in the oxic state of the seawater and the deposition of the deeper water microfacies (MF-2 and MF-3) above a sharp transition. Moreover, an abrupt paleoenvironmental change is represented by the presence of MF2 and MF3 which include dark grey limestones with chert nodules, mud-rich textures, allochthonous skeletal fragments and sponge spiculitics and presence of the planktonic organism, all consistent with deeper water sedimentation. These characteristics can be indicative of an increase in the water depth. The gradual transition to MF-3 with a predominance of mud-rich texture, a decrease in abundance of allochthonous skeletal fragments support the transgressively deepening of the paleoenvironment which continued up to late Albian. It has also been recorded the sponge spicule wackestone with *Microhedbergella rischi* within the most upper part of MF-3 [37,51]. Thus, the presence of the planktonic organism may characterise the deepest part of the paleoenvironment which corresponding to the slope [65]. Further, their REEs data confirm that their oxic conditions remained stable throughout the deeper facies (MF-2 and MF-3), even considering the distal microfacies. The current work, therefore, supports that the shallower part of the paleo-ocean remained relatively less oxic and became suboxic during the Late Aptian-Albian, while deeper facies, which display an overall transgressive motif, were developed in relatively more oxic paleo-oceanic conditions up to the end Albian (Figure 10).

7. Conclusions

The study of the Lower Cretaceous limestone succession, corroborated by detailed microfacies analysis and REE geochemistry, allows us to present new data on ocean paleoredox conditions (Figure 10) and the extent of diagenetic alteration for the studied stratigraphical section in Mescitli area (Eastern Pontides, NE Turkey). Our primary results are as follows:

1. Analyzed micritic limestone samples mainly exhibit a seawater signature including (1) slight LREE depletion relative to the HREEs (ave. 0.72 of Nd/Yb_N and ave. 0.73 of Pr/Yb_N), (2) negative Ce anomalies (Ce^*/Ce = 0.38–0.81; ave. 0.57), (3) positive La anomaly (La^*/La = 0.21–3.02; ave. 1.75) and (4) superchondritic Y/Ho (ave. 46.26).

2. Micritic limestone also shows slight positive Eu^* anomalies (Eu^*/Eu = 1.01–1.65; ave. 1.29) and relatively higher Sm/Yb (1.39–1.26; ave. 2.05) and La/Yb_N (0.68–1.35; 0.96) ratios compared to the modern seawater. This may imply the presence of water-rock interaction between parental seawater and basaltic rocks at elevated temperatures triggered by hydrothermal activity associated with Early Cretaceous basaltic magma generation.

3. The studied sections exhibit negative Ce^* anomalies, varying from 0.38 to 0.81, which may confirm mostly oxic to dysoxic paleoceanographic conditions. Further, dyspoxic (Ce^* = 0.71–0.81) conditions are also recorded in the most upper part of the MF-1 Microfacies during the late Aptian-early Albian.

4. The current work suggests that the shallower part of the paleo-ocean remained relatively less oxic and became suboxic during the Late Aptian-Albian, while deeper facies displaying overall transgressive trend were developed in relatively more oxic paleo-oceanic conditions up to end of Albian.

Author Contributions: Conceptualization, M.Ö.; writing—original draft preparation, M.Ö.; supervision, M.Z.K., I.A.-A., and C.H.; writing—review and editing, I.A-A. and C.H.; review and editing, C.H., R.K., and K.T. All authors have read and agreed to the published version of the manuscript.

Funding: Funding for this study partly originated from Scientific and Technological Research Council of Turkey (TUBITAK-ÇAYDAG, Project no: 115Y005) and Karadeniz Technical University, Turkey (KTU BAP, Project no: FBA-2015-5160; and 7341-FBA-2018-7341).

Acknowledgments: The corresponding author acknowledges the University of Manchester (UK) for facilitating her Post-Doc Research. ISA acknowledges the support from NSERC. Special thanks to Ali Keskin, Sefa İnan, Mert Oğuzhan Dinç, and Aslı Çağla for their help during the sampling period. The anonymous reviewers are also acknowledged for their improvements to the paper.

Conflicts of Interest: The authors declare no conflict of interest.

References

1. Franchi, F. Petrographic and geochemical characterization of the Lower Transvaal Supergroup stromatolitic dolostones (Kanye Basin, Botswana). *Precambrian Res.* **2018**, *310*, 93–113. [CrossRef]

2. Frimmel, H.E. Trace element distribution in Neoproterozoic carbonates as palaeoenvironmental indicator. *Chem. Geol.* **2009**, *258*, 338–353. [CrossRef]

3. Liu, X.M.; Hardisty, D.S.; Lyons, T.W.; Swart, P.K. Evaluating the fidelity of the cerium paleoredox tracer during variable carbonate diagenesis on the Great Bahamas Bank. *Geochim. Et Cosmochim. Acta* **2019**, *248*, 25–42. [CrossRef]

4. Nothdurft, L.D.; Webb, G.E.; Kamber, B.S. Rare earth element geochemistry of Late Devonian reefal carbonates, Canning Basin, Western Australia: Confirmation of a seawater REE proxy in ancient limestones. *Geochim. Et Cosmochim. Acta* **2004**, *68*, 263–283. [CrossRef]

5. Redivo, H.V.; Mizusaki, A.M.; Santana, A.V. REE patterns and trustworthiness of stable carbon isotopes of Salitre Formation, Irecê Basin (Neoproterozoic), São Francisco Craton. *J. S. Am. Earth Sci.* **2019**, *90*, 255–264. [CrossRef]

6. Webb, G.E.; Kamber, B.S. Rare earth elements in Holocene reefal microbialites: A new shallow seawater proxy. *Geochim. Et Cosmochim. Acta* **2000**, *64*, 1557–1565. [CrossRef]

7.   Alibo, D.S.; Nozaki, Y. Rare earth elements in seawater: Particle association, shale-normalization, and Ce oxidation. *Geochim. Et Cosmochim. Acta* **1999**, *63*, 363–372. [CrossRef]
8.   Azmy, K.; Brand, U.; Sylvester, P.; Gleeson, S.A.; Logan, A.; Bitner, M.A. Biogenic and abiogenic low-Mg calcite (bLMC and aLMC): Evaluation of seawater-REE composition, water masses and carbonate diagenesis. *Chem. Geol.* **2011**, *280*, 180–190. [CrossRef]
9.   Nozaki, Y. Rare earth elements and their isotopes. *Encycl. Ocean Sci.* **2001**, *4*, 2354–2366.
10. Piper, D.Z.; Bau, M. Normalized rare earth elements in water, sediments, and wine: Identifying sources and environmental redox conditions. *Am. J. Anal. Chem.* **2013**, *4*, 69–83. [CrossRef]
11. Zhang, J.; Nozaki, Y. Rare earth elements and yttrium in seawater: ICP-MS determinations in the East Caroline, Coral Sea, and South Fiji basins of the western South Pacific Ocean. *Geochim. Et Cosmochim. Acta* **1996**, *60*, 4631–4644. [CrossRef]
12. Zhang, J.; Nozaki, Y. Behavior of rare earth elements in seawater at the ocean margin: A study along the slopes of the Sagami and Nankai troughs near Japan. *Geochim. Et Cosmochim. Acta* **1998**, *62*, 1307–1317. [CrossRef]
13. Sholkovitz, E.R.; Landing, W.M.; Lewis, B.L. Ocean particle chemistry: The fractionation of rare earth elements between suspended particles and seawater. *Geochim. Et Cosmochim. Acta* **1994**, *58*, 1567–1579. [CrossRef]
14. Elderfield, H. The oceanic chemistry of the rare-earth elements. *Philos. Trans. R. Soc. Lond. Ser. A Math. Phys. Sci.* **1988**, *325*, 105–126.
15. Caetano-Filho, S.; Paula-Santos, G.M.; Dias-Brito, D. Carbonate REE + Y signatures from the restricted early marine phase of South Atlantic Ocean (late Aptian–Albian): The influence of early anoxic diagenesis on shale-normalized REE + Y patterns of ancient carbonate rocks. *Palaeogeogr. Palaeoclimatol. Palaeoecol.* **2018**, *500*, 69–83. [CrossRef]
16. Kim, J.H.; Torres, M.E.; Haley, B.A.; Kastner, M.; Pohlman, J.W.; Riedel, M.; Lee, Y.J. The effect of diagenesis and fluid migration on rare earth element distribution in pore fluids of the northern Cascadia accretionary margin. *Chem. Geol.* **2012**, *291*, 152–165. [CrossRef]
17. Li, F.; Webb, G.E.; Algeo, T.J.; Kershaw, S.; Lu, C.; Oehlert, A.M.; Tan, X. Modern carbonate ooids preserve ambient aqueous REE signatures. *Chem. Geol.* **2019**, *509*, 163–177. [CrossRef]
18. Qing, H.; Mountjoy, E.W. Rare earth element geochemistry of dolomites in the Middle Devonian Presqu'ile barrier, Western Canada Sedimentary Basin: Implications for fluid-rock ratios during dolomitization. *Sedimentology* **1994**, *41*, 787–804. [CrossRef]
19. Chen, J.; Algeo, T.J.; Zhao, L.; Chen, Z.Q.; Cao, L.; Zhang, L.; Li, Y. Diagenetic uptake of rare earth elements by bioapatite, with an example from Lower Triassic conodonts of South China. *Earth Sci. Rev.* **2015**, *149*, 181–202. [CrossRef]
20. De Baar, H.J.; German, C.R.; Elderfield, H.; Van Gaans, P. Rare earth element distributions in anoxic waters of the Cariaco Trench. *Geochim. Et Cosmochim. Acta* **1988**, *52*, 1203–1219. [CrossRef]
21. Mongelli, G.; Sinisi, R.; Paternoster, M.; Perri, F. REEs and U distribution in P-rich nodules from Gelasian Apulian Tethyan carbonate: A genetic record. *J. Geochem. Explor.* **2018**, *194*, 19–28. [CrossRef]
22. Sarangi, S.; Mohanty, S.P.; Barik, A. Rare earth element characteristics of Paleoproterozoic cap carbonates pertaining to the Sausar Group, Central India: Implications for ocean paleoredox conditions. *J. Asian Earth Sci.* **2017**, *148*, 31–50. [CrossRef]
23. Özyurt, M.; Kirmaci, M.Z.; Yilmaz, İ.Ö.; Kandemir, R. Sedimentological and Geochemical Records of Lower Cretaceous Carbonate Successions Around Trabzon (NE Turkey). In *Patterns and Mechanisms of Climate, Paleoclimate and Paleoenvironmental Changes from Low-Latitude Regions*; Zhang, Z., Khélif, N., Mezghani, A., Heggy, E., Eds.; Springer: London, UK; Berlin/Heidelberg, Germany, 2019; pp. 19–21.
24. Görür, N. Timing of opening of the Black Sea basin. *Tectonophysics* **1988**, *147*, 247–262. [CrossRef]
25. Eren, M.; Tasli, K. Kilop cretaceous hardground (Kale, Gümüşhane, NE Turkey): Description and origin. *J. Asian Earth Sci.* **2002**, *20*, 433–448. [CrossRef]
26. Kara-Gülbay, R.; Kırmacı, M.Z.; Korkmaz, S. Organic geochemistry and depositional environment of the Aptian bituminous limestone in the Kale Gümüşhane area (NE-Turkey): An example of lacustrine deposits on the platform carbonate sequence. *Org. Geochem.* **2012**, *49*, 6–17. [CrossRef]

27. Kırmacı, M.Z. Sedimatological Investigation of the Upper Jurassic-Lower Cretaceous Berdiga Limestone in the Alucra-Gumushane-Bayburt areas (Eastern Pontide Southern Zone). Ph.D. Thesis, Karadeniz Technical University, Ortahisar/Trabzon, Turkey, 1992; p. 256.

28. Kirmaci, M.Z.; Koch, R.; Bucur, J.I. An Early Cretaceous section in the Kircaova Area (Berdiga Limestone, NE-Turkey) and its correlation with platform carbonates in W-Slovenia. *Facies* **1996**, *34*, 1–21. [CrossRef]

29. Koch, R.; Bucur, I.I.; Kirmaci, M.Z.; Eren, M.; Tasli, K. Upper Jurassic and Lower Cretaceous carbonate rocks of the Berdiga Limestone–Sedimentation on an onbound platform with volcanic and episodic siliciclastic influx. Biostratigraphy, facies and diagenesis (Kircaova, Kale-Gümüşhane area; NE-Turkey). *Neues Jahrb. Für Geol. Paläontol. Abh.* **2008**, *247*, 23–61. [CrossRef]

30. Vincent, S.J.; Guo, I.; Flecker, R.; BouDagher-Fadel, M.K.; Ellam, R.M.; Kandemir, R. Age constraints on intra-formational unconformities in Upper Jurassic-Lower Cretaceous carbonates in northeast Turkey; geodynamic and hydrocarbon implications. *Mar. Petrol. Geol.* **2018**, *91*, 639–657. [CrossRef]

31. Özyurt, M. Origin of Dolomitization in Upper Jurassıc-Lower Cretaceous Platform Carbonates (Berdiga Formation) in Gümüşhane Area. Ph.D. Thesis, Karadeniz Technical University, Trabzon, Turkey, 2019.

32. Özyurt, M.; Kırmacı, M.Z.; Al-Aasm, I.S. Geochemical characteristics of Upper Jurassic–Lower Cretaceous platform carbonates in Hazine Mağara, Gümüşhane (northeast Turkey): Implications for dolomitization and recrystallization. *Can. J. Earth Sci.* **2019**, *56*, 306–320. [CrossRef]

33. Özyurt, M.; Al-Aasm, İ.; Kirmaci, M.Z. Diagenetic Evolution of Upper Jurassic-Lower Cretaceous Berdiga Formation, NE Turkey: Petrographic and Geochemical Evidence. In *Paleobiodiversity and Tectono-Sedimentary Records in the Mediterranean Tethys and Related Eastern Areas*; Boughdiri, M., Bádenas, B., Selden, P., Jaillard, E., Bengtson, P., Granier, B.R.C., Eds.; Springer: London, UK; Berlin/Heidelberg, Germany,, 2019; pp. 175–177.

34. Kırmacı, M.Z.; Yıldız, M.; Kandemir, R.; Eroğlu-Gümrük, T. Multistage dolomitization in Late Jurassic–Early Cretaceous platform carbonates (Berdiga Formation), Başoba Yayla (Trabzon), NE Turkey: Implications of the generation of magmatic arc on dolomitization. *Mar. Petrol. Geol.* **2018**, *89*, 515–529. [CrossRef]

35. Pelin, S. Geological Investigation of Alucra (Giresun) Southeast Region in Terms of Petroleum Opportunities. Ph.D. Thesis, Karadeniz Technical University, Trabzon, Turkey, 1977.

36. Taslı, K. Stratigraphy, Paleogeography and Micropaleontology of Upper Jurassic–Lower Cretaceous Carbonate Sequence in the Gümüshane and Bayburt Areas (NE Turkey). Ph.D. Thesis, Karadeniz Technical University, Trabzon, Turkey, 1991.

37. Taslı, K.; Özer, E.; Yılmaz, C. Biostratigraphic and Environmental Analysis of the Upper Jurassic-Lower Cretaceous Carbonate Sequence in the Başoba Yayla Area (Trabzon, NE Turkey). *Turk. J. Earth Sci.* **2000**, *8*, 125–135.

38. Yildiz, M.; Ziya Kirmaci, M.; Kandemir, R.; Taslı, K. Benthic Foraminiferal Assemblages and Facies Analysis of the Late Jurrasic-Early Cretaceous Platform Carbonate Succession in the Mescitli-İkisu Area (Gümüşhane, NE Turkey). In Proceedings of the ICOCEE-CAPPADOCIA2017 Conference, Nevsehir, Turkey, 8–10 May 2017.

39. Yildiz, M.; Ziya Kirmaci, M.; Kandemir, R. Dolomitization in Late Jurassic-Early Cretaceous Platform Carbonates (Berdiga Formation), Ayralaksa Yayla (Trabzon), NE Turkey. In Proceedings of the 19th EGU General Assembly, Vienna, Austria, 23–28 April 2017; Volume 19, p. 11661.

40. Okay, A.I.; Sahinturk, O. AAPG Memoir 68: Regional and Petroleum Geology of the Black Sea and Surrounding Region. Chapter 15. *Geol. East. Pontides* **1997**, 291–311.

41. Okay, A.I.; Tüysüz, O. *Tethyan Sutures of Northern Turkey*; Geological Society Special Publications: London, UK, 1999; Volume 156, pp. 475–515.

42. Gedikoğlu, A.; Pelin, S.; Özsayar, T. The main lines of geotectonic development of East Pontids in the Mesozoic era: Geocome-I. *Min. Res. Exp. Inst. Geol. Soc. Turk.* **1979**, 551–581.

43. Aydin, F. Geochronology, geochemistry, and petrogenesis of the Maçka subvolcanic intrusions: Implications for the Late Cretaceous magmatic and geodynamic evolution of the eastern part of the Sakarya Zone, northeastern Turkey. *Int. Geol. Rev.* **2014**, *56*, 1246–1275. [CrossRef]

44. Karsli, O.; Dokuz, A.; Uysal, İ.; Aydin, F.; Kandemir, R.; Wijbrans, J. Generation of the Early Cenozoic adakitic volcanism by partial melting of mafic lower crust, Eastern Turkey: Implications for crustal thickening to delamination. *Lithos* **2010**, *114*, 109–120. [CrossRef]

45. Karsli, O.; Dokuz, A.; Uysal, I.; Aydin, F.; Chen, B.; Kandemir, R.; Wijbrans, J. Relative contributions of crust and mantle to generation of Campanian high-K calc-alkaline I-type granitoids in a subduction setting, with special reference to the Harşit Pluton, Eastern Turkey. *Contrib. Mineral. Petrol.* **2010**, *160*, 467–487. [CrossRef]

46. Karsli, O.; Ketenci, M.; Uysal, İ.; Dokuz, A.; Aydin, F.; Chen, B.; Wijbrans, J. Adakite-like granitoid porphyries in the Eastern Pontides, NE Turkey: Potential parental melts and geodynamic implications. *Lithos* **2011**, *127*, 354–372. [CrossRef]

47. Yücel, C.; Arslan, M.; Temizel, I.; Yazar, E.A.; Ruffet, G. Evolution of K-rich magmas derived from a net veined lithospheric mantle in an ongoing extensional setting: Geochronology and geochemistry of Eocene and Miocene volcanic rocks from Eastern Pontides (Turkey). *Gondwana Res.* **2017**, *45*, 65–86. [CrossRef]

48. Yücel, C. Geochronology, geochemistry, and petrology of adakitic Pliocene–Quaternary volcanism in the Şebinkarahisar (Giresun) area, NE Turkey. *Int. Geol. Rev.* **2019**, *61*, 754–777. [CrossRef]

49. Delibaş, O.; Moritz, R.; Ulianov, A.; Chiaradia, M.; Saraç, C.; Revan, K.M.; Göç, D. Cretaceous subduction-related magmatism and associated porphyry-type Cu–Mo prospects in the Eastern Pontides, Turkey: New constraints from geochronology and geochemistry. *Lithos* **2016**, *248*, 119–137. [CrossRef]

50. Yilmaz, C.; Carannante, G.; Kandemir, R. The rift-related Late Cretaceous drowning of the Gümüşhane carbonate platform (NE Turkey). *Boll. Della Soc. Geol. Italiana* **2008**, *127*, 37–50.

51. Özyurt, M.; Kırmacı, M.Z.; Yılmaz, İ.Ö.; Kandemir, R.; Taslı, K. Sedimentological and geochemical approaches for determination of the paleooceanographic and paleoclimatic conditions of Lower Cretaceous marine deposits of the eastern part of Sakarya Zone, NE Turkey. *Int. J. Earth Sci..* Under review.

52. Kırmacı, M.Z.; Akdağ, K. Origin of dolomite in the Late Cretaceous–Paleocene limestone turbidites, eastern Pontides, Turkey. *Sediment. Geol.* **2005**, *181*, 39–57. [CrossRef]

53. Yılmaz, C. Gümüşhane-Bayburt yöresindeki mesozoyik havzalarının tektono-sedimantolojik kayıtları ve kontrol etkenleri. *Türk. Jeol. Bül.* **2002**, *45*, 141–164.

54. Güven, İ. *1/100000 Ölçekli Açınsama Nitelikli Türkiye Jeoloji Haritaları, Trabzon-C28 ve D28 Paftaları*; Jeoloji Etütleri Dairesi, MTA Genel Müdürlüğü: Ankara, Turkey, 1998.

55. Topuz, G.; Altherr, R.; Schwarz, W.H.; Dokuz, A.; Meyer, H.P. Variscan amphibolite-facies rocks from the Kurtoğlu metamorphic complex (Gümüşhane area, Eastern Pontides, Turkey). *Int. J. Earth Sci.* **2007**, *96*, 861. [CrossRef]

56. Dokuz, A. A slab detachment and delamination model for the generation of Carboniferous high-potassium I-type magmatism in the Eastern Pontides, NE Turkey: The Köse composite pluton. *Gondwana Res.* **2011**, *19*, 926–944. [CrossRef]

57. Topuz, G.; Altherr, R.; Siebel, W.; Schwarz, W.H.; Zack, T.; Hasözbek, A.; Şen, C. Carboniferous high-potassium I-type granitoid magmatism in the Eastern Pontides: The Gümüşhane pluton (NE Turkey). *Lithos* **2010**, *116*, 92–110. [CrossRef]

58. Kandemir, R. Sedimentary Characteristics and Depositional Conditions of Lower-Middle Jurassic Şenköy Formation in and Around Gümüşhane. Ph.D. Thesis, Karadeniz Technical University, Trabzon, Turkey, 2004, unpublished work.

59. Kandemir, R.; Yılmaz, C. Lithostratigraphy, facies, and deposition environment of the lower Jurassic Ammonitico Rosso type sediments (ARTS) in the Gümüşhane area, NE Turkey: Implications for the opening of the northern branch of the Neo-Tethys Ocean. *J. Asian Earth Sci.* **2009**, *34*, 586–598. [CrossRef]

60. Eyuboglu, Y. Petrogenesis and U–Pb zircon chronology of felsic tuffs interbedded with turbidites (Eastern Pontides Orogenic Belt, NE Turkey): Implications for Mesozoic geodynamic evolution of the eastern Mediterranean region and accumulation rates of turbidite sequences. *Lithos* **2015**, *212*, 74–92.

61. Yilmaz, C.; Kandemir, R. Sedimentary records of the extensional tectonic regime with temporal cessation: Gumushane Mesozoic Basin (NE Turkey). *Geol. Carpath. Bratisl.* **2006**, *57*, 3.

62. Arslan, M.; Aliyazicioglu, I. Geochemical and petrological characteristics of the Kale (Gümüshane) volcanic rocks: Implications for the Eocene evolution of eastern Pontide arc volcanism, northeast Turkey. *Int. Geol. Rev.* **2001**, *43*, 595–610. [CrossRef]

63. Karsli, O.; Chen, B.; Aydin, F.; Şen, C. Geochemical and Sr–Nd–Pb isotopic compositions of the Eocene Dölek and Sariçiçek Plutons, Eastern Turkey: Implications for magma interaction in the genesis of high-K calc-alkaline granitoids in a post-collision extensional setting. *Lithos* **2007**, *98*, 67–96. [CrossRef]

64. Li, Y.; Zhao, L.; Chen, Z.Q.; Algeo, T.J.; Cao, L.; Wang, X. Oceanic environmental changes on a shallow carbonate platform (Yangou, Jiangxi Province, South China) during the Permian-Triassic transition: Evidence from rare earth elements in conodont bioapatite. *Palaeogeogr. Palaeoclimatol. Palaeoecol.* **2017**, *486*, 6–16. [CrossRef]

65. Premoli Silva, I.; Sliter, W.V. Cretaceous planktonic foraminiferal biostratigraphy and evolutionary trends from the Bottaccione section, Gubbio, Italy. *Paleontogr. Ital.* **1995**, *82*, 1–89.

66. Dunham, R.J. *Classification of Carbonate Rocks According to Depositional Textures*; AAPG: Tulsa, OK, USA, 1962; pp. 108–121.

67. Flügel, E. Microfacies Data: Fabrics. In *Microfacies of Carbonate Rocks*; Springer: Berlin/Heidelberg, Germany, 2004; pp. 177–242.

68. McLennan, S.M.; Taylor, S.R. Archaean Sedimentary Rocks and Their Relation to the Composition of the Archaean Continental Crust. In *Archaean Geochemistry*; Springer: Berlin/Heidelberg, Germany, 1904; pp. 47–72.

69. Bau, M.; Dulski, P. Distribution of yttrium and rare-earth elements in the Penge and Kuruman iron-formations, Transvaal Supergroup, South Africa. *Precambrian Res.* **1996**, *79*, 37–55. [CrossRef]

70. Shields, G.; Stille, P. Diagenetic constraints on the use of cerium anomalies as palaeoseawater redox proxies: An isotopic and REE study of Cambrian phosphorites. *Chem. Geol.* **2001**, *175*, 29–48. [CrossRef]

71. Alexander, B.W.; Bau, M.; Andersson, P.; Dulski, P. Continentally-derived solutes in shallow Archean seawater: Rare earth element and Nd isotope evidence in iron formation from the 2.9 Ga Pongola Supergroup, South Africa. *Geochim. Et Cosmochim. Acta* **2008**, *72*, 378–394. [CrossRef]

72. Tang, Y.; Han, G.; Wu, Q.; Xu, Z. Use of rare earth element patterns to trace the provenance of the atmospheric dust near Beijing, China. *Environ. Earth Sci.* **2013**, *68*, 871–879. [CrossRef]

73. Elderfield, H.; Upstill-Goddard, R.; Sholkovitz, E.R. The rare earth elements in rivers, estuaries and coastal sea waters: Processes affecting crustal input of elements to the ocean and their significance to the composition of sea water. *Geochim. Cosmocim. Acta* **1990**, *55*, 1807–1813.

74. Hu, M.; Ngia, N.R.; Gao, D. Dolomitization and hydrotectonic model of burial dolomitization of the Furongian-Lower Ordovician carbonates in the Tazhong Uplift, central Tarim Basin, NW China: Implications from petrography and geochemistry. *Mar. Petrol. Geol.* **2019**, *106*, 88–115. [CrossRef]

75. Zaky, A.H.; Brand, U.; Azmy, K. A new sample processing protocol for procuring seawater REE signatures in biogenic and abiogenic carbonates. *Chem. Geol.* **2015**, *416*, 36–50. [CrossRef]

76. Gromet, L.P.; Haskin, L.A.; Korotev, R.L.; Dymek, R.F. The "North American shale composite": Its compilation, major and trace element characteristics. *Geochim. Et Cosmochim. Acta* **1984**, *48*, 2469–2482. [CrossRef]

77. Bau, M. Rare-earth element mobility during hydrothermal and metamorphic fluid-rock interaction and the significance of the oxidation state of europium. *Chem. Geol.* **1991**, *93*, 219–230. [CrossRef]

78. Liu, J.; Song, J.; Yuan, H.; Li, X.; Li, N.; Duan, L. Rare earth element and yttrium geochemistry in sinking particles and sediments of the Jiaozhou Bay, North China: Potential proxy assessment for sediment resuspension. *Mar. Pollut. Bull.* **2019**, *144*, 79–91. [CrossRef] [PubMed]

79. Nozaki, Y.; Zhang, J.; Amakawa, H. The fractionation between Y and Ho in the marine environment. *Earth Planet. Sci. Lett.* **1997**, *148*, 329–340. [CrossRef]

80. Bau, M. Controls on the fractionation of isovalent trace elements in magmatic and aqueous systems: Evidence from Y/Ho, Zr/Hf, and lanthanide tetrad effect. *Contrib. Mineral. Petrol.* **1996**, *123*, 323–333. [CrossRef]

81. Banner, J.L.; Hanson, G.N. Calculation of simultaneous isotopic and trace element variations during water-rock interaction with applications to carbonate diagenesis. *Geochim. Et Cosmochim. Acta* **1990**, *54*, 3123–3137. [CrossRef]

82. Banner, J.L.; Hanson, G.N.; Meyers, W.J. Rare earth element and Nd isotopic variations in regionally extensive dolomites from the Burlington-Keokuk Formation (Mississippian); implications for REE mobility during carbonate diagenesis. *J. Sediment. Res.* **1988**, *58*, 415–432.

83. Barrat, J.A.; Boulegue, J.; Tiercelin, J.J.; Lesourd, M. Strontium isotopes and rare-earth element geochemistry of hydrothermal carbonate deposits from Lake Tanganyika, East Africa. *Geochim. Et Cosmochim. Acta* **2000**, *64*, 287–298. [CrossRef]

84. Bau, M. Scavenging of dissolved yttrium and rare earths by precipitating iron oxyhydroxide: Experimental evidence for Ce oxidation, Y-Ho fractionation, and lanthanide tetrad effect. *Geochim. Et Cosmochim. Acta* **1999**, *63*, 67–77. [CrossRef]

85. Li, F.; Yan, J.; Burne, R.V.; Chen, Z.Q.; Algeo, T.J.; Zhang, W.; Xie, S. Paleo-seawater REE compositions and microbial signatures preserved in laminae of Lower Triassic ooids. *Palaeogeogr. Palaeoclimatol. Palaeoecol.* **2017**, *486*, 96–107. [CrossRef]

86. Zhao, M.Y.; Zheng, Y.F. A geochemical framework for retrieving the linked depositional and diagenetic histories of marine carbonates. *Earth Planet. Sci. Lett.* **2017**, *460*, 213–221. [CrossRef]

87. Bau, M.; Alexander, B. Preservation of primary REE patterns without Ce anomaly during dolomitization of Mid-Paleoproterozoic limestone and the potential re-establishment of marine anoxia immediately after the "Great Oxidation Event". *S. Afr. J. Geol.* **2006**, *109*, 81–86. [CrossRef]

88. Bau, M.; Balan, S.; Schmidt, K.; Koschinsky, A. Rare earth elements in mussel shells of the Mytilidae family as tracers for hidden and fossil high-temperature hydrothermal systems. *Earth Planet. Sci. Lett.* **2010**, *299*, 310–316. [CrossRef]

89. Kamber, B.S.; Webb, G.E. The geochemistry of late Archaean microbial carbonate: Implications for ocean chemistry and continental erosion history. *Geochim. Et Cosmochim. Acta* **2001**, *65*, 2509–2525. [CrossRef]

90. Tanaka, K.; Kawabe, I. REE abundances in ancient seawater inferred from marine limestone and experimental REE partition coefficients between calcite and aqueous solution. *Geochem. J.* **2006**, *40*, 425–435. [CrossRef]

91. Shields, G.A.; Webb, G.E. Has the REE composition of seawater changed over geological time? *Chem. Geol.* **2004**, *204*, 103–107. [CrossRef]

92. Bolhar, R.; Van Kranendonk, M.J. A non-marine depositional setting for the northern Fortescue Group, Pilbara Craton, inferred from trace element geochemistry of stromatolitic carbonates. *Precambrian Res.* **2007**, *155*, 229–250. [CrossRef]

93. Bolhar, R.; Kamber, B.S.; Moorbath, S.; Fedo, C.M.; Whitehouse, M.J. Characterisation of early Archaean chemical sediments by trace element signatures. *Earth Planet. Sci. Lett.* **2004**, *222*, 43–60. [CrossRef]

94. MacLeod, K.G.; Irving, A.J. Correlation of cerium anomalies with indicators of paleoenvironment. *J. Sediment. Res.* **1996**, *66*, 948–955.

95. Smrzka, D.; Zwicker, J.; Bach, W.; Feng, D.; Himmler, T.; Chen, D.; Peckmann, J. The behavior of trace elements in seawater, sedimentary pore water, and their incorporation into carbonate minerals: A review. *Facies* **2019**, *65*, 41. [CrossRef]

96. Zhong, S.; Mucci, A. Partitioning of rare earth elements (REEs) between calcite and seawater solutions at 25 C and 1 atm, and high dissolved REE concentrations. *Geochim. Et Cosmochim. Acta* **1995**, *59*, 443–453. [CrossRef]

97. Feng, D.; Chen, D.; Peckmann, J. Rare earth elements in seep carbonates as tracers of variable redox conditions at ancient hydrocarbon seeps. *Terra Nova* **2009**, *21*, 49–56. [CrossRef]

98. Zhu, B.; Ge, L.; Yang, T.; Jiang, S.; Lv, X. Stable isotopes and rare earth element compositions of ancient cold seep carbonates from Enza River, northern Apennines (Italy): Implications for fluids sources and carbonate chimney growth. *Mar. Petrol. Geol.* **2019**, *109*, 434–448. [CrossRef]

99. Auer, G.; Reuter, M.; Hauzenberger, C.A.; Piller, W.E. The impact of transport processes on rare earth element patterns in marine authigenic and biogenic phosphates. *Geochim. Et Cosmochim. Acta* **2017**, *203*, 140–156. [CrossRef]

100. Kato, Y.; Yamaguchi, K.E.; Ohmoto, H. Rare earth elements in Precambrian banded iron formations: Secular changes ofCe andEu anomalies and evolution of atmospheric oxygen. *Evol. Early Earth Atmos. Hydros. Biosph. Constraints Ore Depos.* **2006**, *198*, 269.

101. Tostevin, R.; Shields, G.A.; Tarbuck, G.M.; He, T.; Clarkson, M.O.; Wood, R.A. Effective use of cerium anomalies as a redox proxy in carbonate-dominated marine settings. *Chemical Geology* **2016**, *438*, 146–162. [CrossRef]

102. Lawrence, M.G.; Jupiter, S.D.; Kamber, B.S. Aquatic geochemistry of the rare earth elements and yttrium in the Pioneer River catchment, Australia. *Mar. Freshw. Res.* **2006**, *57*, 725–736. [CrossRef]

103. Parsapoor, A.; Khalili, M.; Mackizadeh, M.A. The behaviour of trace and rare earth elements (REE) during hydrothermal alteration in the Rangan area (Central Iran). *J. Asian Earth Sci.* **2009**, *34*, 123–134. [CrossRef]

104. Klinkhammer, G.P.; Elderfield, H.; Edmond, J.M.; Mitra, A. Geochemical implications of rare earth element patterns in hydrothermal fluids from mid-ocean ridges. *Geochim. Et Cosmochim. Acta* **1994**, *58*, 5105–5113. [CrossRef]

105. Morse, J.W.; Mackenzie, F.T. *Geochemistry of Sedimentary Carbonates*; Elsevier: Amsterdam, The Netherland, 1990.

106. Mresah, M.H. The massive dolomitization of platformal and basinal sequences: Proposed models from the Paleocene, Northeast Sirte Basin, Libya. *Sediment. Geol.* **1998**, *116*, 199–226. [CrossRef]

107. Bischoff, J.L.; Fyfe, W.S. Catalysis, inhibition, and the calcite-aragonite problem; [Part] 1, The aragonite-calcite transformation. *Am. J. Sci.* **1968**, *266*, 65–79. [CrossRef]

108. McManus, K.M. The Aqueous Aragonite to Calcite Transformation: Rate, Mechanisms, and Its Role in the Development of Neomorphic Fabrics. PhD Thesis, Virginia Polytechnic Institute and State University, Blacksburg, VA, USA, 1982.

109. Perdikouri, C.; Piazolo, S.; Kasioptas, A.; Schmidt, B.C.; Putnis, A. Hydrothermal replacement of aragonite by calcite: Interplay between replacement, fracturing and growth. *Eur. J. Mineral.* **2013**, *25*, 123–136. [CrossRef]

110. Hashim, M.S.; Kaczmarek, S.E. Experimental stabilization of carbonate sediments to calcite: Insights into the depositional and diagenetic controls on calcite microcrystal texture. *Earth Planet. Sci. Lett.* **2020**, *538*, 116235. [CrossRef]

111. Moshier, S.O. Microporosity in micritic limestones: A review. *Sediment. Geol.* **1989**, *63*, 191–213. [CrossRef]

112. Lambert, L.; Durlet, C.; Loreau, J.P.; Marnier, G. Burial dissolution of micrite in Middle East carbonate reservoirs (Jurassic–Cretaceous): Keys for recognition and timing. *Mar. Petrol. Geol.* **2006**, *23*, 79–92. [CrossRef]

113. Loucks, R.G.; Lucia, F.J.; Waite, L.E. Origin and description of the micropore network within the Lower Cretaceous Stuart City Trend tight-gas limestone reservoir in Pawnee Field in South Texas. *GCAGS J.* **2013**, *2*, 29–41.

114. Karsli, O.; İlhan, M.; Kandemir, R.; Dokuz, A.; Duygu, L. In Review. Nature of the Early Cretaceous calc-alkaline lamprophyre and high-Nb alkaline basaltic dykes in Sakarya Zone, NE Turkey: Constrains on their linkage to subduction initiation of Neotethyan oceanic Slab. *Contrib. Mineral. Petrol.*. Under review.

115. Reynard, B.; Lécuyer, C.; Grandjean, P. Crystal-chemical controls on rare-earth element concentrations in fossil biogenic apatites and implications for paleoenvironmental reconstructions. *Chem. Geol.* **1999**, *155*, 233–241. [CrossRef]

116. Michard, A.; Albarède, F. The REE content of some hydrothermal fluids. *Chem. Geol.* **1986**, *55*, 51–60. [CrossRef]

117. Olivarez, A.M.; Owen, R.M. REE/Fe variations in hydrothermal sediments: Implications for the REE content of seawater. *Geochim. Et Cosmochim. Acta* **1989**, *53*, 757–762. [CrossRef]

118. Bright, C.A.; Cruse, A.M.; Lyons, T.W.; MacLeod, K.G.; Glascock, M.D.; Ethington, R.L. Seawater rare-earth element patterns preserved in apatite of Pennsylvanian conodonts? *Geochim. Et Cosmochim. Acta* **2009**, *73*, 1609–1624. [CrossRef]

119. German, C.R.; Elderfield, H. Rare earth elements in Saanich Inlet, British Columbia, a seasonally anoxic basin. *Geochim. Et Cosmochim. Acta* **1989**, *53*, 2561–2571. [CrossRef]

120. Koeppenkastrop, D.; Eric, H. Sorption of rare-earth elements from seawater onto synthetic mineral particles: An experimental approach. *Chem. Geol.* **1992**, *95*, 251–263. [CrossRef]

121. Derry, L.A.; Jacobsen, S.B. The chemical evolution of Precambrian seawater: Evidence from REEs in banded iron formations. *Geochim. Et Cosmochim. Acta* **1990**, *54*, 2965–2977. [CrossRef]

122. Jenkyns, H.C. Transient cooling episodes during Cretaceous Oceanic Anoxic Events with special reference to OAE 1a (Early Aptian). *Philos. Trans. Royal Soc. A Math. Phys. Eng. Sci.* **2018**, *376*, 20170073. [CrossRef]

123. Taslı, K.; Özsayar, T. Stratigraphy and paleoenvironmental setting of the Albian–Campanian deposits within the Gümüşßhane province (Eastern Pontides, NE Turkey). *Turk. Assoc. Petrol. Geol. Bull.* **1997**, *9*, 13–29.

Article

Effects of Dolomitization on Porosity during Various Sedimentation-Diagenesis Processes in Carbonate Reservoirs

Leilei Yang [1],*, Linjiao Yu [1], Donghua Chen [1], Keyu Liu [2,3], Peng Yang [2] and Xinwei Li [1]

[1]   Unconventional Petroleum Research Institute, China University of Petroleum, Beijing 102249, China;
2017210109@student.cup.edu.cn (L.Y.); 2018210109@student.cup.edu.cn (D.C.);
2018210513@student.cup.edu.cn (X.L.)
[2]   School of Geosciences, China University of Petroleum (East China), Qingdao 266580, China;
liukeyu@upc.edu.cn (K.L.); yangp@s.upc.edu.cn (P.Y.)
[3]   Qingdao National Laboratory for Marine Science and Technology, Qingdao 266071, China
*   Correspondence: yangleilei@cup.edu.cn

Received: 2 June 2020; Accepted: 23 June 2020; Published: 25 June 2020

Abstract: Carbonate reservoirs, especially dolomite reservoirs, contain large reserves of oil and gas. The complex diagenesis is quite challenging to document the dolomite reservoirs formation and evolution mechanism. Porosity development and evolution in dolomite reservoirs primarily reflect the comprehensive effect of mineral dissolution/precipitation during dolomitization. In this study, multicomponent multiphase flow and solute transport simulation was employed to investigate dolomitization in the deep carbonate strata of the Tarim Basin, Northwest China, where active exploration is currently under way. One- and two-dimensional numerical models with various temperatures, fluid compositions and hydrodynamic characteristics were established to quantificationally study dolomitization and its effect on porosity. After determining the main control factors, detailed petrologic characteristics in the studied area were also analyzed to establish four corresponding diagenetic numerical models under different sedimentary environments. These models enabled a systematic analysis of mineral dissolution/precipitation and a quantitative recovery of porosity evolution during various sedimentation-diagenesis processes. The results allowed for a quantitative evaluation and prediction of reservoir porosity, which would provide a basis for further oil and gas exploration in deep carbonate reservoirs.

Keywords: carbonate reservoirs; sedimentation; diagenesis; dolomitization; porosity

---

1. Introduction

Carbonate reservoirs are widely distributed over the world and crucial for oil and gas exploration because of their proven reserves. However, carbonate reservoirs are usually developed with complicated crack/pores during diagenesis, which make the reservoir prediction difficult and hinder the development of oil and gas exploration [1–7] It is of great importance to understand the effect of diagenesis on the reservoir porosity evolution as it is one of the key factors affecting the reservoir quality.

The formation and evolution of reservoir porosity are the topic of carbonate reservoir research and constitute the premise for the efficiency improvement of oil and gas exploration [5,8–10]. In carbonate reservoirs, reservoir porosity evolution is mainly the result of carbonate minerals dissolution and precipitation, such as dolomitization [11–14]. Fluid–rock reactions are usually the main factor for the porosity evolution because it results in carbonates dissolving to form voids, although in some cases minerals may precipitate to fill such voids [2,15,16]. However, the reservoir condition (e.g., temperature, pressure, mineral species, formation water hydrochemistry and fluid dynamics) is crucial to the porosity evolution during the long process of diagenesis [9,17–22].

As the most important carbonate reservoir, dolomite has received wide attention [13,23–25]. With the continuous exploration and discovery of oil and gas in deep-buried dolomite formations, a string of new issues has arisen. The most important issue is how dolomitization affects the reservoir porosity during various sedimentation–diagenesis processes [26–29]. This, however, has not received notable attention as the majority of previous research works focused on qualitative testing and analysis of dolomitization, which did not provide a systematic description of carbonate minerals under different burial conditions during the whole diagenetic process [30].

In this study, the deep strata of the Tarim Basin in Northwestern China were selected to establish a number of one- and two-dimensional numerical models under various diagenetic conditions. This allowed us to analyze the effect of temperature, fluid and hydrodynamic conditions on dolomitization, and for the quantitative evaluation of the reservoir porosity change induced by dolomitization. Then, quantitative recovery of porosity evolution during various sedimentation–diagenesis processes was successfully implemented. The results allowed for a quantitative evaluation and prediction of reservoir porosity, which would provide a basis for further oil and gas exploration in deep carbonate reservoirs.

2. Geological Background

2.1. Tectonic Locations

Tarim Basin is one of the most important petroliferous basins in China, which contains thick marine carbonate deposits. The most well-developed carbonate sequence within the basin is the Cambrian-Ordovician carbonate, which is also the key intervals for oil and gas exploration [31–33]. The Ordovician carbonate reservoirs in the Shunnan (SN) area, located in the central uplift of the Tarim Basin, are selected in this study, as shown in Figure 1, based on [33,34].

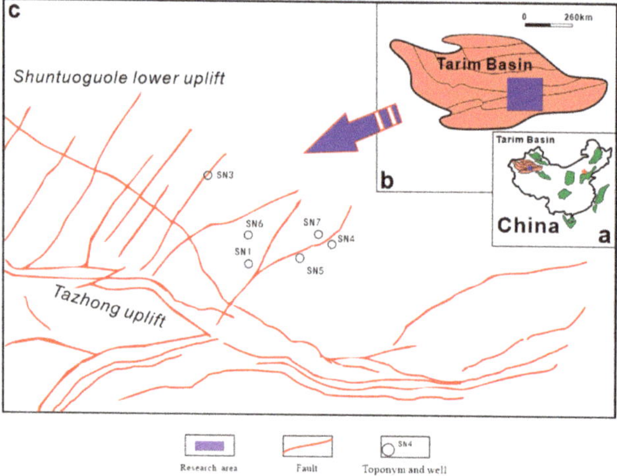

Figure 1. Tectonic locations of the Shunnan (SN) area in the northern slope of the middle Tarim Basin. (**a**) Location of Tarim Basin in China; (**b**) Location of SN in Tarim Basin; (**c**) Tectonic map.

2.2. Diagenesis

A good overview of the typical diagenesis in the studied area has been provided by various previous research, as shown in Figure 2 [32,33,35,36] A series of core and thin sections and other petrological analyses show that the diagenesis mainly includes micritization (Figure 2a), cementation (Figure 2b–g), structural fractures (Figure 2e), dissolution (Figure 2f), dolomitization (Figure 2h–k), and silicification (Figure 2l).

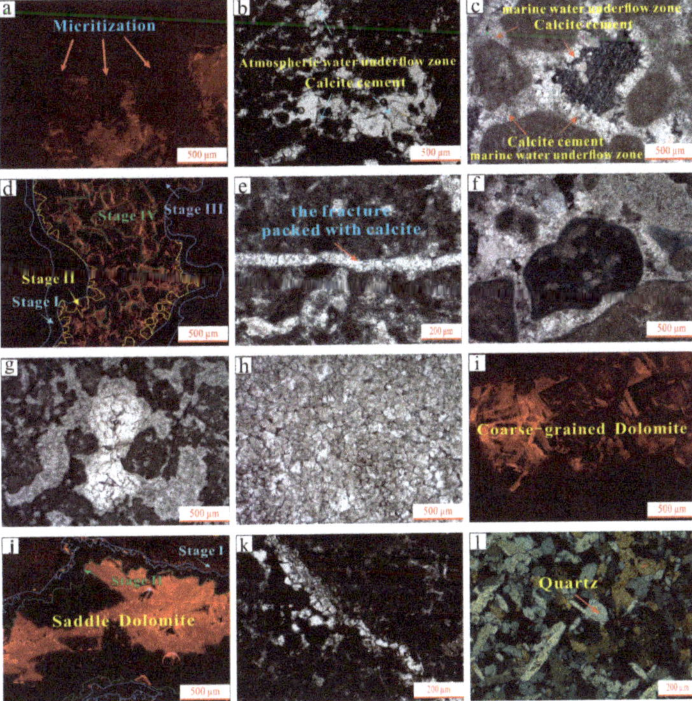

Figure 2. Typical diagenesis types of the Middle-Lower Ordovician carbonate. (**a**) SN6, 7325.50 m, sparry dolomitized calcarenite, small scale, cathodoluminescence; (**b**) SN4, 6361.13 m, micrite calcarenite, plane polarized light; (**c**) SN6, 7325.50 m, sparry dolomitized calcarenite, plane polarized light; (**d**) SN6, 6849.50 m, micrite clastic limestone, cathodoluminescence; (**e**) SN1, 6639.70 m, micrite calcarenite, plane polarized light; (**f**) SN7, 6876.40 m, sparry calcarenite, plane polarized light; (**g**) SN6, 6849.50 m, micrite calcarenite, plane polarized light; (**h**) SN6, 7318.50 m, sand cutting fine crystalline dolomite, plane polarized light; (**i**) SN6, 7319.43 m, fine-medium crystal residual sandy dolomites, cathodoluminescence; (**j**) SN7, 6534.10 m, micrite clastic limestone, cathodoluminescence; (**k**) SN4, 6461.00 m, micrite calcarenite, plane polarized light; (**l**) SN4, 6673.22 m, siliceous limestone, orthogonal polarization.

The diagenetic evolution of the Ordovician reservoirs in the Tarim Basin is complex, which mainly experience submarine diagenesis, meteoric diagenesis and the deep burial diagenetic settings. It is characterized by an early weak cementation, a late compaction and pressure solution, formation of stylolite, dissolution and karstification, mixed dolomitization in the supergene stage and the rupture and formation of multiphase fractures, which are all favorable conditions for developing favorable reservoirs [28,37].

Several studies suggest that the syngenetic karst and inter-stratal karst of the upper highstand system under the Middle-Lower Ordovician III interface are controlled by paleotopography and paleo-fault [38]. Such eogenetic karst reservoirs have a "quasi-layered" development model, and are filled by the subsequent formation of carbonate cement. In the late diagenetic stage, acidic hydrothermal fluids with medium-low temperature, high salinity and rich SiO_2 penetrated along the fault zones, leading to hydrothermal alterations such as hydrothermal dissolution, silica replacement and quartz precipitation to become the main controlling factors for the formation of reservoirs. Liu [39] theorized that the Middle-Lower Ordovician dolomites in the SN area are mainly resulted from shallow burial, burial (transitional environment) and deep burial dolomitizations. The fine-grained dolomite strata formed during the burial stage have the best reservoir properties, causing grain bank metasomatism and sustained dolomitization of the mud-sized dolostone formed under a shallow burial environment [37].

The diagenetic evolution sequence of the reservoirs in the studied area is summarized in Figure 3. According to the diagenetic history, the diagenetic evolution process in the studied reservoir can be categorized into six consecutive stages, including sedimentary-parasyngenetic, parasyngenetic-shallow burial, supergene, shallow burial, middle-deep burial and the deep burial stages.

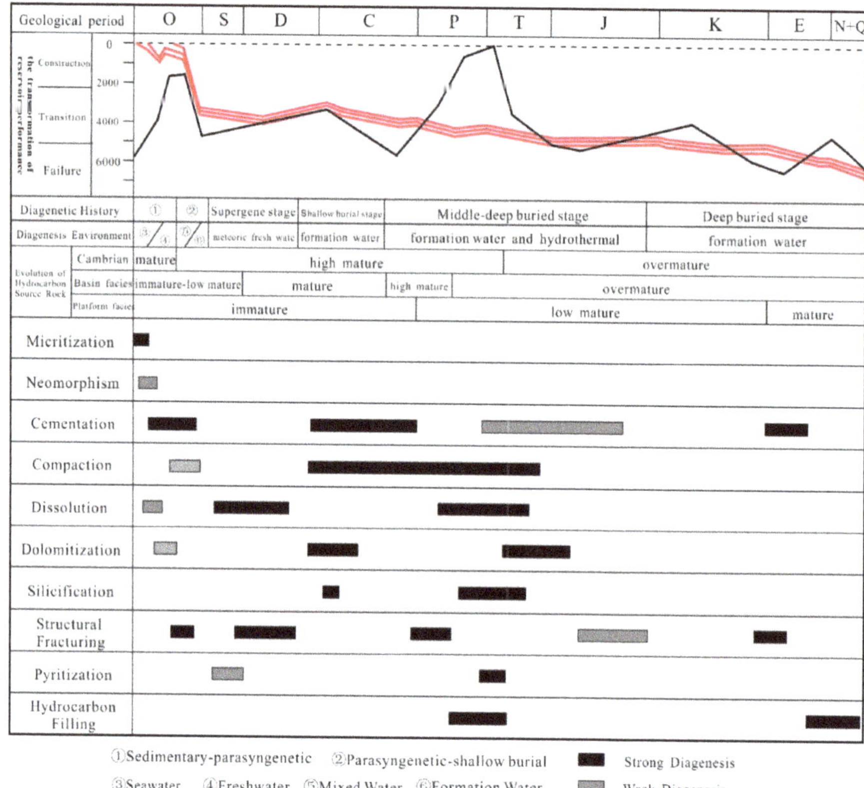

Figure 3. Diagenetic evolution of the Middle-Lower Ordovician carbonate.

2.3. Sedimentary Environments

The diagenetic evolution of the Ordovician carbonate rocks in the SN area occurred under different sedimentary environments. According to the diagenesis and petrological characteristics, the sedimentary environment can be divided into four categories: (1) marine phreatic environment in a subtidal zone, (2) meteoric fresh water and fresh water phreatic environment in a supralittoral zone, (3) seawater and freshwater interaction environment in an intertidal zone and (4) seawater evaporation environment in a shoal setting.

3. Model Setup

3.1. Model Tools

The model tool employed is a multicomponent multiphase flow and solute transport software TOUGHREACT, which introduces reactive chemistry into the multiphase fluid and heat flow code TOUGH. The numerical method for fluid flow and chemical transport simulations is based on the integral finite difference (IFD) method for space discretization. Thermal, hydraulic and chemical fields are coupled to model fluid flow, solute transport and reactions simultaneously [40].

At present, TOUGHREACT has been widely used in diagenesis-related research, enhanced geothermal system, CO_2 storage and other geological study. It can accommodate any number of chemical species present in the liquid, gas and solid phases. Several porosity–permeability models are available to calculate the reservoir porosity and permeability changed by mineral transformation in real time. Mineral dissolution and precipitation are simulated by equilibrium and kinetic equations, which are showed as follows [40].

3.1.1. Equilibrium Mineral Dissolution/Precipitation

The mineral saturation ratio can be expressed as

$$\Omega_m = K_m^{-1} \prod_{j=1}^{N_C} c_j^{v_{mj}} \gamma_j^{v_{mj}} \quad m = 1, \ldots, N_P \tag{1}$$

where m is the equilibrium mineral index, K_m is the corresponding equilibrium constant, c is concentration, γ is thermodynamic activity coefficient, and Nc is aqueous species. At equilibrium, we have

$$SI_m = \log_{10} \Omega_m = 0 \tag{2}$$

where SI_m is called the mineral saturation index.

3.1.2. Kinetic Mineral Dissolution/Precipitation

Kinetic rates could be functions of non-basis species as well. Usually the species appearing in rate laws happen to be basis species.

$$r_n = f(c_1, c_2, \cdots, c_{N_C}) = \pm k_n A_n \left| 1 - \Omega_n^{\theta} \right|^{\eta} \quad n = 1, \ldots, N_q \tag{3}$$

where positive values of r_n indicate dissolution, and negative values precipitation, k_n is the rate constant (moles per unit mineral surface area and unit time), which is temperature dependent, A_n is the specific reactive surface area per kg H_2O and n is the kinetic mineral saturation ratio.

3.2. One-Dimensional Flow

In order to clearly distinguish and understand the influence of any single factor on mineral transformation and porosity evolution, and to avoid multi-factor interference, a simple and typical one-dimensional model was established. The total length of this horizontal direction model was 150 m, divided into 50 equidistant grid cells. Fluid was injected from the left side at a constant rate, which promoted the flow of formation water from the left to the right.

In the Base Case, parameters were set according to the actual physical properties and chemical conditions of the carbonate reservoir studied. The initial temperature was set to 40 °C and the initial pressure was set to equal to the atmospheric pressure. The initial state of the reservoir was assumed to be homogeneous, and the initial porosity was 0.35. At the beginning of the model run, the minerals in the formation were all calcite. The flow rate of fluid in the formation was set to 4 m per year, while the pH of the inflowing external water was 8.5, and the Mg/Ca ratio was 5.25. Thirteen different models, Cases 1–13 were set up to compare the effects of different factors on dolomitization, such as temperature, flow rate, seawater concentration, Mg/Ca ratio, pH and SO_4^{2-} concentration. The specific conditions of these various experiments are listed in Table 1. The values not specified in Table 1 are shown with a dash (-) and indicate same conditions as the Base Case. Table 2 displays four seawater concentrations involved in the simulation.

Table 1. Parameters setting for one-dimensional flow models.

Model No.	T (°C)	Flow Rate (m/yr)	Seawater Index	Mg/Ca	pH	SO_4^{2-} (mmol/L)
Base Case	40	4	1#	5.25	8.5	22.208
Case 1	-		-	-	-	222.08
Case 2	-	2	-	-	-	-
Case 3	-	8	-	-	-	-
Case 4	-	-	2#	-	-	-
Case 5	-	-	3#	-	-	-
Case 6	-	-	4#	-	-	-
Case 7	-	-	-	52.5	-	-
Case 8	-	-	-	10.5	-	-
Case 9	-	-	-	-	6.5	-
Case 10	-	-	-	[illegible]	0	-
Case 11	60		-	-	-	-
Case 12	80	-	-	-	-	-
Case 13	100	-	-	-	-	-

Table 2. Multiple seawater concentrations for one-dimensional flow models.

Seawater Index	Saltness (ppt)	Ca^{2+}	Mg^{2+}	K^+	Na^+	Cl^-	HCO_3^-	SO_4^{2-}
					mmol/L			
1#	27	8.075	42.375	8.051	368.130	428.310	1.836	22.208
2#	15	4.475	23.333	4.436	202.696	235.775	1.016	12.219
3#	42	12.750	66.458	12.641	577.913	672.620	2.754	34.854
4#	32	9.625	50.417	9.564	437.913	509.380	2.180	26.417

3.3. Vertical Profile Flow

Shallow areas, close to the surface were not water-saturated, which inevitably led to unsaturated areas to exit. As different groundwater levels must affect the hydrodynamic conditions in the reservoir, two different groundwater levels (50 m and 100 m) were set to compare their effects on the dolomitization. Except for the different groundwater levels, the simulation conditions in both models were the same.

The initial mineral in the model was calcite, and the initial water type was set as shallow formation water, with a low concentration of K^+, Ca^{2+}, Na^+, Mg^{2+}, Cl^-, SO_4^{2-}, HCO_3^-, etc. During the simulation, external fluids were continuously injected into the reservoir from the upper side of the formation. The external fluid had the same properties as seawater with higher concentrations of Ca^{2+} and Mg^{2+}.

3.4. Diagenesis Evolution

Based on the six successive diagenetic stages categorized in Section 2, six corresponding sub-models were established. The parameters of temperature, pressure and fluid for each diagenetic stage were taken from measured data of the reservoirs studied (Table 3).

In four different sedimentary environments models, the parameter settings were as follows: the simulated temperature and pressure was 25 °C and 0.1 MPa respectively. The seawater component had a salinity of 35 ppt [41]. The initial mineral was calcite, and the mineral content of each sub-model was based on real-time data from the previous stage.

① Marine phreatic environment in the subtidal zone. Under the influence of tides, seawater leaks from the surface. The infiltration rate corresponds to the water exchange rate between seawater and formation.

② Meteoric fresh water and freshwater phreatic environment in the supralittoral zone. The diagenetic process consists of two sub-processes: (1) the atmospheric freshwater leaching process, with the following parameters: infiltration rate referenced to the infiltration rate of rainfall in equatorial regions, fluid composition corresponding to the equatorial region rainwater composition. (2) The shallow layer water flow process, with the fluid defined as mixed atmospheric fresh water and formation water.

③   Seawater evaporation environment in shoal. The infiltration rate corresponds to that at the surface.

④   Seawater and freshwater interaction environment in intertidal zones: The diagenetic process includes two sub-processes: (1) atmospheric freshwater leaching with an infiltration rate corresponding to the annual rainfall in the equatorial region and a fluid composition referring to the rainwater component in the equatorial region. (2) Seawater infiltration process after sea level rise. The infiltration rate corresponds to the water exchange rate between seawater and formation.

Table 3. Parameters of six sub-models corresponding to different diagenetic stages.

Sub-Models No.	Diagenesis Stage	Diagenetic Time (Ma)	Buried Depth (m)	Temperature (°C)	Fluid Composition	Mineral Composition
1#	Sedimentary-parasyngenetic stage	488–465	0–100	25	seawater, meteoric fresh water	micritization, calcite cement
2#	Parasyngenetic-shallow burial stage	465–460	50–600	25–40	mixed water, formation water	calcite cement
3#	Supergene stage	460–455	0–50	25	meteoric fresh water	calcite cement
4#	Shallow burial stage	455–445	50–600	25–40	formation water	calcite cement
5#	Middle-deep buried stage	445–252	600–4600	40–120	formation water, hydrothermal	calcite, dolomite, siliceous cement
6#	Deep buried stage	252–0	4600–7000	120–165	formation water	calcite cement

4. Results

   The main chemical reactions occurring in carbonate reservoirs are shown in reaction formulas (4)–(6). With the intrusion of external fluid, the original balance was disrupted, which made the calcite dissolve and release Ca^{2+} and CO_3^{2-}, which could then combine with Mg^{2+} in the external fluid to generate dolomite. The released Ca^{2+} could combine with SO_4^{2-} to generate gypsum under certain conditions in this process.

$$CaCO_3 \rightarrow Ca^{2+} + CO_3^{2-} \tag{4}$$

$$2CaCO_3 + Mg^{2+} \rightarrow CaMg\,(CO_3)_2 + Ca^{2+} \tag{5}$$

$$Ca^{2+} + SO_4^{2-} \rightarrow CaSO_4 \tag{6}$$

4.1. One-Dimensional Flow

   Relative content of calcite, dolomite and porosity evolution in the Base Case are shown in Figure 4. At the initial model (T = 0 My), the relative content of calcite, dolomite and gypsum were 0.65, 0.00 and 0.00, respectively. Calcite nearer the injection point (left of the model, X = 0 m) dissolved first and gradually transformed into dolomite. All calcite was converted to dolomite within 15 m from the injection point at 0.5 My, while the content of calcite was 0.00 and the content of dolomite was close to 0.60 within 60 m from the injection point at 1 My. The porosity within 15 m from the injection point was up to about 44% at 0.5 My and reached its maximum from the injection point within 60 m at 1 My.

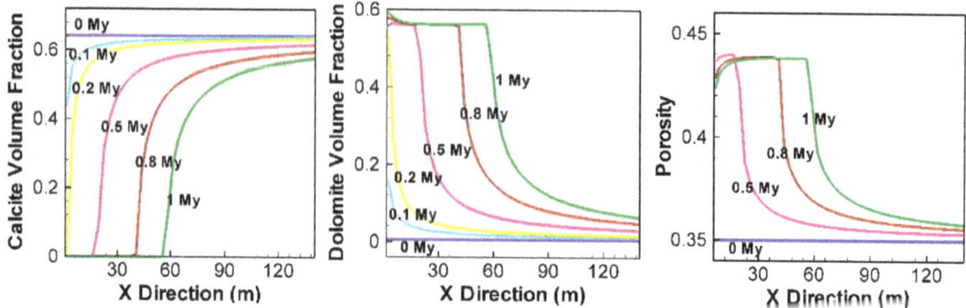

Figure 4. Calcite and dolomite content and porosity evolution (Base Case).

In Case 1, the concentration of SO_4^{2-} in the injected water was 10 times higher than that in the Base Case. Results in Figure 5 show that the calcite nearest to the injection point dissolved first and gradually transformed to dolomite. No calcite was found within 70 m of the injection point after 1 My, when the dolomite content was up to 0.55. During the transformation of calcite to dolomite, the generated Ca^{2+} moved continuously to the right area as the water flows. When Ca^{2+} and SO_4^{2-} concentrations in the system reached the conditions for gypsum precipitation, gypsum was formed. As the curve of gypsum content shows, the content reached the highest value at a distance of 10 m from the injection point after 0.5 My. With continuous injection of external fluids, more and more gypsum precipitated. The gypsum content was close to 0.4 between 20 and 80 m from the injection point after 1 My. The porosity increased to about 0.45 within 20 m of the injection point after 0.5 My, whereas it decreased to about 0.05 between 20 and 80 m from the injection point after 1 My.

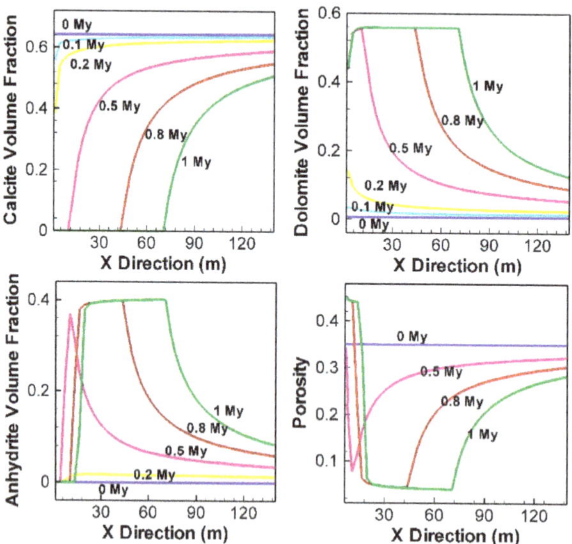

Figure 5. Calcite and dolomite content and porosity evolution (Case 1).

The change of porosity in different models is shown in Figure 6. The main difference between Case 2, Case 3 and the Base Case was hydrodynamic conditions. The fluid velocity in Base Case was 4 m/yr, while that of Case 2 and Case 3 was 2 m/yr and 8 m/yr, respectively. In Case 2, the porosity exhibited a significant increase only within 5 m of the injection point after 0.5 My. It took 0.8 My for the porosity to increase to about 44% in the range of 20 m from the injection point. However, in Case 3, the range where the porosity increased to 44% was close to 40 m from the injection point after 0.5 My and became nearly 90 m after 0.8 My.

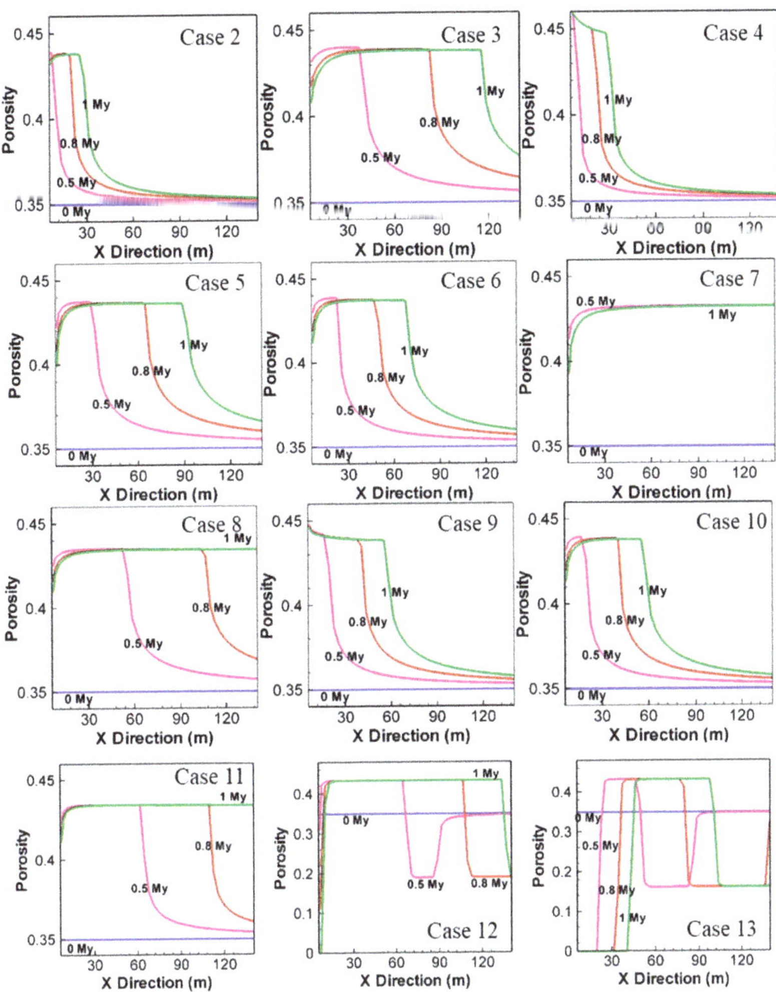

Figure 6. Porosity evolution in each model (Case 2–Case 13).

The difference between Case 4, Case 5, Case 6 and the Base Case was the injected seawater. In the Base Case, Seawater 1# was injected, while Seawater 2#, 3# and 4# were injected to Case 4, Case 5 and Case 6, respectively (Table 2). The salinities of the four types of seawater were 27 ppt, 15 ppt, 32 ppt and 42 ppt, respectively. The porosity in Case 4 was seen to increase significantly and reached a maximum of 46%. The porosity higher than 0.44 at 1 My was within 30 m in Case 4, while within 90 m and 75 m in Case 5 and Case 6, respectively.

The difference between Case 7, Case 8 and the Base Case was the Mg/Ca ratio in the injected fluid. The Mg/Ca ratio in the Base Case was 5.25, while that in Case 7 and Case 8 were 10 times and 2 times that of the Base Case. The porosity in the whole simulation range increased to more than 0.4 after 0.5 My in Case 7 while the range was only about 50 m in Case 8.

Case 9 and Case 10 differed from the Base Case in pH values of the injected fluid, which was 8.5 in the Base Case, against 6.5 and 9 for Case 9 and Case 10, respectively. Within 20 m of the injection point, the porosity in Case 9 was 44% at the injection point, whereas 41% in Case 10.

Case 11, Case 12 and Case 13 differed from the Base Case with regards to the reservoir temperature, which was 40 °C for the Base Case, against 60 °C, 80 °C and 100 °C, for Cases 11, 12 and 13, respectively. These porosity evolution curves were largely different.

4.2. Vertical Profile Flow

The calcite and dolomite content at three key times for groundwater levels of 100 m and 50 m are shown in Figure 7, respectively. Regardless of the level of the groundwater, the external fluid first infiltrated into the saturated zone, after which it infiltrated into the reservoir from the surface, initiating a reaction. The reaction was enhanced at the interface between the unsaturated and the saturated zone. In the unsaturated zone, the reaction was gradually developed in the fluid flow path as a result of the accumulation of ions. For a groundwater level of 100 m, the reaction extended to more than 200 m in the horizontal direction, while for a groundwater level of 50 m, the reaction expanded to 100 m in the horizontal direction after 0.2 My.

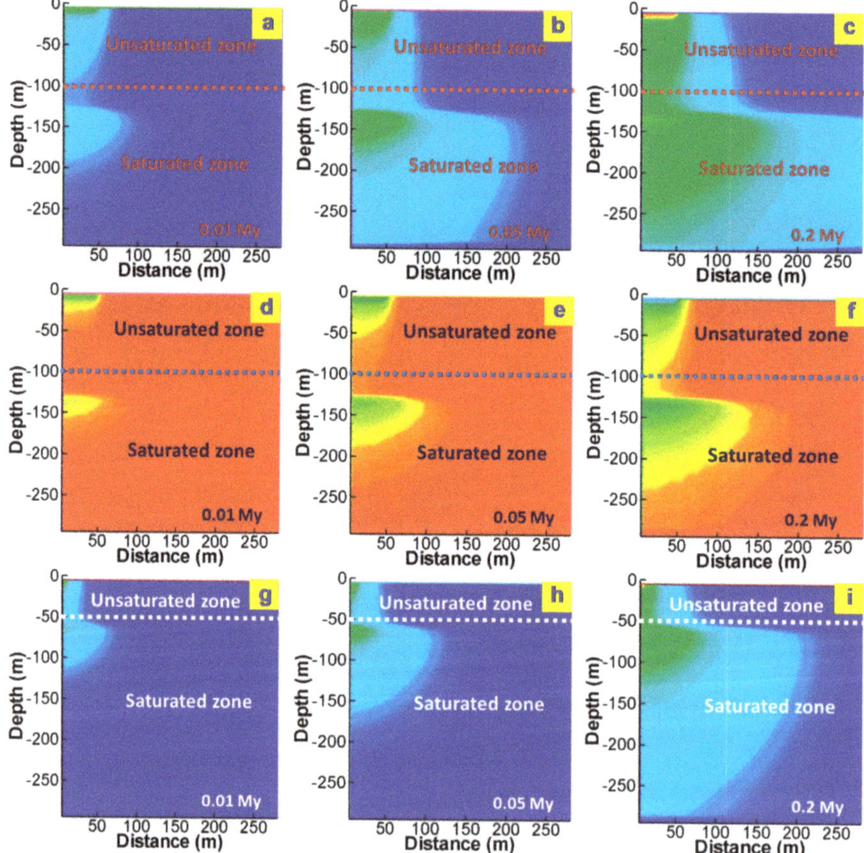

Figure 7. Calcite and dolomite content distribution after 0.01 My, 0.05 My and 0.2 My. (**a–c**) Calcite with groundwater level of 100 m; (**d–f**) Dolomite with groundwater level of 100 m; (**g–i**) Calcite with groundwater level of 50 m.

4.3. Diagenetic Evolution

4.3.1. Marine Phreatic Environment in the Subtidal Zone

Porosity evolution under marine phreatic environment is shown in Figure 8. Strong seawater cementation resulted in a calcite cement content of 5.1% during the sedimentary-parasyngenetic stage. During the parasyngenetic-shallow burial stage, seawater with high concentration of Ca^{2+} and Mg^{2+} was filled, which led to a strong cementation and to a drop of porosity to 20%. Dissolution occurred with the development of numerous secondary pores during the supergene stage. During the shallow burial stage, a large amount of dolomite and calcite precipitated, which decreased porosity to 18%. As a considerable amount of calcite was replaced by dolomite, the porosity decreased to 4.8% during the middle-deep burial stage. Finally, the porosity decreased to 1.6% because of calcite cementation in the deep burial stage.

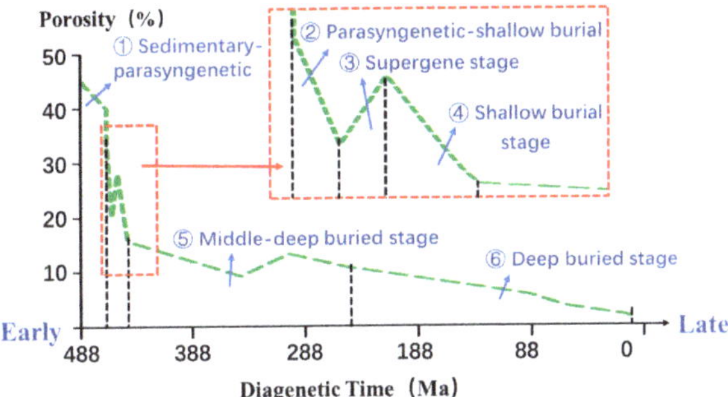

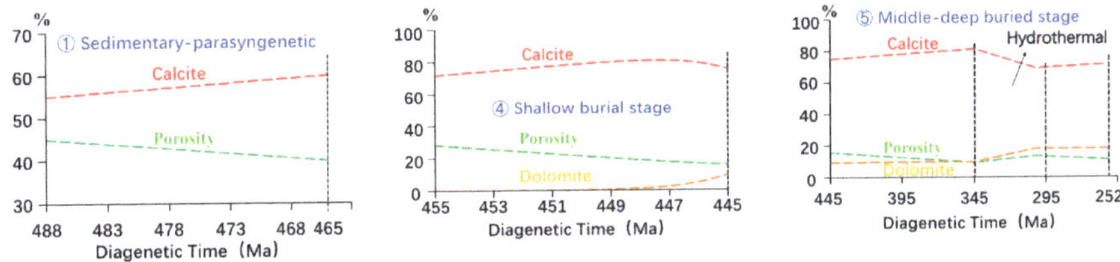

Figure 8. Porosity and typical mineral content evolution under marine phreatic environment.

4.3.2. Meteoric Fresh Water and Freshwater Phreatic Environment in the Supralittoral Zone

Porosity evolution under meteoric fresh water and freshwater phreatic environment is shown in Figure 9. With the synergistic effect of fresh water leaching, freshwater cementation and mechanical compaction, the porosity decreased to 38% during the sedimentary-parasyngenetic stage. As sea levels rose, the overlying seawater seeped into the stratum, and the porosity decreased to 22.2% in the parasyngenetic-shallow burial stage. During the supergene stage, dissolution occurred in the freshwater leaching environment, and the porosity increased to 33.9%, while during the burial stage, calcite cementation and dolomitization made the porosity decrease to 14.5%. As high salinity acidic hydrothermal fluid acted as diagenetic fluid, contributing to cementation, dissolution and dolomitization, porosity was almost constant throughout the middle-deep buried stage. Poor fluidity of formation water led to the weak cementation of calcite and the porosity reduced to 4.2% during the deep burial stage.

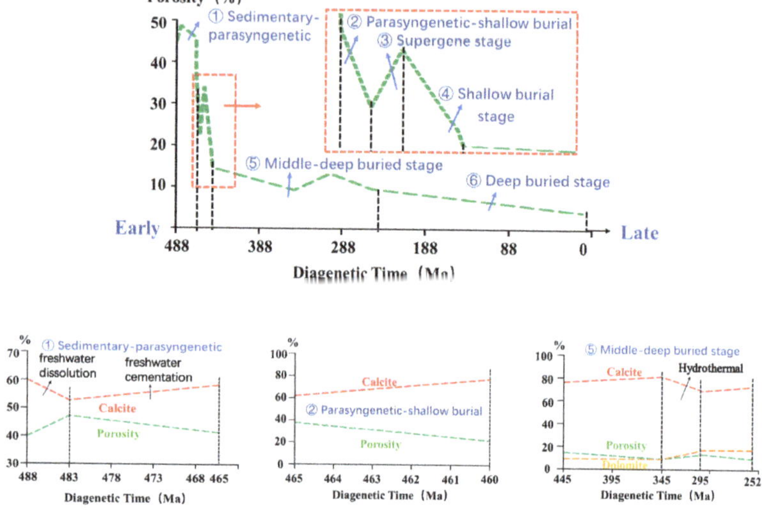

Figure 9. Porosity and typical mineral content evolution under meteoric fresh water and freshwater phreatic environment.

4.3.3. Seawater Evaporation Environment in Shoal

Porosity evolution under seawater evaporation environment is shown in Figure 10. During the sedimentary-parasyngenetic stage, cementation, dolomitization and mechanical compaction were the main reactions and porosity decreased to 36%. In the parasyngenetic-shallow burial stage, poor fluidity of the infiltration fluid caused a large amount of calcite cement, resulting in a decrease in porosity. The porosity increased to 30.5% during the supergene stage. Due to calcite cementation and dolomitization, porosity decreased to 18.2% during the shallow burial stage. During the middle-deep buried stage, a large amount of calcite was converted into dolomite and a small amount of siliceous cement was formed, resulting in porosity decrease. Lastly, porosity decreased to 10.8% during the deep burial stage.

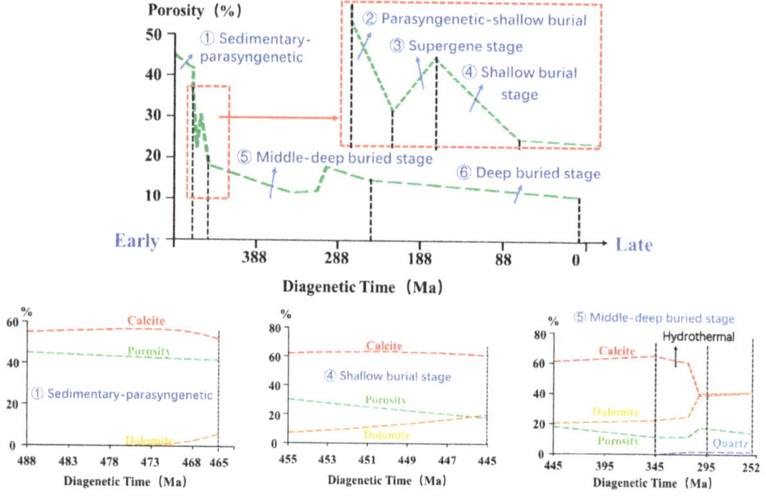

Figure 10. Porosity and typical mineral content evolution under seawater evaporation environment.

4.3.4. Seawater and Freshwater Interaction Environment in Intertidal Zones

Porosity evolution under seawater and freshwater interaction environment is shown in Figure 11. Dissolution, cementation and mechanical compaction were the main diagenetic processes during the sedimentary-parasyngenetic stage, and the porosity decreased to 35%. The porosity decreased to 19.7% because of calcite cement during the shallow burial stage and increased to 29% under the action of atmospheric freshwater leaching during the surface stage. In the shallow burial stage, the pore water gradually evolved into dolomitized fluid, causing calcite cementation and dolomitization and the porosity decreased to 13.9%. During the middle-deep buried stage, acidic hydrothermal fluid with medium-low temperature, high salinity and rich SiO_2 penetrated along the fault zone, and the porosity decreased to 9.3%. However, the porosity increased to 19.0% after hydrothermal alteration. The poor fluidity of the formation water and low porosity-permeability resulted in a weak cementation and a porosity decrease of 17.1% during the deep burial stage.

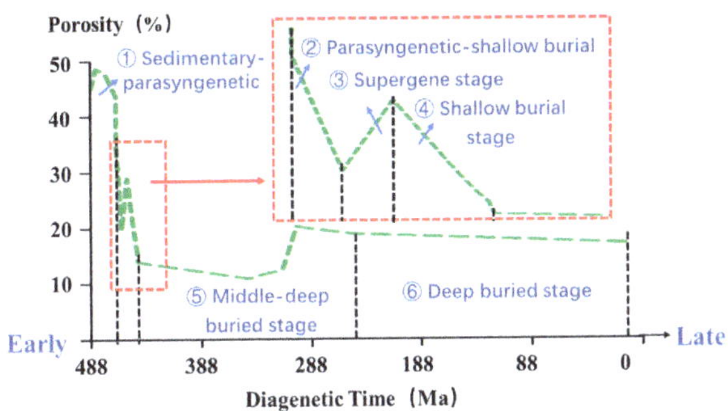

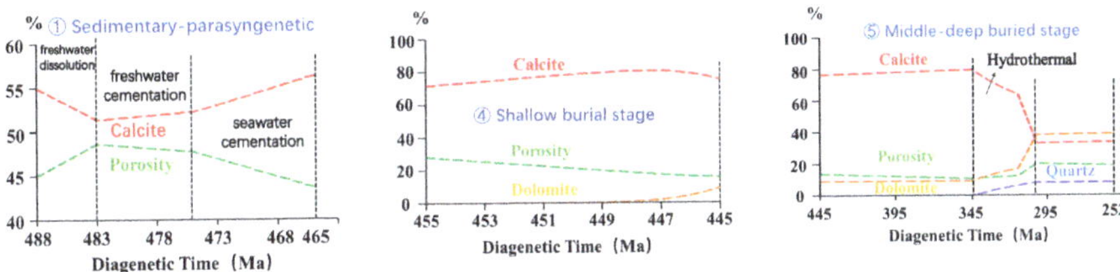

Figure 11. Porosity and typical mineral content evolution under seawater and freshwater interaction environment.

5. Discussion

5.1. Effect of Geological Factors on Dolomitization

The effect of the temperature, flow rate, fluid properties and other geological factors on dolomitization varied [23,42,43]. Results of one-dimensional models indicated that if the flow rate of the external fluid, i.e., the hydrodynamic condition, was different, the degree of dolomitization varied. The porosity change extent was consistent under various flow rate conditions, but different amounts of time were needed for the porosity to reach the same level. In short, the better the hydrodynamic conditions, the faster the fluid migration, which then pushed the reaction and expands the scope of dolomitization.

In two-dimensional models, comparing different groundwater surfaces, it was found that when the groundwater level was low, a 100 m unsaturated zone exist in the shallow layer, which contained much more gas. When external fluid entered the formation, it took a longer path to reach saturation, while more pores need to be filled in the meantime. Therefore, the reaction occurred later, and the degree of reaction was also weak due to the large number of ions being diluted during the porous flow process.

Different solution properties also lead to various levels of dolomitization [13,43]. The higher the Mg/Ca ratio in the fluid, the better the precipitation of dolomite, due to the large amount of Mg needed in the dolomitization process. However, the pH value of the fluid only had a limited influence compared with other factors.

Dolomitization is a reaction that involves a volume decrease, i.e., the volume of minerals is reduced with the conversion of calcite to dolomite, so that the number of pores within the reservoir increases [13,23]. However, if a large amount of gypsum that can occupy some pores is formed in this process, then the porosity will decrease. The results of this study demonstrated that the amount and the degree of precipitation gypsum were various under different diagenetic temperatures, which led to a large disparity in reservoir porosity.

5.2. Comparison of Model and Test Results during Successive Diagenetic Stages

Diagenetic models' results were consistent with test data both in mineral transformation and porosity evolution. In models, calcite and dolomite contents increased and the porosity decreased from 45% to about 35% throughout the sedimentary-parasyngenetic stage (488–465 Ma), meanwhile, an observation also shows that the main diagenesis was carbonate cementation. It is worth noting that the porosity evolution varied under different sedimentary environments, which was determined by the complex fluid composition.

The depth ranges in the parasyngenetic-shallow burial stage (465–460 Ma) was 50–600 m. As sea levels rose, the overlying seawater seeped into the formation and replenished Ca^{2+} and HCO_3^- in the pore water. Both test and model results indicate that increased temperature and pressure with the burial depth was beneficial to calcite cementation.

During the supergene stage (460–455 Ma), deeply buried rocks were uplifted below the diving surface due to tectonic movements and underwent short-term erosion and weathering. With the leaching and dissolution of meteoric fresh water, fresh water with low ion concentration infiltrated into the formation, thus dissolving calcite, which led to the development of a large number of secondary pores.

During the shallow burial stage, porosity and mineral contents evolution show that the main diagenetic reactions were cementation and dolomitization, associated with a porosity decrease. As the burial depth increased, a gradual increase of formation temperature favored the overcoming of the dynamic obstacles to dolomite formation, which were dominant at the beginning of the shallow burial stage [44]. High-salinity concentrated seawater is the diagenetic fluid at this stage, which can easily flow downward under the effect of gravity and cause convection with low-density seawater at the bottom. Following the penetration of the high-salinity seawater in the stratum, calcite would begin to cement, causing the Ca^{2+} concentration in the fluid to decrease, and the Mg^{2+}/Ca^{2+} to increase, which is beneficial for the fluid to overcome the kinetic obstacles. Calcite can then be replaced by dolomite, hence resulting in the decrease in calcite content with burial time. Previous studies have also shown that massive dolomitization of sediments is the main diagenesis in the shallow burial stage [45,46].

During the middle-deep burial stage, the diagenetic fluids were medium-high salinity, SiO_2-rich and CO_2-rich deep hydrothermal fluids. SiO_2 had a relatively high saturation in the hot liquid, whose temperature was about 200 °C after entering the formation (the formation temperature was about 100 °C). As the fluid flowed upward, reactions became different because of the different crack and fracture system in the formation. Strong hydrothermal dissolution happened in the fault zone of SN4 and SN5, with a developed fracture system.

During the deep burial stage (252–0 Ma), the temperature and pressure were high, which was conducive to the cementation and dolomitization. Calcite cementation occurred when high Ca^{2+} fluid entered the formation from the Lower Cambrian source rock, leading the content of calcite to increase and the porosity to decrease. With the high concentration of Ca^{2+}, the Mg/Ca continuously diminished, which is unfavorable for the dolomite precipitation.

5.3. Implications for Reservoir Evolution under Various Sedimentary Environments

The carbonate rock in the studied area experienced complex diagenesis under different sedimentary environments [43]. The various diagenetic phenomena observed were in the present results from the comprehensive effects of the sedimentary environments [47].

In the marine phreatic environment, cementation and compaction were the main diagenesis. As the carbonate deposits were all located below the sea level, the diagenetic fluid was seawater with a high concentration of Ca^{2+} and Mg^{2+}, which favored the formation of calcium carbonate but not that of dolomite. A large number of thick ring-like cements and fibrous calcite cements could be seen in the thin slices of the SN6, which is a typical single well representative of sedimentary diagenesis in the seawater subsurface flow environment.

In the meteoric water and freshwater phreatic environment, dissolution, cementation and compaction were the main diagenesis. With the sea level decline, some carbonate sediments became exposed. Atmospheric fresh water leach causing calcite dissolution porosity increased. With continuous infiltration, the saturation increased and calcite cement was formed, leading the porosity to decrease. A large number of fenestrae secondary pores and few equiaxed-granular freshwater cements filling the slab hole could be seen in the thin slices of the SN7, which is a typical single well of sedimentary diagenesis in the meteoric fresh water and freshwater phreatic environment.

In the seawater evaporative environment, dolomitization and compaction were the main diagenesis. Under dry climatic conditions, strong evaporation caused seawater to concentrate. High salinity seawater flowed into the formation, resulting in an increase in the concentration of Mg^{2+} and Ca^{2+} in the pore water. This hypersaline water with a high concentration of Mg^{2+} increased the probability of effective collision between Mg^{2+} and $CO_3{}^{2-}$, increasing the dolomite content. SN3 and SN5 are typical wells of the sedimentary diagenesis in the seawater evaporation environment.

In the seawater and freshwater interaction environment, cementation, dissolution and compaction were the main diagenesis. Controlled by secondary sedimentary cycles and the sea level decline temporarily, carbonate deposits in the intertidal zone were exposed to the atmosphere. Afterwards, as the sea level rose, carbonate deposits were submerged again by seawater. A large number of intragranular and intergranular dissolved pores and casting pores were formed in SN4, and the pores were filled with equiaxed-granular freshwater cements and fibrous calcite seawater cements formed during the parasyngenetic stage.

6. Conclusions

A series of test analysis and numerical simulation were combined to study the effects of various factors (such as temperature, flow rate, seawater concentration, Mg/Ca ratio, pH and $SO_4{}^S$ concentration) on dolomitization during diagenesis in carbonate reservoirs. The degree of dolomitization varied with the flow rate and other hydrodynamic conditions of the external fluid. The better the hydrodynamic conditions, the faster the fluid migration, which then pushed the reaction and expanded the scope of dolomitization. Different solution properties and minerals also led to various levels of dolomitization. During successive diagenetic stages under different sedimentary environments with various temperatures, minerals, solutions and other conditions, reservoir experienced complicated fluid–rock reactions. The diagenetic process and porosity evolution curves in four different sedimentary environments were re-established in the studied area. The main controlling factors of dolomitization were identified, which is helpful to clarify the genesis and distribution of carbonate reservoirs. As the fluid–rock interaction mechanism in carbonate reservoirs is a complex process, further experimental and numerical simulation studies are planned to elaborate and enhance our understanding.

Author Contributions: Conceptualization, L.Y. (Leilei Yang) and K.L.; Methodology, L.Y. (Leilei Yang), and K.L.; Software, L.Y. (Leilei Yang); Validation, L.Y. (Linjiao Yu) and D.C.; Investigation, X.L.; Resources, P.Y.;

Writing—original draft preparation, L.Y. (Leilei Yang); Writing—review and editing, L.Y. (Leilei Yang) and D.C.; Visualization, X.L.; Project administration, L.Y. (Leilei Yang); Funding acquisition, L.Y. (Leilei Yang) and K.L. All authors have read and agreed to the published version of the manuscript.

Funding: This research is supported by the Natural Science Foundation of China (No. 41702249) and the Strategic Priority Research Program of the Chinese Academy of Sciences, Grant No. XDA14010401. The data that support the findings of this study are available on request from the corresponding author. Please contact Leilei Yang, yangleilei@cup.edu.cn, for data used in this study.

Conflicts of Interest: The authors declare no conflicts of interest.

References

1. Alvarado, V.; Manrique, E. Enhanced Oil Recovery. An Update Review. *Energies* **2010**, *3*, 1529–1575. [CrossRef]

2. Araujo, T.P.; Leite, M.G.P. Flow simulation with reactive transport applied to carbonate rock diagenesis. *Mar. Pet. Geol.* **2017**, *88*, 94–106. [CrossRef]

3. Ehrenberg, S.N.; Walderhaug, O.; Bjorlykke, K. Carbonate porosity creation by mesogenetic dissolution: Reality or illusion? *AAPG Bull.* **2012**, *96*, 217–233. [CrossRef]

4. Kang, Y. Reservoir rock characteristics of paleozoic marine facies carbonate rock in the Tarim Basin. *Pet. Geol. Exp.* **2007**, *29*, 217–223.

5. Rashid, F.; Glover, P.W.J.; Lorinczi, P.; Hussein, D.; Collier, R.; Lawrence, J. Permeability prediction in tight carbonate rocks using capillary pressure measurements. *Mar. Pet. Geol.* **2015**, *68*, 536–550. [CrossRef]

6. Teichert, B.M.A.; Johnson, J.E.; Solomon, E.A.; Giosan, L.; Rose, K.; Kocherla, M.; Connolly, E.C.; Torres, M.E. Composition and origin of authigenic carbonates in the Krishna–Godavari and Mahanadi Basins, eastern continental margin of India. *Mar. Pet. Geol.* **2014**, *58*, 438–460. [CrossRef]

7. Zou, C.; Zhu, R.; Liu, K.; Su, L.; Bai, B.; Zhang, X.; Yuan, X.; Wang, J. Tight gas sandstone reservoirs in China: Characteristics and recognition criteria. *J. Pet. Sci. Eng.* **2012**, *88*, 82–91. [CrossRef]

8. Javanbakht, M.; Wanas, H.A.; Jafarian, A.; Shahsavan, N.; Sahraeyan, M. Carbonate diagenesis in the Barremian-Aptian Tirgan Formation (Kopet-Dagh Basin, NE Iran): Petrographic, geochemical and reservoir quality constraints. *J. Afr. Earth Sci.* **2018**, *144*, 122–135. [CrossRef]

9. Ronchi, P.; Ortenzi, A.; Borromeo, O.; Claps, M.; Zempolich, W.G. Depositional setting and diagenetic processes and their impact on the reservoir quality in the late Visean-Bashkirian Kashagan carbonate platform (Pre-Caspian Basin, Kazakhstan). *AAPG Bull.* **2010**, *94*, 1313–1348. [CrossRef]

10. Yasuda, E.Y.; dos Santos, R.G.; Trevisan, O.V. Kinetics of carbonate dissolution and its effects on the porosity and permeability of consolidated porous media. *J. Pet. Sci. Eng.* **2013**, *112*, 284–289. [CrossRef]

11. Li, J.; Ma, Y.; Huang, K.; Zhang, Y.; Wang, W.; Liu, J.; Li, Z.; Lu, S. Quantitative characterization of organic acid generation, decarboxylation, and dissolution in a shale reservoir and the corresponding applications—A case study of the Bohai Bay Basin. *Fuel* **2018**, *214*, 538–545. [CrossRef]

12. Li, Y.; Yang, W.; Wang, Q.; Song, Y.; Jiang, Z.; Guo, L.; Zhang, Y.; Wang, J. Influence of the actively migrated diagenetic elements on the hydrocarbon generation potential in tuffaceous shale. *Fuel* **2019**, *256*, 115795. [CrossRef]

13. Read, J.F.; Husinec, A.; Cangialosi, M.; Loehn, C.W.; Prtoljan, B. Climate controlled, fabric destructive, reflux dolomitization and stabilization via marine- and synorogenic mixed fluids: An example from a large Mesozoic, calcite-sea platform, Croatia. *Palaeogeogr. Palaeoclimatol. Palaeoecol.* **2016**, *449*, 108–126. [CrossRef]

14. Zhao, H.; Ning, Z.; Zhao, T.; Zhang, R.; Wang, Q. Effects of mineralogy on petrophysical properties and permeability estimation of the Upper Triassic Yanchang tight oil sandstones in Ordos Basin, Northern China. *Fuel* **2016**, *186*, 328–338. [CrossRef]

15. Miller, K.; Vanorio, T.; Keehm, Y. Evolution of permeability and microstructure of tight carbonates due to numerical simulation of calcite dissolution. *J. Geophys. Res. Solid Earth* **2017**, *122*, 4460–4474. [CrossRef]

16. Pearce, J.K.; Golab, A.; Dawson, G.K.W.; Knuefing, L.; Goodwin, C.; Golding, S.D. Mineralogical controls on porosity and water chemistry during O_2-SO_2-CO_2 reaction of CO_2 storage reservoir and cap-rock core. *Appl. Geochem.* **2016**, *75*, 152–168. [CrossRef]

17. Adams, A.; Diamond, L.W. Early diagenesis driven by widespread meteoric infiltration of a Central European carbonate ramp: A reinterpretation of the Upper Muschelkalk. *Sediment. Geol.* **2017**, *362*, 37–52. [CrossRef]

18. Berkowitz, B.; Singurindy, O.; Lowell, R.P. Mixing-driven diagenesis and mineral deposition: $CaCO_3$ precipitation in salt water-fresh water mixing zones. *Geophys. Res. Lett.* **2003**, *30*, 4. [CrossRef]

19. Dumariska-Slowik, M.; Powolny, T.; Sikorska-Jaworowska, M.; Heflik, W.; Morgun, V.; Xuan, B.T. Mineralogical and geochemical constraints on the origin and evolution of albitites from Dmytrivka at the Oktiabrski complex, Southeast Ukraine. *Lithos* **2019**, *334*, 231–244. [CrossRef]

20. Jones, G.D.; Xiao, Y.T. Geothermal convection in the Tengiz carbonate platform, Kazakhstan: Reactive transport models of diagenesis and reservoir quality. *AAPG Bull.* **2006**, *90*, 1251–1272. [CrossRef]

21. Jons, N.; Kahl, W.A.; Bach, W. Reaction-induced porosity and onset of low-temperature carbonation in abyssal peridotites: Insights from 3D high-resolution microtomography. *Lithos* **2017**, *268*, 274–284. [CrossRef]

22. Putnis, A.; Hinrichs, R.; Putnis, C.V.; Golla-Schindler, U.; Collins, L.G. Hematite in porous red-clouded feldspars: Evidence of large-scale crustal fluid rock interaction. *Lithos* **2007**, *95*, 10–18. [CrossRef]

23. Iannace, A.; Capuano, M.; Galluccio, L. "Dolomites and dolomites" in Mesozoic platform carbonates of the Southern Apennines: Geometric distribution, petrography and geochemistry. *Palaeogeogr. Palaeoclimatol. Palaeoecol.* **2011**, *310*, 324–339. [CrossRef]

24. Li, Q.; Zhang, Y.; Dong, L.; Guo, Z. Oligocene syndepositional lacustrine dolomite: A study from the southern Junggar Basin, NW China. *Palaeogeogr. Palaeoclimatol. Palaeoecol.* **2018**, *503*, 69–80. [CrossRef]

25. Tosti, F.; Mastandrea, A.; Guido, A.; Demasi, F.; Russo, F.; Riding, R. Biogeochemical and redox record of mid–late Triassic reef evolution in the Italian Dolomites. *Palaeogeogr. Palaeoclimatol. Palaeoecol.* **2014**, *399*, 52–66. [CrossRef]

26. Ebrahimi, P.; Vilcaez, J. Effect of brine salinity and guar gum on the transport of barium through dolomite rocks: Implications for unconventional oil and gas wastewater disposal. *J. Environ. Manag.* **2018**, *214*, 370–378. [CrossRef]

27. Giorgioni, M.; Iannace, A.; D'Amore, M.; Dati, F.; Galluccio, L.; Guerriero, V.; Mazzoli, S.; Parente, M.; Strauss, C.; Vitale, S. Impact of early dolomitization on multi-scale petrophysical heterogeneities and fracture intensity of low-porosity platform carbonates (Albian-Cenomanian, southern Apennines, Italy). *Mar. Pet. Geol.* **2016**, *73*, 462–478. [CrossRef]

28. Huang, S.; Gong, Y.; Huang, K.; Tong, H. The Influence of Burial History on Carbonate Dissolution and Precipitation A Case Study from Feixianguan Formation of Triassic, NE Sichuan and Ordovician Carbonate of Northern Tarim Basin. *Adv. Earth Sci.* **2010**, *25*, 381–390.

29. Vincent, B.; Waters, J.; Witkowski, F.; Daniau, G.; Oxtoby, N.; Crowley, S.; Ellam, R. Diagenesis of Rotliegend sandstone reservoirs (offshore Netherlands): The origin and impact of dolomite cements. *Sediment. Geol.* **2018**, *373*, 272–291. [CrossRef]

30. Jasionowski, M.; Peryt, T.M.; Durakiewicz, T. Polyphase dolomitisation of the Wuchiapingian Zechstein Limestone (Ca1) isolated reefs (Wolsztyn Palaeo-Ridge, Fore-Sudetic Monocline, SW Poland). *Geol. Q.* **2014**, *58*, 493–510.

31. Chen, X.; Yi, W.; Lu, W. The Paleokarst Reservoirs of Oil/Gas FIelds in China. *Acta Sedimentol. Sin.* **2004**, *22*, 244–253.

32. Ni, X.; Zhang, L.; Shen, A.; Qiao, Z.; Han, L. Diagenesis and pore evolution of the Ordovician karst reservoir in Yengimahalla-Hanilcatame region of Tarim Basin. *J. Palaeogeogr.* **2010**, *12*, 467–479.

33. Wang, T.; Song, D.; Li, M.; Yang, C.; Ni, Z. Natural gas source and deep gas exploration potential of the Ordovician Yingshan Formation in the Shunnan-Gucheng region, Tarim Basin. *Oil Gas Geol.* **2014**, *35*, 753–762.

34. Yu, Z.; Liu, K.; Zhao, M.; Liu, S.; Zhuo, Q.; Lu, X. Characterization of Diagenesis and the Petroleum Charge in Kela 2 Gas Field, Kuqa Depression, Tarim Basin. *Earth Sci.* **2016**, *41*, 533–545.

35. Jiao, C.; He, Z.; Xing, X.; Qing, H.; He, B.; Li, C. Tectonic hydrothermal dolomite and its significance of reservoirs in Tarim basin. *Acta Pet. Sin.* **2011**, *27*, 277–284.

36. You, D.; Han, J.; Hu, W.; Qian, Y.; Cao, Z.; Chen, Q.; Li, H. Characteristics and genesis of pores and micro-pores in ultra—deep limestones: A case study of Yijianfang Formation limestones from Shunnan-7 and Shuntuo-1 wells in Tarim Basin. *Oil Gas Geol.* **2017**, *38*, 693–702.

37. Li, Z.; Li, J.; Zhang, P.; Yu, J.; Liu, J.; Yang, L. Key Structural-Fluid Evolution and Reservoir Diagenesis of Deep-buried Carbonates: An Example from the Ordovician Yingshan Formation in Tazhong, Tarim Basin. *Bull. Mineral. Petrol. Geochem.* **2016**, *35*, 827–838.

38. Lu, Z.; Chen, H.; Qing, H.; Chi, G.; You, D.; Hang, Y.; Zhang, S. Petrography, fluid inclusion and isotope studies in Ordovician carbonate reservoirs in the Shunnan area, Tarim basin, NW China: Implications for the nature and timing of silicification. *Sediment. Geol.* **2017**, *359*, 29–43. [CrossRef]
39. Liu, W.; Huang, Q.; Wang, K.; Shi, S.; Jiang, H. Characteristics of hydrothermal activity in the Tarim Basin and its reworking effect on carbonate reservoirs. *Nat. Gas Ind.* **2016**, *36*, 14–21. [CrossRef]
40. Xu, T.; Sonnenthal, E.; Spycher, N.; Pruess, K. TOUGHREACT—A simulation program for non-isothermal multiphase reactive geochemical transport in variably saturated geologic media: Applications to geothermal injectivity and CO_2 geological sequestration. *Comput. Geosci.* **2006**, *32*, 145–165. [CrossRef]
41. Ma, J.; Li, C.; Li, B. Distribution and changes of current, temperature and salinity in the equatorial region of the western Pacific Ocean. *Acta Oceanol. Sin.* **1985**, *7*, 131–142.
42. Helpa, V.; Rybacki, E.; Ahart, R.; Muralco, L.F.G.; Rhede, D.; Jeřábek, P.; Dresen, G. Reaction kinetics of dolomite rim growth. *Contrib. Mineral. Petrol.* **2014**, *167*, 1001. [CrossRef]
43. Montes-Hernandez, G.; Findling, N.; Renard, F. Dissolution-precipitation reactions controlling fast formation of dolomite under hydrothermal conditions. *Appl. Geochem.* **2016**, *73*, 169–177. [CrossRef]
44. Machel, H.G. Concepts and models of dolomitization: A critical reappraisal. *Geol. Soc. Lond. Spec. Publ.* **2004**, *235*, 7–63. [CrossRef]
45. Huang, Q.; Liu, D.; Ye, N.; Li, Y. Reservoir characteristics and diagenesis of the Cambrian dolomite in the Tarim Basin. *J. Northeast Pet. Univ.* **2013**, *37*, 63–74.
46. Li, P.; Chen, G.; Zeng, Q.; Yi, J.; Hu, G. Genesis of Lower Ordovician Dolomite in Central Tarim Basin. *Acta Sedimentol. Sin.* **2011**, *29*, 842–856.
47. Hu, M.; Hu, Z.; Li, S.; Wang, Y. Geochemical Characteristics and Genetic Mechanism of the Ordovician Dolostone in the Tazhong Area, Tarim Basin. *Acta Geol. Sin.* **2011**, *85*, 2060–2069.

MDPI
St. Alban-Anlage 66
4052 Basel
Switzerland
Tel. +41 61 683 77 34
Fax +41 61 302 89 18
www.mdpi.com

Minerals Editorial Office
E-mail: minerals@mdpi.com
www.mdpi.com/journal/minerals